Redox Imbalance and Mitochondrial Abnormalities in Kidney Disease

Redox Imbalance and Mitochondrial Abnormalities in Kidney Disease

Editor

Liang-Jun Yan

MDPI • Basel • Beijing • Wuhan • Barcelona • Belgrade • Manchester • Tokyo • Cluj • Tianjin

Editor
Liang-Jun Yan
University of North Texas Health Science Center
USA

Editorial Office
MDPI
St. Alban-Anlage 66
4052 Basel, Switzerland

This is a reprint of articles from the Special Issue published online in the open access journal *Biomolecules* (ISSN 2218-273X) (available at: https://www.mdpi.com/journal/biomolecules/special_issues/Redox_Imbalance).

For citation purposes, cite each article independently as indicated on the article page online and as indicated below:

LastName, A.A.; LastName, B.B.; LastName, C.C. Article Title. *Journal Name* **Year**, *Volume Number*, Page Range.

ISBN 978-3-0365-3757-3 (Hbk)
ISBN 978-3-0365-3758-0 (PDF)

Contents

About the Editor

Liang-Jun Yan is a professor in the Department of Pharmaceutical Sciences, College of Pharmacy at the University of North Texas Health Science Center. He completed his B.S. in Biochemistry at Peking University and earned his Ph.D. in Molecular and Cell Biology from the University of California at Berkeley. His research has centered on redox imbalance and mitochondrial oxidative stress in aging and age-related metabolic disease. His lab is currently working on kidney disease, in particular, ischemia- and drug-induced acute kidney injury and chronic diabetic kidney disease. The objective of these investigations is to elucidate the mechanisms of mitochondrial oxidative stress and redox dysregulation in kidney diseases and to identify novel redox-associated targets for developing potential therapeutics. Relevant to these studies, his group is also interested in studying the therapeutic values of natural products derived from plants and herbs that can be used to fight kidney disease.

biomolecules

MDPI

Editorial

Redox Imbalance and Mitochondrial Abnormalities in Kidney Disease

Liang-Jun Yan

Department of Pharmaceutical Sciences, College of Pharmacy, University of North Texas Health Science Center, Fort Worth, TX 76107, USA; liang-jun.yan@unthsc.edu; Tel.: +1-817-735-2386; Fax: +1-817-735-2603

Citation: Yan, L.-J. Redox Imbalance and Mitochondrial Abnormalities in Kidney Disease. *Biomolecules* **2022**, *12*, 476. https://doi.org/10.3390/biom12030476

Received: 10 March 2022
Accepted: 17 March 2022
Published: 21 March 2022

The kidneys carry out fundamental life-sustaining functions by removing waste substances, controlling salt and water balance, retaining substances vital to the body such as glucose and proteins, and maintaining blood pH balance [1]. By performing these functions, the kidneys are heavily dependent on mitochondrial ATP production that not only consumes a large amount of oxygen, but also exposes each kidney to a variety of damages, such as diabetic hyperglycemia, drug toxicity, ischemic injury, heavy metal toxicity, and sepsis [2]. All these damages are thought to be associated with cellular redox imbalance and mitochondrial oxidative stress, as well as mitochondrial abnormalities. The clinical symptoms of these insults can be anywhere from acute kidney injury (AKI) to chronic kidney disease (CKD) that, if left untreated, can eventually lead to end-stage kidney failure [3].

To fend off the abovementioned insults, it is necessary to understand the underlying mechanisms of redox imbalance, oxidative stress, and mitochondrial abnormalities involved in kidney disease. In this Special Issue of *Biomolecules* entitled "Redox Imbalance and Mitochondrial Abnormalities in Kidney Disease", we have collected 11 papers that cover a variety of topics focusing on oxidative stress, mitochondrial dysfunction, and antioxidation enhancement implicated in kidney disease or kidney transplantation.

Among the 11 papers, seven are review articles, and four are original research articles. Yan reviewed the sources of NAD^+/NADH redox imbalance and their roles in diabetic kidney disease (DKD) [3]. The gist of the paper is that NAD^+/NADH redox imbalance leads to the disruption of mitochondrial homeostasis, and an increase in nephron oxidative damage that culminates as diabetic nephropathy reflected by deteriorating kidney functions. While mitochondrial oxidative stress is a central event in DKD, it is also hypothesized by Yan and Allen to be a unifying mechanism underlying cadmium-induced kidney toxicity [2] that mainly involves oxidative damage to proximal tubule epithelial cells within the nephrons. Moreover, Zhu et al. also stressed in their review article that mitochondrial oxidative stress-induced cell death is the major mechanism of podocytopathies involving the dysregulation of p53, PI3K/Akt, P38/MAPK, unfolded protein response, and endoplasmic reticulum that may all be the downstream targets of oxidative stress and redox imbalance [4].

In the article "Glucose- and non-glucose-induced mitochondrial dysfunction in diabetic kidney disease" [5], Ito et al. summarized the effects of toxic metabolites of glucose, such as advanced glycation end-products and the detrimental effects of non-glucose events, such as lipotoxicity, hypoxia, and endothelin-1 receptor signaling on the pathogenesis of DKD [4]. It should be noted that lipotoxicity can also be caused by the accumulation of acetyl-CoA, which is not only a substrate for lipid biosynthesis, but can also lead to an increase in protein acetylation [6]. Both processes are thought to contribute to the pathogenesis of kidney disease, along with the potential contributions of Krebs cycle metabolites, such as citrate, succinate, and fumarate [6].

Given that mitochondrial redox imbalance and oxidative stress are the central themes of kidney disease, it is natural to explore therapeutic approaches that can counteract kidney injury by mitigating oxidative stress and mitochondrial dysfunction. Indeed, several papers in this Special Issue have just explored such approaches. For example, the role of SGLT2 inhibitors as antioxidants in the kidney was reviewed by Liorens-Cebria et al. [7]. Additionally, Wongmekiat et al. provided evidence that purple rice husk is nephoprotective in DKD, with an underlying mechanism involving PGC-1α, Sirt3, and SOD2 [8]. Interestingly, renalase, as a flavin-dependent detoxifying enzyme, may be regarded as a risk factor for death in patients with CKD [9], though further studies are needed to confirm the link between renalase levels and death. Approaches that can prevent mitochondrial uncoupling and preserve renal function in transplanted kidneys have also been discussed in this Special Issue. These include the controlled oxygenated rewarming of kidney grafts [10] and the suppression of inflammation-induced kidney damage by the use of a free radical scavenger Prc-210 [11]. The underlying protective mechanisms of these approaches likely involve the mitigation of mitochondrial oxidative stress and the restoration of redox balance.

In summary, mitochondrial oxidative stress and redox imbalance drive renal mitochondrial abnormalities, which can be attributed to the over-production of mitochondrial reactive oxygen species. As summarized by Aranda-Rivera et al. and Ito et al. in their review articles published in this Special Issue [5,12], numerous processes and signaling pathways are involved, and these can collectively contribute to the pathogenesis of a given kidney disease. These pathways include posttranslational modifications of proteins, mitochondrial fusion and fission, mitophagy, biogenesis, Warburg effects, Krebs cycle, electron transport chain, oxidative phosphorylation, and mitochondrial uncoupling. Further dissection of each of these processes or signaling pathways may provide novel insights into designing additional therapeutic approaches to fight kidney disease.

Conflicts of Interest: The author declares no conflict of interest.

References

1. Yan, L.J. Folic acid-induced animal model of kidney disease. *Anim. Model. Exp. Med.* **2021**, *4*, 329–342. [CrossRef] [PubMed]
2. Yan, L.J.; Allen, D.C. Cadmium-induced kidney injury: Oxidative damage as a unifying mechanism. *Biomolecules* **2021**, *11*, 1575. [CrossRef] [PubMed]
3. Yan, L.J. Nadh/nad(+) redox imbalance and diabetic kidney disease. *Biomolecules* **2021**, *11*, 730. [CrossRef] [PubMed]
4. Zhu, Y.-T.; Wan, C.; Lin, J.-H.; Hammes, H.-P.; Zhang, C. Mitochondrial oxidative stress and cell death in podocytopathies. *Biomolecules* **2022**, *12*, 403. [CrossRef]
5. Ito, M.; Gurumani, M.Z.; Merscher, S.; Fornoni, A. Glucose- and non-glucose-induced mitochondrial dysfunction in diabetic kidney disease. *Biomolecules* **2022**, *12*, 351. [CrossRef]
6. Jiménez-Uribe, A.P.; Hernández-Cruz, E.Y.; Ramírez-Magaña, K.J.; Pedraza-Chaverri, J. Involvement of tricarboxylic acid cycle metabolites in kidney diseases. *Biomolecules* **2021**, *11*, 1259. [CrossRef] [PubMed]
7. Llorens-Cebrià, C.; Molina-Van den Bosch, M.; Vergara, A.; Jacobs-Cachá, C.; Soler, M.J. Antioxidant roles of sglt2 inhibitors in the kidney. *Biomolecules* **2022**, *12*, 143. [CrossRef] [PubMed]
8. Wongmekiat, O.; Lailerd, N.; Kobroob, A.; Peerapanyasut, W. Protective effects of purple rice husk against diabetic nephropathy by modulating pgc-1α/sirt3/sod2 signaling and maintaining mitochondrial redox equilibrium in rats. *Biomolecules* **2021**, *11*, 1224. [CrossRef] [PubMed]
9. Knop, W.; Serwin, N.M.; Cecerska-Heryć, E.; Grygorcewicz, B.; Dołęgowska, B.; Gomółka, A.; Wiśniewska, M.; Ciechanowski, K. Elevated levels of renalase, the β-nad(p)h isomerase, can be used as risk factors of major adverse cardiovascular events and all-cause death in patients with chronic kidney disease. *Biomolecules* **2021**, *11*, 1514. [CrossRef] [PubMed]
10. Zlatev, H.; von Horn, C.; Minor, T. Preservation of mitochondrial coupling and renal function by controlled oxygenated rewarming of porcine kidney grafts. *Biomolecules* **2021**, *11*, 1880. [CrossRef] [PubMed]
11. Goesch, T.R.; Wilson, N.A.; Zeng, W.; Verhoven, B.M.; Zhong, W.; Coumbe Gitter, M.M.; Fahl, W.E. Suppression of inflammation-associated kidney damage post-transplant using the new prc-210 free radical scavenger in rats. *Biomolecules* **2021**, *11*, 1054. [CrossRef] [PubMed]
12. Aranda-Rivera, A.K.; Cruz-Gregorio, A.; Aparicio-Trejo, O.E.; Pedraza-Chaverri, J. Mitochondrial redox signaling and oxidative stress in kidney diseases. *Biomolecules* **2021**, *11*, 1144. [CrossRef] [PubMed]

biomolecules

MDPI

Review

Cadmium-Induced Kidney Injury: Oxidative Damage as a Unifying Mechanism

Liang-Jun Yan * and Daniel C. Allen

Department of Pharmaceutical Sciences, College of Pharmacy, University of North Texas Health Science Center, Fort Worth, TX 76107, USA; danielallen@my.unthsc.edu
* Correspondence: liang-jun.yan@unthsc.edu; Tel.: +1-817-735-2386; Fax: +1-817-735-2603

Abstract: Cadmium is a nonessential metal that has heavily polluted the environment due to human activities. It can be absorbed into the human body via the gastrointestinal tract, respiratory tract, and the skin, and can cause chronic damage to the kidneys. The main site where cadmium accumulates and causes damage within the nephrons is the proximal tubule. This accumulation can induce dysfunction of the mitochondrial electron transport chain, leading to electron leakage and production of reactive oxygen species (ROS). Cadmium may also impair the function of NADPH oxidase, resulting in another source of ROS. These ROS together can cause oxidative damage to DNA, proteins, and lipids, triggering epithelial cell death and a decline in kidney function. In this article, we also reviewed evidence that the antioxidant power of plant extracts, herbal medicines, and pharmacological agents could ameliorate cadmium-induced kidney injury. Finally, a model of cadmium-induced kidney injury, centering on the notion that oxidative damage is a unifying mechanism of cadmium renal toxicity, is also presented. Given that cadmium exposure is inevitable, further studies using animal models are warranted for a detailed understanding of the mechanism underlying cadmium induced ROS production, and for the identification of more therapeutic targets.

Keywords: cadmium; kidney injury; renal toxicity; mitochondria; oxidative damage; proximal tubule

Citation: Yan, L.-J.; Allen, D.C. Cadmium-Induced Kidney Injury: Oxidative Damage as a Unifying Mechanism. *Biomolecules* **2021**, *11*, 1575. https://doi.org/10.3390/biom11111575

Academic Editor: Theodoros Eleftheriadis

Received: 16 September 2021
Accepted: 20 October 2021
Published: 23 October 2021

1. Introduction

The kidney is a vital organ that performs critical physiological functions by actively filtering excess fluid and secreting waste products including urea, uric acid, and creatinine [1,2]. It is through the process of filtration and reabsorption that the kidneys maintain homeostasis of water, acid-base and, electrolytes [3]. Moreover, the kidney also secretes hormones that participate in the control and regulation of hemodynamics, red blood cell production, and vitamin D maturation [3]. Under abnormal conditions such as fasting and insulin resistance, the kidney can also make glucose via the gluconeogenic pathway [4–6] using noncarbohydrate precursors such as pyruvate, alanine, lactate, and glycerol [7].

The kidney is also vulnerable to injuries caused by numerous challenges such as ischemia [8–12], drug toxicity [13–19], environmental heavy metal exposure [20–27], hypertension [28–30], immune injury [31,32], and diabetes [33–36]. In terms of environmental risk factors, human kidney disease caused by environmental pollutants and occupational-linked toxins is a major public health issue [37]. Cadmium is a toxic heavy metal mainly derived from chemical stabilizers, pigments, nickel-cadmium batteries, and metal coatings and alloys [38]. It is also a toxic element in cigarettes [38]. Accordingly, contaminated soil, air, drinking water, food chains [39,40], and cigarettes, as well as children's plastic toys [41], are the major sources of human cadmium exposure. Numerous studies focusing on cadmium toxicity have established that the kidney is a primary organ site for cadmium accumulation [42,43]. Indeed, cadmium exposure has been tightly associated with renal dysfunction and kidney damage, causing polyuria and proteinuria [23,24]. The proximal tubule is the major site of cadmium deposition, accumulation, and damage because of the development of proximal tubular epithelial cell hypertrophy with occurrence of polyuria and proteinuria [44–46]. Therefore, it is important to counteract cadmium-induced kidney injury to safeguard kidney function.

2. Cadmium Absorption, Transportation, and Accumulation in the Kidney

Cadmium has a high affinity toward thiol groups and can selectively form complexes with proteins and peptides whose cysteine residues are available for cadmium binding [47,48]. After ingestion of cadmium-contaminated water, food, and/or cigarette smoking, cadmium can be absorbed into circulation via the gastrointestinal tract, respiratory tract, or the skin [49–51]. Once in the blood, cadmium binds to albumin and other cysteine-containing proteins and peptides such as glutathione [37] and gets transported via many avenues to the liver [37] whereby the heavy metal is then released and induces the expression of metallothionein that then binds tightly to cadmium [52,53]. This binding serves the purpose of detoxification as the cadmium-metallothionein complex is usually considered nontoxic [54]. The cadmium-metallothionein complex can be released into the bloodstream and is then filtered at the glomerulus and reabsorbed by the proximal tubular epithelial cells [55]. This is followed by release of cadmium from the degradation of the cadmium-metallothionein complex [55]. The free form of cadmium in the proximal tubular region of the nephron can then bind to pre-existing renal metallothionein and induce further renal expression of metallothionein [50]. When renal metallothionein is exhausted [56,57], the nonmetallothionein bound cadmium accumulates and induces nephrotoxicity [49–51,58,59], primarily in the proximal tubular region (Figure 1) via generation of oxygen free radicals [60–62]. As up to 50% of the body's cadmium pool can deposit in the kidney [37] and the half-life of cadmium in the kidney is approximately 45 years [63–67], cadmium-caused renal toxicity can pose a major threat to human health, particularly in countries where environmental control and regulation are lacking. It should be noted that while the binding of cadmium to metallothionein is a well-established mechanism, other thiol-containing proteins and peptides such as albumin and glutathione can also bind cadmium, leading to functional impairment of these cadmium bound target proteins and peptides [50].

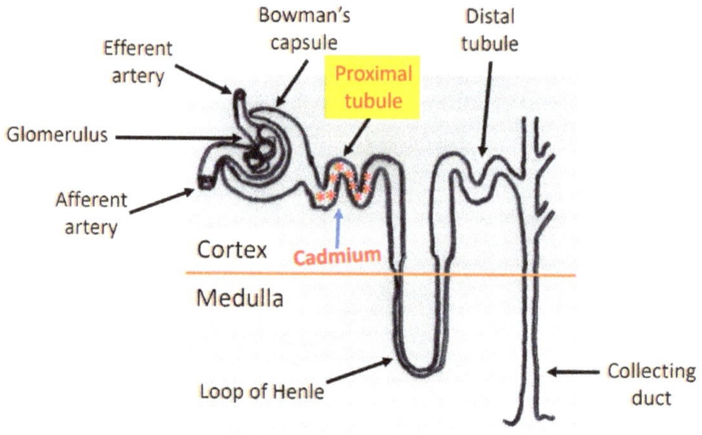

Figure 1. Diagram showing the proximal convoluted tubule as the major site of cadmium accumulation and toxicity in the nephrons "*".

3. Cadmium-Induced Animal Models of Kidney Injury

Given the fact that human cadmium exposure is a chronic process at a very low level, any investigation of cadmium renal toxicity would require many years of monitoring and follow-up studies. Therefore, animal models using mice or rats have been widely used to replicate the pathophysiological mechanisms of cadmium renal toxicity [39,40,42,68]. In numerous cases, high doses of cadmium were applied in these animal models to shorten the duration of the studies and facilitate the process of obtaining insights into the mechanisms of cadmium renal toxicity. As mentioned above, studies using rodent models as well as results from human subjects have established that the primary target of cadmium in the nephron is the proximal tubule, whereby cadmium causes overall dysfunction of the epithelial cells [51,69,70], resulting in polyuria and proteinuria [50,51]. There is also an increase in urinary excretion of amino acids, glucose, and electrolytes such as Na^+. K^+, and Ca^{2+} [50,51]. Increasing evidence also indicates that a variety of risk factors such as aging [71], malnutrition [72], obesity [73–75], and diabetes [27,76] can further superimpose on cadmium renal toxicity and aggravate cadmium-induced renal dysfunction.

It should be stressed that in animal model studies of cadmium renal injury, a variety of doses, routes, and duration of exposures have been performed. The purpose of all these approaches is to try to replicate or recapitulate the toxico-kinetics and underlying mechanisms of long-term, low-level exposure that commonly occur in humans [50].

4. Mechanisms of Cadmium-Induced Renal Toxicity

What is the proposed mechanism of cadmium-induced kidney injury? Based on numerous studies, all injurious pathways converge on ROS production and culminate in oxidative stress [77–81], which suggests that oxidative damage is a unifying mechanism of cadmium-induced renal toxicity and injury. We also think that the major sources of ROS causing oxidative damage in this context are mitochondria and NADPH oxidase, described as follows.

5. Sources of Reactive Oxygen Species

5.1. Mitochondria

Mitochondria are well known as the intracellular site of ROS production [82–85]. Among the electron transport chain components complexes I, II and III have all been established as major sites of ROS production [86–89]. These sites are not perfect even under normal conditions and can leak electrons out of the transport chain [90,91] (Figure 2). The leaked electrons can then partially reduce oxygen to form superoxide anion, which is the precursor of all other reactive oxygen species including H_2O_2, hydroxyl radical, and peroxynitrite [92,93] (Figure 3). Additionally, dihydrolipoamide dehydrogenase involved enzyme complexes such as pyruvate dehydrogenase, α-ketoglutarate dehydrogenase, and branched chain amino acid dehydrogenase can also produce superoxide anion in a variety of experimental and pathological conditions [94–97].

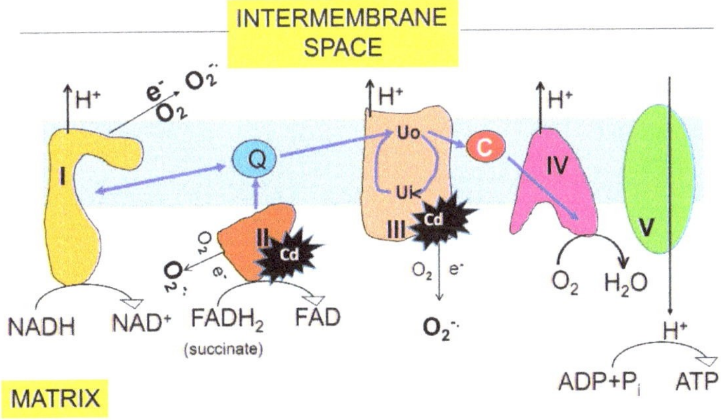

Figure 2. Diagram showing mitochondrial electron transport chain and oxidative phosphorylation. Complexes I, II, and III all can generate superoxide anion. This process can be enhanced by pathophysiological conditions such as cadmium exposure and accumulation. Note that it has been suggested that complexes II and III are the likely sites interacting with cadmium [98].

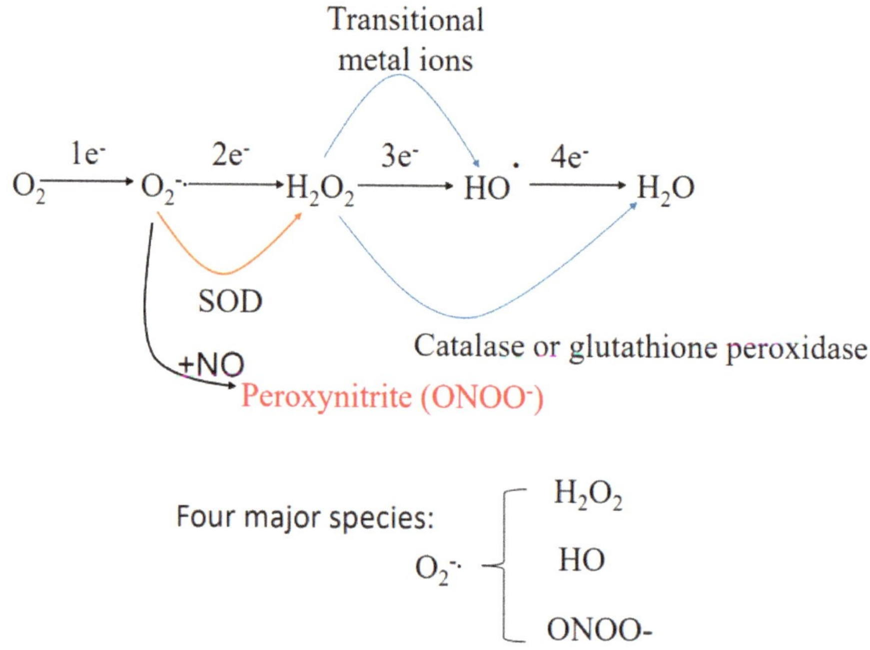

Figure 3. Production of other reactive oxygen species and reactive nitrogen species from the initial species superoxide. Superoxide can be dismutated by superoxide dismutase to form H_2O_2, which can be further detoxified by catalase. In the presence of metal ions such as iron, H_2O_2 can also generate very reactive species hydroxyl radical. Additionally, superoxide can react with nitric oxide to form peroxynitrite that is also very reactive toward macromolecules.

5.2. NADPH Oxidase

NADPH oxidase (NOX) can generate superoxide anion using NADPH as its reducing agent [99]. So far, seven NOXs have been identified (NOX 1-5, Duox1 and DuoX2) [99]. These isoforms differ in many aspects including catalytic oxidase subunit, tissue distribution, intra-cellular location, and mechanisms of regulation [100,101]. All NOXs are composed of multiple subunits. Upon stimulation, these subunits will come together and assemble to form a membrane-associated complex to generate superoxide at the expense of NADPH [102]. Figure 4 shows a representative diagram of NOX assembly upon stimulation whereby the major site of ROS production is the gp91phox subunit with other proteins being the ancillary units required for the regulation and functioning of the whole enzyme complex. It should be noted that Figure 4 only shows the assembly of NOX2. The structural and compositional variations of other NOX isoforms [103] and their potential interaction with cadmium may also play a role in cadmium induced renal toxicity. Under normal conditions, these NOXs function in a beneficial way by regulating kidney metabolism and homeostasis including glucose transport, gluconeogenesis, renal hemodynamics, and electrolyte transport and balance [99]. Under pathophysiological conditions, these NOXs, in particular NOX2 and NOX4 in the kidney, can overgenerate ROS that are damaging to cellular components including DNA, proteins, and lipids, causing cell death and kidney injury [99,104–106]. It has been reported that cadmium exposure can increase the expression of NOX1 subunits, leading to increased ROS production from the enzyme [107]. Nevertheless, it is not known exactly which subunit in the NADPH oxidase physically interacts with cadmium at the present time. It should be noted that xanthine oxidase [108,109] and nitric oxide synthase [110–112], although not a major source of ROS in the kidney, may also contribute to renal oxidative stress under a variety of pathological and experimental conditions including cadmium exposure. It should also be pointed out that comprehensive evaluations of the roles of NADPH oxidases, xanthine oxidase, and nitric oxide synthase in cadmium-induced kidney injury remain to be conducted.

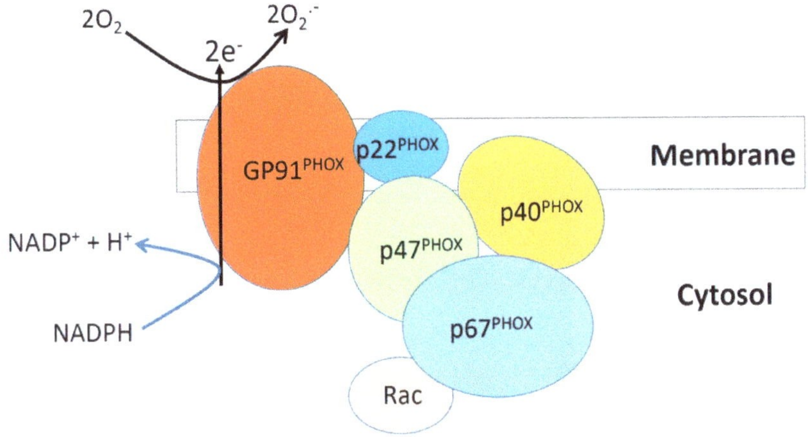

Figure 4. NADPH oxidase assembly and superoxide production at the expense of NADPH. Upon stimulation, each individual subunit of the enzyme is recruited to the membrane and form a membrane-associated complex. Only one subunit GP91 catalyzes partial reduction of oxygen. This figure is adapted from reference [102]. Please not that shown here is NOX2 assembly. For structures and components of other NXO isoforms, please refer reference [103].

6. Effects of Cadmium on Mitochondrial Function

Cadmium can enter mitochondria and accumulate therein [113,114]. This is likely facilitated by mitochondrial membrane channels, and solute molecule carriers and receptors [114]. Once inside the mitochondria, cadmium can bind thiol-containing proteins and impair the corresponding protein function [80]. Studies have demonstrated that upon exposure to cadmium, kidney mitochondria displayed deformation, swelling, and vaculation, concurrent with increased SOD1 expression and decreased SOD2 and catalase expression [115]. Additionally, the anti-apoptotic protein BCL-2 was also found decreased by cadmium exposure [115]; so was the ratio between reduced glutathione and oxidized glutathione [60]. All these could be a generalized cadmium mitochondrial toxicity and the ultimate outcome would be reflected by overproduction of mitochondrial ROS, disruption of mitochondrial metabolic pathways, and impairment of mitochondrial pores, membrane channels and transporters [80]. It has been reported that complex II and complex III may be the major sites impaired by cadmium in the nephrons [98] while the effects of cadmium on proximal tubular mitochondrial complex I (NADH-ubiquinone oxidoreductase) remain unclear. Disruption of all these processes by cadmium would increase mitochondrial ROS production and eventually lead to cell death and kidney injury [80,114,116–119]. An outline of cadmium induced ROS production, oxidative damage to macromolecules, cell death, and kidney injury is shown in Figure 5, highlighting the concept that oxidative damage is a unifying mechanism of cadmium-induced kidney injury.

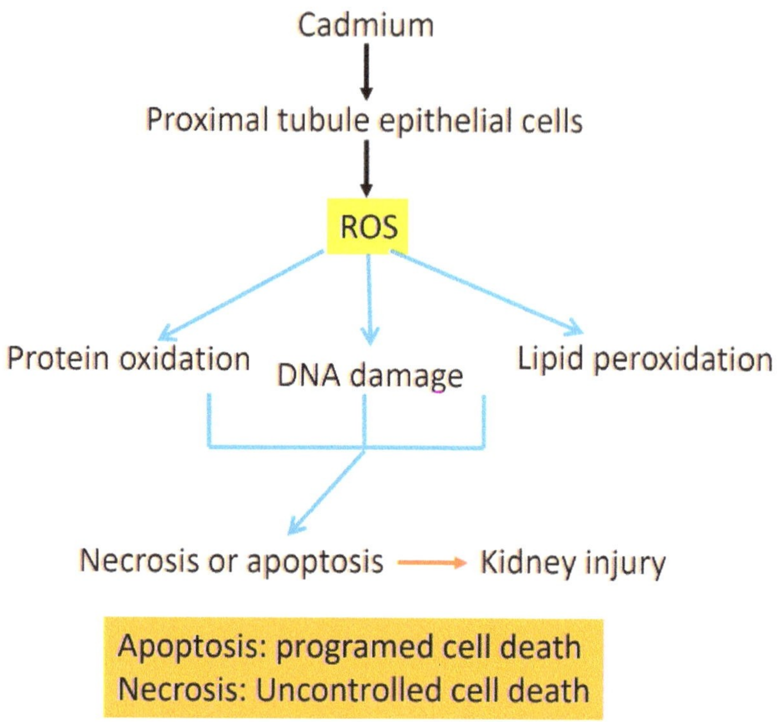

Figure 5. ROS can damage DNA, proteins, and lipids. Damage of the molecules impairs the biological function of each molecule, leading to cell death and kidney injury. Cell death may include both necrosis and apoptosis.

7. Counteracting Effects of Natural Products, Chemicals and Pharmacological Agents on Cadmium-Induced Kidney Injury

In further support of the notion that oxidative stress and oxidative damage are the universal mechanisms underlying renal toxicity by cadmium, we herein tabulate evidence that numerous natural products such as plant extracts and herbal medicines have been used to counteract the oxidative, deleterious effects of cadmium on the kidney. Many of these studies used cadmium-induced animal models of kidney injury as a platform [120]. Table 1 selectively shows some of the reported plant extracts and herbs as well as exogenous chemicals and pharmacological agents that can attenuate cadmium-induced oxidative stress involved in kidney injury. Additionally, many of these approaches can also induce the activation of endogenous cellular defense systems such as Nrf2, superoxide dismutase, glutathione peroxidase, and catalase [74,121–124]. Some studies using kidney cell lines such as HEK293 are also included in the table. It should be pointed out that among all the compounds and chemicals listed in Table 1, it is very difficult to identify which one would be the most efficient in terms of combating confirmed cadmium-induced kidney injury as cross examination and comparison of these natural products on a same platform under exactly the same experimental conditions have not been conducted. In addition, whether administration of these natural antioxidants could increase the efflux of cadmium out of the body remains to be investigated.

Table 1. Counteracting effects of exogenous compounds such as plant extracts, herbs, chemicals and pharmacological agents on cadmium induced renal toxicity.

Plant/Extract/Chemical	Rodent Model	Mechanism	Reference
Allium hirtifolium boiss	Rats	Anti-oxidative stress	[125]
Apple juice	Rats	Anti-oxidative stress	[126]
Arctium lappa	Rats	Anti-oxidative stress	[127]
Carnosic acid	Mice and cells	Anti-oxidative damage	[128]
Catechin	Rats	Anti-oxidative damage	[129]
Caffeic acid phenethyl ester	Rats	Anti-oxidative stress	[130,131]
Chelidonium majus leaves	Rats	Antidiuretic	[132]
Chorella pyrenoidosa	Rats	Antihyperglycemic	[133]
Cleistocalyx nervosum var. paniala	Rats	Increasing antioxidation power	[134]
Coriandrum sativum leaf	Mice	Anti-oxidative stress	[135]
Curcumin	Rats	Anti-oxidative stress	[136]
Edaravone	Mice/Cells	Inhibiting oxidative stress	[137]
Elderberry	Rats	Increasing antioxidant enzymes	[138]
Epigallocatechin-3-gallate	Rats	Increasing antioxidant defense	[139]
Eucommia ulmoides bark	Rats	Anti-oxidative damage	[140]
Ferulic acid	Rats	Anti-oxidative stress	[78]
Fragaria ananassa	Rats	Anti-oxidative stress	[141]
Ginger	Rats	Decrease lipid peroxidation	[142]
Glutathione	Rats	Anti-oxidative stress	[143]
Glycyrrhiza glabra	Rats	Anti-oxidative stress	[144]
Grape seed procyanidin	Mice	Antioxidants	[145]
Grape skin/purple carrot	Rats	Anti-oxidative damage	[146]
Green/black/red/white tea	Rats	Anti-oxidative damage	[147]
Green olive leaf	Renal cells (MCD4)	Anti-oxidative stress	[148]
Herbal adaptogens	Chicken	Anti-oxidative damage	[21]
Ipomoea aquatic/Enhydra fluctuans	Mice	Anti-oxidation/anti-apoptosis	[149]
Irvingia gabonesis stem bark	Rats	Increasing antioxidant defense	[150]
Licorice	Rats	Anti-oxidative damage	[151] **
Ligustrazine	Rats	Restoring renal function	[152]
Lipoic acid	Rats	Anti-apoptosis	[68,153]
Onion/garlic	Rats	Anti-oxidative stress	[154]
Origanum majorana L.	Rats	Anti-oxidative damage	[155]
Persea americana seeds	Rats	Mitigating oxidative stress	[156]
Physalis peruviana L	Rats	Anti-oxidation/Anti-apoptosis	[157]
Picroliv	Rats	Anti-oxidative stress	[158]
Plantamajoside	Rats	Decrease oxidative damage	[159]
Pleurotus ostreatus	Rats (female)	Mitigating oxidative damage	[160]
Potentilla anserine	Mice and cells	Anti-oxidative stress	[161]
Puerarin	Rat proximal tubule cells	Restoring mitochondrial function	[162]
Quercetin	Rats	Suppressing ER stress	[163]
Resveratrol	Chickens	Anti-oxidative stress	[164]
Roflumilast	Rats	Increasing antioxidant defense	[165]
Rosmarinic acid	Mice	Anti-oxidative damage	[166]
Royal jelly	Mice (male)	Antioxidation/Nrf2 activation	[39]
Rutin	Rats	Inhibiting oxidative stress	[167]
Salvia officinalis	Rats	Anti-oxidative damage	[168]
Salvia miltiorrhiza	Rats	Anti-oxidative injury	[169]
Sana Makki	Rats	Anti-oxidative stress/Nrf2	[170]
Selenium yeast	Chicken	Mitigating necroptosis	[171]
Sesamol	Rats	Inhibiting oxidative stress	[172]
SInapic acid	Rats	Inhibiting oxidative stress	[173]
Solanum torvum Swartz	Rats	Anti-oxidative stress	[174]
Spinacia oleracea polysaccharides	HEK293 cells	Anti-oxidative stress	[175]
Telmisartan	Mice	Suppressing oxidative stress	[176]
Tetrahydrobiopterin	Rats	Maintaining mitochondria integrity	[177]
Thunbergia laurifolia leaf	Kidney cells	Increasing antioxidant enzymes	[178]

Table 1. *Cont.*

Plant/Extract/Chemical	Rodent Model	Mechanism	Reference
Thymus serrulatus essential oil	Rats	Anti-oxidative stress	[179]
Thymoquinone	Rats	Increasing glutathione	[180]
Tinospora cordifolia	Rats	Anti-oxidative stress	[181]
Trehalose	Rats	Inhibiting oxidative stress	[182]
Tribulus terrestris linn	Rats	Anti-oxidation	[183]
Vitamin C	Rabbits	Anti-oxidative stress	[184]
Vitamin E	Rats	Enhancing antioxidant defense	[185]

** Please note that licorice could also pose renal toxicity under certain conditions [186].

8. Other Potential Interventional Approaches

In addition to the plant extracts, herbs and pharmacological agents as shown in Table 1, there are other approaches that have also been applied to counteract cadmium-induced kidney injury. For example, caloric restriction as an established interventional approach for aging and age-related diseases [187–190] has been demonstrated to mitigate cadmium-induced renal toxicity and kidney dysfunction [191]. Dietary restriction of calcium intake has also been shown to enhance cadmium-induced expression of metallothionein, which could minimize cadmium toxicity [192]. The protective effects of preconditioning and postconditioning observed in numerous studies [193–195], if any, elicited by a variety of approaches including ischemia, hypoxia, chemicals or pharmacological agents are yet to be investigated. Additionally, metal chelation using specific chelating agents may also be considered as an interventional approach [196]. Recent findings that persulfide and polysulfide can bind to cadmium thereby decreasing cadmium toxicity [197–199] may also provide potential approaches for counteracting cadmium-induced kidney injury.

9. Postulated Model of Cadmium-Induced Proximal Tubule Lesion

Prozialeck and Edwards proposed a model of proximal convoluted tubular cell injury in 2012 [50] that we think elaborates very well the mechanism of cadmium-induced kidney injury in terms of oxidative damage as a unifying mechanism. This model is similar to what has been proposed to explain the mechanisms of ischemic acute kidney injury [8,10,200,201]. Essentially, as diagramed in Figure 6, under healthy conditions and in the absence of cadmium deposit and accumulation, epithelial cells in the proximal tubule are closely associated with each other via specialized junctional structures. These epithelial cells align orderly and tightly on the tubular basement membrane via local adhesion molecules to collectively achieve filtration and reabsorption. In the presence of cadmium, which can accumulate in the cytosolic and mitochondrial compartments, cadmium binds thiol-containing proteins and peptides, leading to functional impairment of these cadmium-bound proteins and peptides. Consequently, such impairments cause mitochondrial electron leakage or NADPH oxidase dysfunction, resulting in enhanced production of ROS and elevated levels of oxidative stress. If the oxidative stress is mild, the tubular cells can repair themselves and resume normal function. It should be noted that this self-repair is likely achieved by de-differentiated tubular epithelial cells instead of differentiated and fixed tubular progenitor cells [202,203]. However, if the oxidative stress is severe and overwhelms cellular repair capacity, an irreversible damage process occurs and cells die by means of apoptosis, necrosis or both [204,205], leading to cell-cell and cell-basement membrane dissociations. This would lead to proteinuria, polyuria, and a progressive decline in kidney function. This functional decline, however, may be intervened and halted by the antioxidative approaches shown in Table 1 if applied appropriately. It should be pointed out that in order to distinguish cadmium-induced proteinuria from primary glomerular lesion, the magnitude of proteinuria and a cadmium concentration dependent manner will need to be characterized.

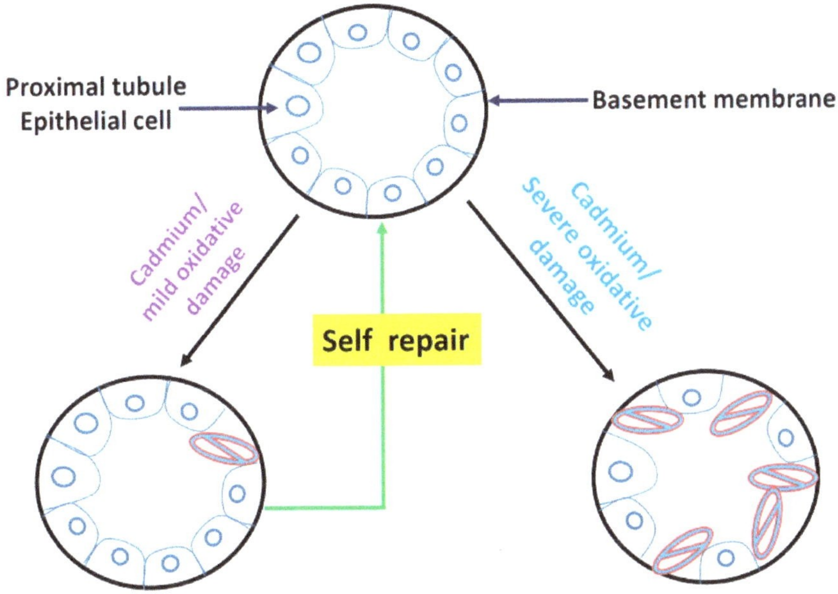

Figure 6. Schematic diagram depicting cadmium-induced injury to proximal tubular epithelial cells. When oxidative damage is mild, the cells can mobilize their repair defense system and self-repair, leading to maintenance of cellular function. When oxidative damage is severe, cells lose their self-repair capacity and die, leading to widespread cell death and kidney injury. (Adapted from reference [50]).

10. Diagnosis of Cadmium-Induced Kidney Injury

While diagnosis of cadmium-induced kidney injury is complicated by factors such as dosage of exposure, duration of exposure, early stage injury or late stage irreversible injury as well as whether there does any exist underlying disease, a series of parameters could be combined to indicate whether a kidney injury is caused by cadmium exposure. These parameters include measurements of blood and urine cadmium, urinary metallothionein, urinary β2-microglobulin and *N*-acetyl-β-glucosaminidase. In addition, kidney injury molecule-1 (Kim-1) could also be used to indicate early stage of cadmium-induced proximal tubular injury [50]. Moreover, severe cadmium poisoning could also cause pains in the spine and joints [24]. Collectively, measurements of these biomarkers or indices should provide good evidence that a cadmium-caused kidney injury has occurred. It should be noted that once these biomarkers appear and are detectable, cadmium induced kidney injury might be at an advanced stage that is irreversible. Therefore, novel biomarkers of cadmium-induced early stage kidney injury remain to be explored.

11. Summary

Cadmium exposure and cadmium-induced kidney disease are major public health issues. Mitochondria and NADPH oxidase can be impaired by cadmium that accumulates in the proximal tubular site of nephrons, and thus, are the major sources of ROS. Therefore, the main underlying mechanism of cadmium renal toxicity is enhanced oxidative stress and associated damage to DNA, proteins, and lipids, eventually leading to cell death, kidney injury, and decline in kidney function (Figure 5). Given that cadmium exposure is inevitable in the foreseeable future, cadmium-induced animal models of kidney disease should continue to play an important role in investigating the etiological, pathological, pharmacogenetic, pharmacological, and therapeutic aspects of cadmium-induced kidney disorders. Finally, in addition to elucidating the detailed mechanisms of cadmium-caused mitochondrial oxidative stress and redox imbalance, future studies should also explore novel biomarkers that can be used to diagnose early kidney injury by cadmium exposure. Moreover, whether administration of natural products such as those listed in Table 1 could increase the efflux of cadmium out of the body remains to be investigated.

Author Contributions: Conceptulization, L.-J.Y.; original draft preparation, L.-J.Y.; review and editing, D.C.A. and L.-J.Y. All authors have read and agreed to the published version of the manuscript.

Funding: This research received no external funding.

Institutional Review Board Statement: Not applicable.

Informed Consent Statement: Not applicable.

Conflicts of Interest: The authors declare no conflict of interest.

References

1. Koeppen, B.M.; Stanton, B.A. *Renal Physiology*, 5th ed.; Elsevier: Philadelphia, PA, USA, 2013.
2. Rennke, H.G.; Denker, B.M. *Renal Pathology: The Essentials*, 5th ed.; Wolters Kluwer: New York, NY, USA, 2020.
3. Dipiro, J.T.; Talbet, R.L.; Yee, G.C.; Matzke, G.R.; Wells, B.G.; Posey, L.M. *Pharmacotherapy: A Pathophysiological Approach*, 9th ed.; McGraw-Hill Education: New York, NY, USA, 2014.
4. Sumida, K.D.; Garrett, J.H.; McJilton, W.T.; Hevener, A.L.; Donovan, C.M. Effect of endurance training and fasting on renal gluconeogenic enzymes in the rat. *Int. J. Sport Nutr. Exerc. Metab.* **2004**, *14*, 323–332. [CrossRef]
5. Kaneko, K.; Soty, M.; Zitoun, C.; Duchampt, A.; Silva, M.; Philippe, E.; Gautier-Stein, A.; Rajas, F.; Mithieux, G. The role of kidney in the inter-organ coordination of endogenous glucose production during fasting. *Mol. Metab.* **2018**, *16*, 203–212. [CrossRef]
6. Sharma, R.; Tiwari, S. Renal gluconeogenesis in insulin resistance: A culprit for hyperglycemia in diabetes. *World J. Diabetes* **2021**, *12*, 556–568. [CrossRef] [PubMed]
7. Lieberman, M.; Marks, A.D. *Marks' Basic Medical Biochemistry: A Clinical Approach*, 4th ed.; Wolters Kluwer/Lippincott Williams & Wilkins: Philadelphia, PA, USA, 2013.
8. Molitoris, B.A.; Marrs, J. The role of cell adhesion molecules in ischemic acute renal failure. *Am. J. Med.* **1999**, *106*, 583–592. [CrossRef]
9. Scholz, H.; Boivin, F.J.; Schmidt-Ott, K.M.; Bachmann, S.; Eckardt, K.U.; Scholl, U.I.; Persson, P.B. Kidney physiology and susceptibility to acute kidney injury: Implications for renoprotection. *Nat. Rev. Nephrol.* **2021**, *17*, 335–349. [CrossRef]
10. Zhuang, S.; Lu, B.; Daubert, R.A.; Chavin, K.D.; Wang, L.; Schnellmann, R.G. Suramin promotes recovery from renal ischemia/reperfusion injury in mice. *Kidney Int.* **2009**, *75*, 304–311. [CrossRef] [PubMed]
11. Tran, M.; Tam, D.; Bardia, A.; Bhasin, M.; Rowe, G.C.; Kher, A.; Zsengeller, Z.K.; Akhavan-Sharif, M.R.; Khankin, E.V.; Saintgeniez, M.; et al. Pgc-1alpha promotes recovery after acute kidney injury during systemic inflammation in mice. *J. Clin. Investig.* **2011**, *121*, 4003–4014. [CrossRef]
12. Dare, A.J.; Bolton, E.A.; Pettigrew, G.J.; Bradley, J.A.; Saeb-Parsy, K.; Murphy, M.P. Protection against renal ischemia-reperfusion injury in vivo by the mitochondria targeted antioxidant mitoq. *Redox Biol.* **2015**, *5*, 163–168. [CrossRef]
13. Perazella, M.A. Renal vulnerability to drug toxicity. *Clin. J. Am. Soc. Nephrol.* **2009**, *4*, 1275–1283. [CrossRef]
14. Cohen, A.; Ioannidis, K.; Ehrlich, A.; Regenbaum, S.; Cohen, M.; Ayyash, M.; Tikva, S.S.; Nahmias, Y. Mechanism and reversal of drug-induced nephrotoxicity on a chip. *Sci. Transl. Med.* **2021**, *13*, eabd6299. [CrossRef]
15. Hosohata, K. Role of oxidative stress in drug-induced kidney injury. *Int. J. Mol. Sci.* **2016**, *17*, 1826. [CrossRef] [PubMed]
16. Osman, A.T.; Sharkawi, S.M.Z.; Hassan, M.I.A.; Abo-Youssef, A.M.; Hemeida, R.A.M. Empagli fl ozin and neohesperidin protect against methotrexate-induced renal toxicity via suppression of oxidative stress and inflammation in male rats. *Food Chem. Toxicol.* **2021**, *155*, 112406. [CrossRef] [PubMed]
17. Haider, L.; Sharif, S.; Hasan, A.; McFarlane, I.M. Low-dose methotrexate toxicity in the setting of vancomycin-induced acute kidney injury. *Am. J. Med. Case Rep.* **2020**, *8*, 206–209. [CrossRef] [PubMed]
18. Hall, A.M.; Trepiccione, F.; Unwin, R.J. Drug toxicity in the proximal tubule: New models, methods and mechanisms. *Pediatr. Nephrol.* **2021**. [CrossRef] [PubMed]
19. Wen, J.; Zeng, M.; Shu, Y.; Guo, D.; Sun, Y.; Guo, Z.; Wang, Y.; Liu, Z.; Zhou, H.; Zhang, W. Aging increases the susceptibility of cisplatin-induced nephrotoxicity. *Age* **2015**, *37*, 112. [CrossRef]
20. Lash, L.H. Diverse roles of mitochondria in renal injury from environmental toxicants and therapeutic drugs. *Int. J. Mol. Sci.* **2021**, *22*, 4172. [CrossRef]
21. Bharavi, K.; Reddy, A.G.; Rao, G.S.; Reddy, A.R.; Rao, S.V. Reversal of cadmium-induced oxidative stress in chicken by herbal adaptogens withania somnifera and ocimum sanctum. *Toxicol. Int.* **2010**, *17*, 59–63. [CrossRef]
22. Kataria, A.; Trasande, L.; Trachtman, H. The effects of environmental chemicals on renal function. *Nat. Rev. Nephrol.* **2015**, *11*, 610–625. [CrossRef]
23. Johri, N.; Jacquillet, G.; Unwin, R. Heavy metal poisoning: The effects of cadmium on the kidney. *Biometals* **2010**, *23*, 783–792. [CrossRef]

24. Vervaet, B.A.; D'Haese, P.C.; Verhulst, A. Environmental toxin-induced acute kidney injury. *Clin. Kidney J.* **2017**, *10*, 747–758. [CrossRef]
25. Rizwan, S.; Naqshbandi, A.; Farooqui, Z.; Khan, A.A.; Khan, F. Protective effect of dietary flaxseed oil on arsenic-induced nephrotoxicity and oxidative damage in rat kidney. *Food Chem. Toxicol.* **2014**, *68*, 99–107. [CrossRef]
26. Kimura, A.; Ishida, Y.; Hayashi, T.; Wada, T.; Yokoyama, H.; Sugaya, T.; Mukaida, N.; Kondo, T. Interferon-gamma plays protective roles in sodium arsenite-induced renal injury by up-regulating intrarenal multidrug resistance-associated protein 1 expression. *Am. J. Pathol.* **2006**, *169*, 1118–1128. [CrossRef] [PubMed]
27. Barregard, L.; Bergstrom, G.; Fagerberg, B. Cadmium, type 2 diabetes, and kidney damage in a cohort of middle-aged women. *Environ. Res.* **2014**, *135*, 311–316. [CrossRef] [PubMed]
28. Gangadhariah, M.H.; Luther, J.M.; Garcia, V.; Paueksakon, P.; Zhang, M.Z.; Hayward, S.W.; Love, H.D.; Falck, J.R.; Manthati, V.L.; Imig, J.D.; et al. Hypertension is a major contributor to 20-hydroxyeicosatetraenoic acid-mediated kidney injury in diabetic nephropathy. *J. Am. Soc. Nephrol.* **2015**, *26*, 597–610. [CrossRef] [PubMed]
29. Wang, Z.; do Carmo, J.M.; da Silva, A.A.; Fu, Y.; Hall, J.E. Mechanisms of synergistic interactions of diabetes and hypertension in chronic kidney disease: Role of mitochondrial dysfunction and er stress. *Curr. Hypertens Rep.* **2020**, *22*, 15. [CrossRef] [PubMed]
30. Ware, K.; Yildiz, V.; Xiao, M.; Medipally, A.; Hemminger, J.; Scarl, R.; Satoskar, A.A.; Hebert, L.; Ivanov, I.; Biederman, L.; et al. Hypertension and the kidney: Reduced kidney mass is bad for both normotensive and hypertensive rats. *Am. J. Hypertens* **2021**. [CrossRef] [PubMed]
31. Narsipur, S.S.; Peterson, O.W.; Smith, R.; Bigby, T.D.; Parthasarathy, S.; Gabbai, F.B.; Wilson, C.B.; Blantz, R.C. Mechanisms of glomerular immune injury: Effects of antioxidant treatment. *J. Am. Soc. Nephrol.* **2003**, *14*, 1748–1755. [CrossRef]
32. Duann, P.; Lianos, E.A. Mechanisms of ho-1 mediated attenuation of renal immune injury: A gene profiling study. *Transl. Res.* **2011**, *158*, 249–261. [CrossRef]
33. Persson, F.; Rossing, P. Diagnosis of diabetic kidney disease: State of the art and future perspective. *Kidney Int. Suppl.* **2018**, *8*, 2–7. [CrossRef]
34. Hallan, S.; Sharma, K. The role of mitochondria in diabetic kidney disease. *Curr. Diabetes Rep.* **2016**, *16*, 61. [CrossRef] [PubMed]
35. Wei, P.Z.; Szeto, C.C. Mitochondrial dysfunction in diabetic kidney disease. *Clin. Chim. Acta* **2019**, *496*, 108–116. [CrossRef]
36. Yan, L.J. Nadh/nad(+) redox imbalance and diabetic kidney disease. *Biomolecules* **2021**, *11*, 730. [CrossRef]
37. Orr, S.E.; Bridges, C.C. Chronic kidney disease and exposure to nephrotoxic metals. *Int. J. Mol. Sci.* **2017**, *18*, 1039. [CrossRef]
38. Edwards, J.R.; Prozialeck, W.C. Cadmium, diabetes and chronic kidney disease. *Toxicol. Appl. Pharmacol.* **2009**, *238*, 289–293. [CrossRef]
39. Almeer, R.S.; AlBasher, G.I.; Alarifi, S.; Alkahtani, S.; Ali, D.; Abdel Moneim, A.E. Royal jelly attenuates cadmium-induced nephrotoxicity in male mice. *Sci. Rep.* **2019**, *9*, 5825. [CrossRef]
40. Almeer, R.S.; Kassab, R.B.; AlBasher, G.I.; Alarifi, S.; Alkahtani, S.; Ali, D.; Abdel Moneim, A.E. Royal jelly mitigates cadmium-induced neuronal damage in mouse cortex. *Mol. Biol. Rep.* **2019**, *46*, 119–131. [CrossRef] [PubMed]
41. Aurisano, N.; Huang, L.; Mila, I.C.L.; Jolliet, O.; Fantke, P. Chemicals of concern in plastic toys. *Environ. Int.* **2021**, *146*, 106194. [CrossRef] [PubMed]
42. Chen, C.; Han, X.; Wang, G.; Liu, D.; Bao, L.; Jiao, C.; Luan, J.; Hou, Y.; Xu, Y.; Wang, H.; et al. Nrf2 deficiency aggravates the kidney injury induced by subacute cadmium exposure in mice. *Arch. Toxicol.* **2021**, *95*, 883–893. [CrossRef] [PubMed]
43. Satarug, S.; Nishijo, M.; Lasker, J.M.; Edwards, R.J.; Moore, M.R. Kidney dysfunction and hypertension: Role for cadmium, p450 and heme oxygenases? *Tohoku J. Exp. Med.* **2006**, *208*, 179–202. [CrossRef] [PubMed]
44. Hao, R.; Song, X.; Sun-Waterhouse, D.; Tan, X.; Li, F.; Li, D. Mir-34a/sirt1/p53 signaling pathway contributes to cadmium-induced nephrotoxicity: A preclinical study in mice. *Environ. Pollut* **2021**, *282*, 117029. [CrossRef]
45. Kim, K.S.; Lim, H.J.; Lim, J.S.; Son, J.Y.; Lee, J.; Lee, B.M.; Chang, S.C.; Kim, H.S. Curcumin ameliorates cadmium-induced nephrotoxicity in sprague-dawley rats. *Food Chem. Toxicol.* **2018**, *114*, 34–40. [CrossRef]
46. Tripathi, S.; Srivastav, A.K. Cytoarchitectural alterations in kidney of wistar rat after oral exposure to cadmium chloride. *Tissue Cell* **2011**, *43*, 131–136. [CrossRef]
47. Rikans, L.E.; Yamano, T. Mechanisms of cadmium-mediated acute hepatotoxicity. *J. Biochem. Mol. Toxicol.* **2000**, *14*, 110–117. [CrossRef]
48. Kim, S.C.; Cho, M.K.; Kim, S.G. Cadmium-induced non-apoptotic cell death mediated by oxidative stress under the condition of sulfhydryl deficiency. *Toxicol. Lett.* **2003**, *144*, 325–336. [CrossRef]
49. Godt, J.; Scheidig, F.; Grosse-Siestrup, C.; Esche, V.; Brandenburg, P.; Reich, A.; Groneberg, D.A. The toxicity of cadmium and resulting hazards for human health. *J. Occup. Med. Toxicol.* **2006**, *1*, 22. [CrossRef] [PubMed]
50. Prozialeck, W.C.; Edwards, J.R. Mechanisms of cadmium-induced proximal tubule injury: New insights with implications for biomonitoring and therapeutic interventions. *J. Pharmacol. Exp. Ther.* **2012**, *343*, 2–12. [CrossRef] [PubMed]
51. Prozialeck, W.C.; Vaidya, V.S.; Liu, J.; Waalkes, M.P.; Edwards, J.R.; Lamar, P.C.; Bernard, A.M.; Dumont, X.; Bonventre, J.V. Kidney injury molecule-1 is an early biomarker of cadmium nephrotoxicity. *Kidney Int.* **2007**, *72*, 985–993. [CrossRef]
52. Shaikh, Z.A.; Lucis, O.J. Cadmium and zinc binding in mammalian liver and kidneys. *Arch. Environ. Health* **1972**, *24*, 419–425. [CrossRef]
53. Shaikh, Z.A.; Lucis, O.J. Biological differences in cadmium and zinc turnover. *Arch. Environ. Health* **1972**, *24*, 410–418. [CrossRef]

54. Liu, J.; Liu, Y.; Habeebu, S.S.; Klaassen, C.D. Susceptibility of mt-null mice to chronic cdcl2-induced nephrotoxicity indicates that renal injury is not mediated by the cdmt complex. *Toxicol. Sci.* **1998**, *46*, 197–203.
55. Klaassen, C.D.; Liu, J. Role of metallothionein in cadmium-induced hepatotoxicity and nephrotoxicity. *Drug Metab. Rev.* **1997**, *29*, 79–102. [CrossRef]
56. Nomiyama, K.; Nomiyama, H. Tissue metallothioneins in rabbits chronically exposed to cadmium, with special reference to the critical concentration of cadmium in the renal cortex. *Dev. Toxicol. Environ. Sci.* **1982**, *9*, 47–67. [PubMed]
57. Nomiyama, K.; Nomiyama, H. Critical concentration of 'unbound' cadmium in the rabbit renal cortex. *Experientia* **1986**, *42*, 149. [CrossRef] [PubMed]
58. Prozialeck, W.C.; Edwards, J.R.; Lamar, P.C.; Liu, J.; Vaidya, V.S.; Bonventre, J.V. Expression of kidney injury molecule-1 (kim-1) in relation to necrosis and apoptosis during the early stages of cd-induced proximal tubule injury. *Toxicol. Appl. Pharmacol.* **2009**, *238*, 306–314. [CrossRef] [PubMed]
59. Fels, J.; Scharner, B.; Zarbock, R.; Zavala Guevara, I.P.; Lee, W.K.; Barbier, O.C.; Thevenod, F. Cadmium complexed with beta2-microglubulin, albumin and lipocalin-2 rather than metallothionein cause megalin:Cubilin dependent toxicity of the renal proximal tubule. *Int. J. Mol. Sci.* **2019**, *20*, 2379. [CrossRef] [PubMed]
60. Nair, A.R.; Lee, W.K.; Smeets, K.; Swennen, Q.; Sanchez, A.; Thevenod, F.; Cuypers, A. Glutathione and mitochondria determine acute defense responses and adaptive processes in cadmium-induced oxidative stress and toxicity of the kidney. *Arch. Toxicol.* **2015**, *89*, 2273–2289. [CrossRef]
61. Bagchi, D.; Bagchi, M.; Hassoun, E.A.; Stohs, S.J. Cadmium-induced excretion of urinary lipid metabolites, DNA damage, glutathione depletion, and hepatic lipid peroxidation in sprague-dawley rats. *Biol. Trace Elem. Res.* **1996**, *52*, 143–154. [CrossRef]
62. Hassoun, E.A.; Stohs, S.J. Cadmium-induced production of superoxide anion and nitric oxide, DNA single strand breaks and lactate dehydrogenase leakage in j774a.1 cell cultures. *Toxicology* **1996**, *112*, 219–226. [CrossRef]
63. Sotomayor, C.G.; Groothof, D.; Vodegel, J.J.; Eisenga, M.F.; Knobbe, T.J.; IJmker, J.; Lammerts, R.G.M.; de Borst, M.H.; Berger, S.P.; Nolte, I.M.; et al. Plasma cadmium is associated with increased risk of long-term kidney graft failure. *Kidney Int.* **2021**, *99*, 1213–1224. [CrossRef]
64. Fransson, M.N.; Barregard, L.; Sallsten, G.; Akerstrom, M.; Johanson, G. Physiologically-based toxicokinetic model for cadmium using markov-chain monte carlo analysis of concentrations in blood, urine, and kidney cortex from living kidney donors. *Toxicol. Sci.* **2014**, *141*, 365–376. [CrossRef]
65. Jarup, L. Cadmium overload and toxicity. *Nephrol. Dial. Transplant.* **2002**, *17*, 35–39. [CrossRef]
66. Jarup, L.; Persson, B.; Elinder, C.G. Decreased glomerular filtration rate in solderers exposed to cadmium. *Occup. Environ. Med.* **1995**, *52*, 818–822. [CrossRef]
67. Kazantzis, G. Renal tubular dysfunction and abnormalities of calcium metabolism in cadmium workers. *Environ. Health Perspect.* **1979**, *28*, 155–159. [CrossRef]
68. Chen, S.; Liu, G.; Long, M.; Zou, H.; Cui, H. Alpha lipoic acid attenuates cadmium-induced nephrotoxicity via the mitochondrial apoptotic pathways in rat. *J. Inorg. Biochem.* **2018**, *184*, 19–26. [CrossRef]
69. Prozialeck, W.C.; Edwards, J.R.; Vaidya, V.S.; Bonventre, J.V. Preclinical evaluation of novel urinary biomarkers of cadmium nephrotoxicity. *Toxicol. Appl. Pharmacol.* **2009**, *238*, 301–305. [CrossRef]
70. Sarma, S.N.; Saleem, A.; Lee, J.Y.; Tokumoto, M.; Hwang, G.W.; Man Chan, H.; Satoh, M. Effects of long-term cadmium exposure on urinary metabolite profiles in mice. *J. Toxicol. Sci.* **2018**, *43*, 89–100. [CrossRef] [PubMed]
71. Porter, G.A. Risk factors for toxic nephropathies. *Toxicol. Lett.* **1989**, *46*, 269–279. [CrossRef]
72. Nath, R.; Prasad, R.; Palinal, V.K.; Chopra, R.K. Molecular basis of cadmium toxicity. *Prog. Food Nutr. Sci.* **1984**, *8*, 109–163.
73. Gong, P.; Wang, M.; Yang, W.; Chang, X.; Wang, L.; Chen, F. Integrated metabolomics coupled with pattern recognition and pathway analysis to reveal molecular mechanism of cadmium-induced diabetic nephropathy. *Toxicol. Res.* **2021**, *10*, 777–791. [CrossRef]
74. Gong, P.; Chang, X.; Chen, X.; Bai, X.; Wen, H.; Pi, S.; Yang, W.; Wang, L.; Chen, F. Metabolomics study of cadmium-induced diabetic nephropathy and protective effect of caffeic acid phenethyl ester using uplc-q-tof-ms combined with pattern recognition. *Environ. Toxicol. Pharmacol.* **2017**, *54*, 80–92. [CrossRef] [PubMed]
75. Zhou, B.; Gentry, A.; Xu, Q.; Young, J.L.; Yan, X.; Pagidas, K.; Yang, Y.; Watson, W.H.; Kong, M.; Cai, L.; et al. Effects of cadmium and high-fat diet on essential metal concentration in the mouse testis. *Toxicol. Rep.* **2021**, *8*, 718–723. [CrossRef] [PubMed]
76. Hagedoorn, I.J.M.; Gant, C.M.; Huizen, S.V.; Maatman, R.; Navis, G.; Bakker, S.J.L.; Laverman, G.D. Lifestyle-related exposure to cadmium and lead is associated with diabetic kidney disease. *J. Clin. Med.* **2020**, *9*, 2432. [CrossRef] [PubMed]
77. Shah, S.; Iqbal, M.; Karam, J.; Salifu, M.; McFarlane, S.I. Oxidative stress, glucose metabolism, and the prevention of type 2 diabetes: Pathophysiological insights. *Antioxid. Redox Signal.* **2007**, *9*, 911–929. [CrossRef] [PubMed]
78. Sanjeev, S.; Bidanchi, R.M.; Murthy, M.K.; Gurusubramanian, G.; Roy, V.K. Influence of ferulic acid consumption in ameliorating the cadmium-induced liver and renal oxidative damage in rats. *Environ. Sci Pollut Res. Int.* **2019**, *26*, 20631–20653. [CrossRef] [PubMed]
79. Zhang, H.; Reynolds, M. Cadmium exposure in living organisms: A short review. *Sci. Total Environ.* **2019**, *678*, 761–767. [CrossRef]
80. Cannino, G.; Ferruggia, E.; Luparello, C.; Rinaldi, A.M. Cadmium and mitochondria. *Mitochondrion* **2009**, *9*, 377–384. [CrossRef]
81. Genchi, G.; Sinicropi, M.S.; Lauria, G.; Carocci, A.; Catalano, A. The effects of cadmium toxicity. *Int. J. Environ. Res. Public Health* **2020**, *17*, 3782. [CrossRef] [PubMed]

82. Murphy, M.P. How mitochondria produce reactive oxygen species. *Biochem. J.* **2009**, *417*, 1–13. [CrossRef]
83. Rosca, M.G.; Vazquez, E.J.; Chen, Q.; Kerner, J.; Kern, T.S.; Hoppel, C.L. Oxidation of fatty acids is the source of increased mitochondrial reactive oxygen species production in kidney cortical tubules in early diabetes. *Diabetes* **2012**, *61*, 2074–2083. [CrossRef]
84. Sena, L.A.; Chandel, N.S. Physiological roles of mitochondrial reactive oxygen species. *Mol. Cell* **2012**, *48*, 158–167. [CrossRef]
85. Bhargava, P.; Schnellmann, R.G. Mitochondrial energetics in the kidney. *Nat. Rev. Nephrol.* **2017**, *13*, 629–646. [CrossRef] [PubMed]
86. Trewin, A.J.; Bahr, L.L.; Almast, A.; Berry, B.J.; Wei, A.Y.; Foster, T.H.; Wojtovich, A.P. Mitochondrial reactive oxygen species generated at the complex-ii matrix or intermembrane space microdomain have distinct effects on redox signaling and stress sensitivity in caenorhabditis elegans. *Antioxid. Redox Signal.* **2019**, *31*, 594–607. [CrossRef]
87. Ralph, S.J.; Moreno-Sanchez, R.; Neuzil, J.; Rodriguez-Enriquez, S. Inhibitors of succinate: Quinone reductase/complex ii regulate production of mitochondrial reactive oxygen species and protect normal cells from ischemic damage but induce specific cancer cell death. *Pharm. Res.* **2011**, *28*, 2695–2730. [CrossRef] [PubMed]
88. Brand, M.D. Riding the tiger—Physiological and pathological effects of superoxide and hydrogen peroxide generated in the mitochondrial matrix. *Crit. Rev. Biochem. Mol. Biol.* **2020**, *55*, 592–661. [CrossRef]
89. Brand, M.D.; Goncalves, R.L.; Orr, A.L.; Vargas, L.; Gerencser, A.A.; Borch Jensen, M.; Wang, Y.T.; Melov, S.; Turk, C.N.; Matzen, J.T.; et al. Suppressors of superoxide-h2o2 production at site iq of mitochondrial complex i protect against stem cell hyperplasia and ischemia-reperfusion injury. *Cell Metab.* **2016**, *24*, 582–592. [CrossRef]
90. Ames, B.N.; Shigenaga, M.K. Oxidants are a major contributor to aging. *Ann. N. Y. Acad. Sci.* **1992**, *663*, 85–96. [CrossRef]
91. Yan, L.J. Pathogenesis of chronic hyperglycemia: From reductive stress to oxidative stress. *J. Diabetes Res.* **2014**, *2014*, 137919. [CrossRef]
92. Salvemini, D.; Doyle, T.M.; Cuzzocrea, S. Superoxide, peroxynitrite and oxidative/nitrative stress in inflammation. *Biochem. Soc. Trans.* **2006**, *34*, 965–970. [CrossRef]
93. Kassab, A.; Piwowar, A. Cell oxidant stress delivery and cell dysfunction onset in type 2 diabetes. *Biochimie* **2012**, *94*, 1837–1848. [CrossRef] [PubMed]
94. Klivenyi, P.; Starkov, A.A.; Calingasan, N.Y.; Gardian, G.; Browne, S.E.; Yang, L.; Bubber, P.; Gibson, G.E.; Patel, M.S.; Beal, M.F. Mice deficient in dihydrolipoamide dehydrogenase show increased vulnerability to mptp, malonate and 3-nitropropionic acid neurotoxicity. *J. Neurochem.* **2004**, *88*, 1352–1360. [CrossRef] [PubMed]
95. Vaubel, R.A.; Rustin, P.; Isaya, G. Mutations in the dimer interface of dihydrolipoamide dehydrogenase promote site-specific oxidative damages in yeast and human cells. *J. Biol. Chem.* **2011**, *286*, 40232–40245. [CrossRef]
96. Ambrus, A.; Torocsik, B.; Tretter, L.; Ozohanics, O.; Adam-Vizi, V. Stimulation of reactive oxygen species generation by disease-causing mutations of lipoamide dehydrogenase. *Hum. Mol. Genet.* **2011**, *20*, 2984–2995. [CrossRef]
97. Bunik, V.I.; Brand, M.D. Generation of superoxide and hydrogen peroxide by side reactions of mitochondrial 2-oxoacid dehydrogenase complexes in isolation and in cells. *Biol. Chem.* **2018**, *399*, 407–420. [CrossRef]
98. Wang, Y.; Fang, J.; Leonard, S.S.; Rao, K.M. Cadmium inhibits the electron transfer chain and induces reactive oxygen species. *Free Radic. Biol. Med.* **2004**, *36*, 1434–1443. [CrossRef]
99. Sedeek, M.; Nasrallah, R.; Touyz, R.M.; Hebert, R.L. Nadph oxidases, reactive oxygen species, and the kidney: Friend and foe. *J. Am. Soc. Nephrol.* **2013**, *24*, 1512–1518. [CrossRef]
100. Suh, Y.A.; Arnold, R.S.; Lassegue, B.; Shi, J.; Xu, X.; Sorescu, D.; Chung, A.B.; Griendling, K.K.; Lambeth, J.D. Cell transformation by the superoxide-generating oxidase mox1. *Nature* **1999**, *401*, 79–82. [CrossRef] [PubMed]
101. Cheng, G.; Cao, Z.; Xu, X.; van Meir, E.G.; Lambeth, J.D. Homologs of gp91phox: Cloning and tissue expression of nox3, nox4, and nox5. *Gene* **2001**, *269*, 131–140. [CrossRef]
102. Gardiner, G.J.; Deffit, S.N.; McLetchie, S.; Perez, L.; Walline, C.C.; Blum, J.S. A role for nadph oxidase in antigen presentation. *Front. Immunol* **2013**, *4*, 295. [CrossRef]
103. Drummond, G.R.; Selemidis, S.; Griendling, K.K.; Sobey, C.G. Combating oxidative stress in vascular disease: Nadph oxidases as therapeutic targets. *Nat. Rev. Drug Discov.* **2011**, *10*, 453–471. [CrossRef] [PubMed]
104. Cho, S.; Yu, S.L.; Kang, J.; Jeong, B.Y.; Lee, H.Y.; Park, C.G.; Yu, Y.B.; Jin, D.C.; Hwang, W.M.; Yun, S.R.; et al. Nadph oxidase 4 mediates tgf-beta1/smad signaling pathway induced acute kidney injury in hypoxia. *PLoS ONE* **2019**, *14*, e0219483.
105. Liu, Q.; Liang, X.; Liang, M.; Qin, R.; Qin, F.; Wang, X. Ellagic acid ameliorates renal ischemic-reperfusion injury through nox4/jak/stat signaling pathway. *Inflammation* **2020**, *43*, 298–309. [CrossRef]
106. Lima, N.K.S.; Farias, W.R.A.; Cirilo, M.A.S.; Oliveira, A.G.; Farias, J.S.; Aires, R.S.; Muzi-Filho, H.; Paixao, A.D.O.; Vieira, L.D. Renal ischemia-reperfusion leads to hypertension and changes in proximal tubule na(+) transport and renin-angiotensin-aldosterone system: Role of nadph oxidase. *Life Sci.* **2021**, *266*, 118879. [CrossRef]
107. Pinheiro Junior, J.E.G.; Moraes, P.Z.; Rodriguez, M.D.; Simoes, M.R.; Cibin, F.; Pinton, S.; Barbosa Junior, F.; Pecanha, F.M.; Vassallo, D.V.; Miguel, M.; et al. Cadmium exposure activates nadph oxidase, renin-angiotensin system and cyclooxygenase 2 pathways in arteries, inducing hypertension and vascular damage. *Toxicol. Lett.* **2020**, *333*, 80–89. [CrossRef]
108. Asagba, S.O. Alteration in the activity of oxidative enzymes in the tissues of male wistar albino rats exposed to cadmium. *Int. J. Occup. Med. Environ. Health* **2010**, *23*, 55–62. [CrossRef] [PubMed]
109. Greene, E.L.; Paller, M.S. Xanthine oxidase produces o2-. In posthypoxic injury of renal epithelial cells. *Am. J. Physiol.* **1992**, *263*, F251–F255. [CrossRef] [PubMed]

110. Ratliff, B.B.; Abdulmahdi, W.; Pawar, R.; Wolin, M.S. Oxidant mechanisms in renal injury and disease. *Antioxid. Redox Signal.* **2016**, *25*, 119–146. [CrossRef]
111. Noris, M.; Remuzzi, G. Physiology and pathophysiology of nitric oxide in chronic renal disease. *Proc. Assoc. Am. Physicians* **1999**, *111*, 602–610. [CrossRef]
112. Soyupek, S.; Oksay, T.; Sutcu, R.; Armagan, A.; Gokalp, O.; Perk, H.; Delibas, N. The effect of cadmium toxicity on renal nitric oxide synthase isoenzymes. *Toxicol. Ind. Health* **2012**, *28*, 624–628. [CrossRef] [PubMed]
113. Pavon, N.; Buelna-Chontal, M.; Macias-Lopez, A.; Correa, F.; Uribe-Alvarez, C.; Hernandez-Esquivel, L.; Chavez, E. On the oxidative damage by cadmium to kidney mitochondrial functions. *Biochem. Cell Biol.* **2019**, *97*, 187–192. [CrossRef]
114. Thevenod, F.; Lee, W.K.; Garrick, M.D. Iron and cadmium entry into renal mitochondria: Physiological and toxicological implications. *Front. Cell Dev. Biol* **2020**, *8*, 848. [CrossRef]
115. Liu, Q.; Zhang, R.; Wang, X.; Shen, X.; Wang, P.; Sun, N.; Li, X.; Li, X.; Hai, C. Effects of sub-chronic, low-dose cadmium exposure on kidney damage and potential mechanisms. *Ann. Transl. Med.* **2019**, *7*, 177. [CrossRef]
116. Ghasemi, H.; Postampour, F.; Ranjbar, A. The role of oxidative stress in metals toxicity/mitochondrial dysfunction as a key player. *GMJ* **2014**, *3*, 2–13.
117. Barnett, L.M.A.; Cummings, B.S. Nephrotoxicity and renal pathophysiology: A contemporary perspective. *Toxicol. Sci.* **2018**, *164*, 379–390. [CrossRef] [PubMed]
118. Gobe, G.; Crane, D. Mitochondria, reactive oxygen species and cadmium toxicity in the kidney. *Toxicol. Lett.* **2010**, *198*, 49–55. [CrossRef]
119. Ge, J.; Zhang, C.; Sun, Y.C.; Zhang, Q.; Lv, M.W.; Guo, K.; Li, J.L. Cadmium exposure triggers mitochondrial dysfunction and oxidative stress in chicken (*Gallus gallus*) kidney via mitochondrial upr inhibition and nrf2-mediated antioxidant defense activation. *Sci. Total Environ.* **2019**, *689*, 1160–1171. [CrossRef] [PubMed]
120. Sandbichler, A.M.; Hockner, M. Cadmium protection strategies—A hidden trade-off? *Int. J. Mol. Sci* **2016**, *17*, 139. [CrossRef]
121. Li, X.; Zhang, H.; Sun, F. Cdse/zns quantum dots exhibited nephrotoxicity through mediating oxidative damage and inflammatory response. *Aging* **2020**, *13*, 12194–12206. [CrossRef]
122. Wang, X.Y.; Wang, Z.Y.; Zhu, Y.S.; Zhu, S.M.; Fan, R.F.; Wang, L. Alleviation of cadmium-induced oxidative stress by trehalose via inhibiting the nrf2-keap1 signaling pathway in primary rat proximal tubular cells. *J. Biochem. Mol. Toxicol.* **2018**, *32*, e22011. [CrossRef]
123. Ashrafizadeh, M.; Ahmadi, Z.; Farkhondeh, T.; Samarghandian, S. Back to nucleus: Combating with cadmium toxicity using nrf2 signaling pathway as a promising therapeutic target. *Biol. Trace Elem. Res.* **2020**, *197*, 52–62. [CrossRef] [PubMed]
124. Choudhury, C.; Mazumder, R.; Kumar, R.; Dhar, B.; Sengupta, M. Cadmium induced oxystress alters nrf2-keap1 signaling and triggers apoptosis in piscine head kidney macrophages. *Aquat. Toxicol.* **2021**, *231*, 105739. [CrossRef] [PubMed]
125. Dastan, D.; Karimi, S.; Larki-Harchegani, A.; Nili-Ahmadabadi, A. Protective effects of allium hirtifolium boiss extract on cadmium-induced renal failure in rats. *Environ. Sci. Pollut. Res. Int.* **2019**, *26*, 18886–18892. [CrossRef] [PubMed]
126. Handan, B.A.; De Moura, C.F.G.; Cardoso, C.M.; Santamarina, A.B.; Pisani, L.P.; Ribeiro, D.A. Protective effect of grape and apple juices against cadmium intoxication in the kidney of rats. *Drug Res.* **2020**, *70*, 503–511. [CrossRef] [PubMed]
127. Suliman Al-Gebaly, A. Ameliorative effect of arctium lappa against cadmium genotoxicity and histopathology in kidney of wistar rat. *Pak. J. Biol. Sci* **2017**, *20*, 314–319. [CrossRef] [PubMed]
128. Das, S.; Dewanjee, S.; Dua, T.K.; Joardar, S.; Chakraborty, P.; Bhowmick, S.; Saha, A.; Bhattacharjee, S.; De Feo, V. Carnosic acid attenuates cadmium induced nephrotoxicity by inhibiting oxidative stress, promoting nrf2/ho-1 signalling and impairing tgf-beta1/smad/collagen iv signalling. *Molecules* **2019**, *24*, 4176. [CrossRef] [PubMed]
129. Wongmekiat, O.; Peerapanyasut, W.; Kobroob, A. Catechin supplementation prevents kidney damage in rats repeatedly exposed to cadmium through mitochondrial protection. *Naunyn. Schmiedebergs Arch. Pharmacol.* **2018**, *391*, 385–394. [CrossRef] [PubMed]
130. Gong, P.; Chen, F.; Liu, X.; Gong, X.; Wang, J.; Ma, Y. Protective effect of caffeic acid phenethyl ester against cadmium-induced renal damage in mice. *J. Toxicol. Sci.* **2012**, *37*, 415–425. [CrossRef] [PubMed]
131. Kobroob, A.; Chattipakorn, N.; Wongmekiat, O. Caffeic acid phenethyl ester ameliorates cadmium-induced kidney mitochondrial injury. *Chem. Biol. Interact.* **2012**, *200*, 21–27. [CrossRef]
132. Koriem, K.M.; Arbid, M.S.; Asaad, G.F. Chelidonium majus leaves methanol extract and its chelidonine alkaloid ingredient reduce cadmium-induced nephrotoxicity in rats. *J. Nat. Med.* **2013**, *67*, 159–167. [CrossRef] [PubMed]
133. Senthilkumar, T.; Sangeetha, N.; Ashokkumar, N. Antihyperglycemic, antihyperlipidemic, and renoprotective effects of chlorella pyrenoidosa in diabetic rats exposed to cadmium. *Toxicol. Mech. Methods* **2012**, *22*, 617–624. [CrossRef]
134. Poontawee, W.; Natakankitkul, S.; Wongmekiat, O. Protective effect of *Cleistocalyx nervosum* var. Paniala fruit extract against oxidative renal damage caused by cadmium. *Molecules* **2016**, *21*, 133. [CrossRef]
135. Nishio, R.; Tamano, H.; Morioka, H.; Takeuchi, A.; Takeda, A. Intake of heated leaf extract of coriandrum sativum contributes to resistance to oxidative stress via decreases in heavy metal concentrations in the kidney. *Plant. Foods Hum. Nutr* **2019**, *74*, 204–209. [CrossRef]
136. Tubsakul, A.; Sangartit, W.; Pakdeechote, P.; Kukongviriyapan, V.; Apaijit, K.; Kukongviriyapan, U. Curcumin mitigates hypertension, endothelial dysfunction and oxidative stress in rats with chronic exposure to lead and cadmium. *Tohoku J. Exp. Med.* **2021**, *253*, 69–76. [CrossRef]

137. Fan, S.R.; Ren, T.T.; Yun, M.Y.; Lan, R.; Qin, X.Y. Edaravone attenuates cadmium-induced toxicity by inhibiting oxidative stress and inflammation in icr mice. *Neurotoxicology* **2021**, *86*, 1–9. [CrossRef]
138. Kopec, A.; Sikora, E.; Piatkowska, E.; Borczak, B.; Czech, T. Possible protective role of elderberry fruit lyophilizate against selected effects of cadmium and lead intoxication in wistar rats. *Environ. Sci. Pollut Res. Int.* **2016**, *23*, 8837–8848. [CrossRef] [PubMed]
139. Chen, J.; Du, L.; Li, J.; Song, H. Epigallocatechin-3-gallate attenuates cadmium-induced chronic renal injury and fibrosis. *Food Chem. Toxicol.* **2016**, *96*, 70–78. [CrossRef] [PubMed]
140. Liu, E.; Han, L.; Wang, J.; He, W.; Shang, H.; Gao, X.; Wang, T. Eucommia ulmoides bark protects against renal injury in cadmium-challenged rats. *J. Med. Food* **2012**, *15*, 307–314. [CrossRef] [PubMed]
141. Elkhadragy, M.F.; Al-Olayan, E.M.; Al-Amiery, A.A.; Abdel Moneim, A.E. Protective effects of fragaria ananassa extract against cadmium chloride-induced acute renal toxicity in rats. *Biol. Trace Elem. Res.* **2018**, *181*, 378–387. [CrossRef]
142. Onwuka, F.C.; Erhabor, O.; Eteng, M.U.; Umoh, I.B. Protective effects of ginger toward cadmium-induced testes and kidney lipid peroxidation and hematological impairment in albino rats. *J. Med. Food* **2011**, *14*, 817–821. [CrossRef]
143. Veljkovic, A.R.; Nikolic, R.S.; Kocic, G.M.; Pavlovic, D.D.; Cvetkovic, T.P.; Sokolovic, D.T.; Jevtovic, T.M.; Basic, J.T.; Laketic, D.M.; Marinkovic, M.R.; et al. Protective effects of glutathione and lipoic acid against cadmium-induced oxidative stress in rat's kidney. *Ren Fail.* **2012**, *34*, 1281–1287. [CrossRef]
144. Mohamed, N.E. Effect of aqueous extract of glycyrrhiza glabra on the biochemical changes induced by cadmium chloride in rats. *Biol. Trace Elem. Res.* **2019**, *190*, 87–94. [CrossRef]
145. Chen, Q.; Zhang, R.; Li, W.M.; Niu, Y.J.; Guo, H.C.; Liu, X.H.; Hou, Y.C.; Zhao, L.J. The protective effect of grape seed procyanidin extract against cadmium-induced renal oxidative damage in mice. *Environ. Toxicol. Pharmacol.* **2013**, *36*, 759–768. [CrossRef]
146. Claudio, S.R.; Pidone Ribeiro, F.A.; De Lima, E.C.; Santamarina, A.B.; Pisani, L.P.; Pereira, C.S.D.; Fujiyama Oshima, C.T.; Ribeiro, D.A. The protective effect of grape skin or purple carrot extracts against cadmium intoxication in kidney of rats. *Pathophysiology* **2019**, *26*, 263–269. [CrossRef] [PubMed]
147. Winiarska-Mieczan, A. The potential protective effect of green, black, red and white tea infusions against adverse effect of cadmium and lead during chronic exposure—A rat model study. *Regul. Toxicol. Pharmacol.* **2015**, *73*, 521–529. [CrossRef]
148. Ranieri, M.; Di Mise, A.; Difonzo, G.; Centrone, M.; Venneri, M.; Pellegrino, T.; Russo, A.; Mastrodonato, M.; Caponio, F.; Valenti, G.; et al. Green olive leaf extract (ole) provides cytoprotection in renal cells exposed to low doses of cadmium. *PLoS ONE* **2019**, *14*, e0214159. [CrossRef] [PubMed]
149. Dua, T.K.; Dewanjee, S.; Khanra, R.; Bhattacharya, N.; Bhaskar, B.; Zia-Ul-Haq, M.; De Feo, V. The effects of two common edible herbs, ipomoea aquatica and enhydra fluctuans, on cadmium-induced pathophysiology: A focus on oxidative defence and anti-apoptotic mechanism. *J. Transl. Med.* **2015**, *13*, 245. [CrossRef] [PubMed]
150. Ojo, O.A.; Ajiboye, B.O.; Oyinloye, B.E.; Ojo, A.B.; Olarewaju, O.I. Protective effect of irvingia gabonensis stem bark extract on cadmium-induced nephrotoxicity in rats. *Interdiscip. Toxicol.* **2014**, *7*, 208–214. [CrossRef]
151. Aksoy, N.; Dogan, Y.; Iriadam, M.; Bitiren, M.; Uzer, E.; Ozgonul, A.; Aksoy, S. Protective and therapeutic effects of licorice in rats with acute tubular necrosis. *J. Ren. Nutr.* **2012**, *22*, 336–343. [CrossRef]
152. Lan, Z.; Bi, K.S.; Chen, X.H. Ligustrazine attenuates elevated levels of indoxyl sulfate, kidney injury molecule-1 and clusterin in rats exposed to cadmium. *Food Chem Toxicol.* **2014**, *63*, 62–68. [CrossRef]
153. Luo, T.; Liu, G.; Long, M.; Yang, J.; Song, R.; Wang, Y.; Yuan, Y.; Bian, J.; Liu, X.; Gu, J.; et al. Treatment of cadmium-induced renal oxidative damage in rats by administration of alpha-lipoic acid. *Environ. Sci. Pollut. Res. Int.* **2017**, *24*, 1832–1844. [CrossRef]
154. Suru, S.M. Onion and garlic extracts lessen cadmium-induced nephrotoxicity in rats. *Biometals* **2008**, *21*, 623–633. [CrossRef]
155. Shati, A.A. Effects of origanum majorana l. On cadmium induced hepatotoxicity and nephrotoxicity in albino rats. *Saudi Med. J.* **2011**, *32*, 797–805.
156. Osukoya, O.A.; Oyinloye, B.E.; Ajiboye, B.O.; Olokode, K.A.; Adeola, H.A. Nephroprotective and anti-inflammatory potential of aqueous extract from persea americana seeds against cadmium-induced nephrotoxicity in wistar rats. *Biometals* **2021**, *34*, 1141–1153. [CrossRef]
157. Dkhil, M.A.; Al-Quraishy, S.; Diab, M.M.; Othman, M.S.; Aref, A.M.; Abdel Moneim, A.E. The potential protective role of physalis peruviana l. Fruit in cadmium-induced hepatotoxicity and nephrotoxicity. *Food Chem Toxicol.* **2014**, *74*, 98–106. [CrossRef]
158. Yadav, N.; Khandelwal, S. Effect of picroliv on cadmium-induced hepatic and renal damage in the rat. *Hum. Exp. Toxicol.* **2006**, *25*, 581–591. [CrossRef]
159. Jung, H.Y.; Seo, D.W.; Hong, C.O.; Kim, J.Y.; Yang, S.Y.; Lee, K.W. Nephroprotection of plantamajoside in rats treated with cadmium. *Environ. Toxicol. Pharmacol.* **2015**, *39*, 125–136. [CrossRef]
160. Dkhil, M.A.; Diab, M.S.M.; Lokman, M.S.; El-Sayed, H.; Bauomy, A.A.; Al-Shaebi, E.M.; Al-Quraishy, S. Nephroprotective effect of pleurotus ostreatus extract against cadmium chloride toxicity in rats. *An. Acad. Bras. Cienc.* **2020**, *92*, e20191121. [CrossRef] [PubMed]
161. Shen, R.; Liu, D.; Hou, C.; Liu, D.; Zhao, L.; Cheng, J.; Wang, D.; Bai, D. Protective effect of potentilla anserina polysaccharide on cadmium-induced nephrotoxicity in vitro and in vivo. *Food Funct.* **2017**, *8*, 3636–3646. [CrossRef]
162. Song, X.B.; Liu, G.; Wang, Z.Y.; Wang, L. Puerarin protects against cadmium-induced proximal tubular cell apoptosis by restoring mitochondrial function. *Chem. Biol. Interact.* **2016**, *260*, 219–231. [CrossRef] [PubMed]

163. Alshammari, G.M.; Al-Qahtani, W.H.; AlFaris, N.A.; Albekairi, N.A.; Alqahtani, S.; Eid, R.; Yagoub, A.E.A.; Al-Harbi, L.N.; Yahya, M.A. Quercetin alleviates cadmium chloride-induced renal damage in rats by suppressing endoplasmic reticulum stress through sirt1-dependent deacetylation of xbp-1s and eif2alpha. *Biomed. Pharmacother* **2021**, *141*, 111862. [CrossRef]
164. Zhang, Q.; Zhang, C.; Ge, J.; Lv, M.W.; Talukder, M.; Guo, K.; Li, Y.H.; Li, J.L. Ameliorative effects of resveratrol against cadmium-induced nephrotoxicity via modulating nuclear xenobiotic receptor response and pink1/parkin-mediated mitophagy. *Food Funct.* **2020**, *11*, 1856–1868. [CrossRef] [PubMed]
165. Ansari, M.N.; Aloliet, R.I.; Ganaie, M.A.; Khan, T.H.; Najeeb Ur, R.; Imam, F.; Hamad, A.M. Roflumilast, a phosphodiesterase 4 inhibitor, attenuates cadmium-induced renal toxicity via modulation of nf-kappab activation and induction of nqo1 in rats. *Hum. Exp. Toxicol.* **2019**, *38*, 588–597. [CrossRef]
166. Joardar, S.; Dewanjee, S.; Bhowmick, S.; Dua, T.K.; Das, S.; Saha, A.; De Feo, V. Rosmarinic acid attenuates cadmium-induced nephrotoxicity via inhibition of oxidative stress, apoptosis, inflammation and fibrosis. *Int. J. Mol. Sci.* **2019**, *20*, 2027. [CrossRef]
167. Abarikwu, S.O.; Njoku, R.C.; Lawrence, C.J.; Charles, I.A.; Ikewuchi, J.C. Rutin ameliorates oxidative stress and preserves hepatic and renal functions following exposure to cadmium and ethanol. *Pharm. Biol.* **2017**, *55*, 2161–2169. [CrossRef] [PubMed]
168. Rashwan, H.M.; Mohammed, H.E.; El-Nekeety, A.A.; Hamza, Z.K.; Abdel-Aziem, S.H.; Hassan, N.S.; Abdel-Wahhab, M.A. Bioactive phytochemicals from salvia officinalis attenuate cadmium-induced oxidative damage and genotoxicity in rats. *Environ. Sci. Pollut Res. Int.* **2021**. [CrossRef]
169. He, L.; Zhang, Q.Q.; Lu, X.Y.; Li, X.X. [effects of water extract of salvia miltiorrhiza aganst renal injury on rats exposed to cadmium]. *Zhonghua Yi Xue Za Zhi* **2017**, *97*, 57–61. [PubMed]
170. Albasher, G.; Albrahim, T.; Aljarba, N.; Alharbi, R.I.; Alsultan, N.; Alsaiari, J.; Rizwana, H. Involvement of redox status and the nuclear-related factor 2 in protecting against cadmium-induced renal injury with sana makki (cassia senna l.) pre-treatment in male rats. *An. Acad. Bras. Cienc.* **2020**, *92*, e20191237. [CrossRef] [PubMed]
171. Chen, H.; Li, P.; Shen, Z.; Wang, J.; Diao, L. Protective effects of selenium yeast against cadmium-induced necroptosis through mir-26a-5p/pten/pi3k/akt signaling pathway in chicken kidney. *Ecotoxicol. Environ. Saf.* **2021**, *220*, 112387. [CrossRef]
172. Bas, H.; Apaydin, F.G.; Kalender, S.; Kalender, Y. Lead nitrate and cadmium chloride induced hepatotoxicity and nephrotoxicity: Protective effects of sesamol on biochemical indices and pathological changes. *J. Food Biochem.* **2021**, *45*, e13769. [CrossRef]
173. Ansari, M.A.; Raish, M.; Ahmad, A.; Alkharfy, K.M.; Ahmad, S.F.; Attia, S.M.; Alsaad, A.M.S.; Bakheet, S.A. Sinapic acid ameliorate cadmium-induced nephrotoxicity: In vivo possible involvement of oxidative stress, apoptosis, and inflammation via nf-kappab downregulation. *Environ. Toxicol. Pharmacol.* **2017**, *51*, 100–107. [CrossRef]
174. Ramamurthy, C.H.; Subastri, A.; Suyavaran, A.; Subbaiah, K.C.; Valluru, L.; Thirunavukkarasu, C. Solanum torvum swartz. Fruit attenuates cadmium-induced liver and kidney damage through modulation of oxidative stress and glycosylation. *Environ. Sci. Pollut. Res. Int.* **2016**, *23*, 7919–7929. [CrossRef]
175. Mzoughi, Z.; Souid, G.; Timoumi, R.; Le Cerf, D.; Majdoub, H. Partial characterization of the edible spinacia oleracea polysaccharides: Cytoprotective and antioxidant potentials against cd induced toxicity in hct116 and hek293 cells. *Int. J. Biol. Macromol.* **2019**, *136*, 332–340. [CrossRef]
176. Fouad, A.A.; Jresat, I. Protective effect of telmisartan against cadmium-induced nephrotoxicity in mice. *Life Sci.* **2011**, *89*, 29–35. [CrossRef]
177. Mostafa, D.G.; Ahmed, S.F.; Hussein, O.A. Protective effect of tetrahydrobiopterin on hepatic and renal damage after acute cadmium exposure in male rats. *Ultrastruct. Pathol.* **2018**, *42*, 516–531. [CrossRef] [PubMed]
178. Junsi, M.; Takahashi Yupanqui, C.; Usawakesmanee, W.; Slusarenko, A.; Siripongvutikorn, S. Thunbergia laurifolia leaf extract increased levels of antioxidant enzymes and protected human cell-lines in vitro against cadmium. *Antioxidants* **2020**, *9*, 47. [CrossRef] [PubMed]
179. Ansari, M.N.; Rehman, N.U.; Karim, A.; Imam, F.; Hamad, A.M. Protective effect of thymus serrulatus essential oil on cadmium-induced nephrotoxicity in rats, through suppression of oxidative stress and downregulation of nf-kappab, inos, and smad2 mrna expression. *Molecules* **2021**, *26*, 1252. [CrossRef] [PubMed]
180. Fouad, A.A.; Alwadaani, H.A.; Jresat, I. Protective effect of thymoquinone against nephrotoxicity induced by cadmium in rats. *Intenrational Sch. Sci. Res. Innov.* **2016**, *10*, 73–76.
181. Padma, V.V.; Baskaran, R.; Divya, S.; Priya, L.B.; Saranya, S. Modulatory effect of tinospora cordifolia extract on cd-induced oxidative stress in wistar rats. *Integr. Med. Res.* **2016**, *5*, 48–55. [CrossRef]
182. Fan, R.F.; Li, Z.F.; Zhang, D.; Wang, Z.Y. Involvement of nrf2 and mitochondrial apoptotic signaling in trehalose protection against cadmium-induced kidney injury. *Metallomics* **2020**, *12*, 2098–2107. [CrossRef] [PubMed]
183. Lakshmi, G.D.; Kumar, P.R.; Bharavi, K.; Annapurna, P.; Rajendar, B.; Patel, P.T.; Kumar, C.S.; Rao, G.S. Protective effect of tribulus terrestris linn on liver and kidney in cadmium intoxicated rats. *Indian J. Exp. Biol* **2012**, *50*, 141–146.
184. Ali, S.; Hussain, S.; Khan, R.; Mumtaz, S.; Ashraf, N.; Andleeb, S.; Shakir, H.A.; Tahir, H.M.; Khan, M.K.A.; Ulhaq, M. Renal toxicity of heavy metals (cadmium and mercury) and their amelioration with ascorbic acid in rabbits. *Environ. Sci. Pollut. Res. Int.* **2019**, *26*, 3909–3920. [CrossRef]
185. Fang, J.; Xie, S.; Chen, Z.; Wang, F.; Chen, K.; Zuo, Z.; Cui, H.; Guo, H.; Ouyang, P.; Chen, Z.; et al. Protective effect of vitamin e on cadmium-induced renal oxidative damage and apoptosis in rats. *Biol. Trace Elem. Res.* **2021**, *199*, 4675–4687. [CrossRef]
186. Yu, J.; Liu, Y.; Guo, J.; Tao, W.; Chen, Y.; Fan, X.; Shen, J.; Duan, J.A. Health risk of licorice-yuanhua combination through induction of colonic h2s metabolism. *J. Ethnopharmacol.* **2019**, *236*, 136–146. [CrossRef] [PubMed]

187. Van Cauwenberghe, C.; Vandendriessche, C.; Libert, C.; Vandenbroucke, R.E. Caloric restriction: Beneficial effects on brain aging and alzheimer's disease. *Mamm. Genome* **2016**, *27*, 300–319. [CrossRef]
188. Estrela, G.R.; Wasinski, F.; Batista, R.O.; Hiyane, M.I.; Felizardo, R.J.; Cunha, F.; de Almeida, D.C.; Malheiros, D.M.; Camara, N.O.; Barros, C.C.; et al. Caloric restriction is more efficient than physical exercise to protect from cisplatin nephrotoxicity via ppar-alpha activation. *Front. Physiol.* **2017**, *8*, 116. [CrossRef] [PubMed]
189. Sohal, R.S.; Forster, M.J. Caloric restriction and the aging process: A critique. *Free Radic. Biol. Med.* **2014**, *73*, 366–382. [CrossRef]
190. Sohal, R.S.; Weindruch, R. Oxidative stress, caloric restriction, and aging. *Science* **1996**, *273*, 59–63. [CrossRef]
191. Shaikh, Z.A.; Jordan, S.A.; Tang, W. Protection against chronic cadmium toxicity by caloric restriction. *Toxicology* **1999**, *133*, 93–103. [CrossRef]
192. Felley-Bosco, E.; Diezi, J. Dietary calcium restriction enhances cadmium-induced metallothionein synthesis in rats. *Toxicol. Lett.* **1992**, *60*, 139–144. [CrossRef]
193. Jin, Z.; Wu, J.; Yan, L.J. Chemical conditioning as an approach to ischemic stroke tolerance: Mitochondria as the target. *Int. J. Mol. Sci.* **2016**, *17*, 351. [CrossRef] [PubMed]
194. Wu, J.; Li, R.; Li, W.; Ren, M.; Thangthaeng, N.; Sumien, N.; Liu, R.; Yang, S.; Simpkins, J.W.; Forster, M.J.; et al. Administration of 5-methoxyindole-2-carboxylic acid that potentially targets mitochondrial dihydrolipoamide dehydrogenase confers cerebral preconditioning against ischemic stroke injury. *Free Radic. Biol. Med.* **2017**, *113*, 244–254. [CrossRef]
195. Wu, J.; Jin, Z.; Yang, X.; Yan, L.J. Post-ischemic administration of 5-methoxyindole-2-carboxylic acid at the onset of reperfusion affords neuroprotection against stroke injury by preserving mitochondrial function and attenuating oxidative stress. *Biochem. Biophys. Res. Commun.* **2018**, *497*, 444–450. [CrossRef]
196. Kim, J.J.; Kim, Y.S.; Kumar, V. Heavy metal toxicity: An update of chelating therapeutic strategies. *J. Trace Elem. Med. Biol.* **2019**, *54*, 226–231. [PubMed]
197. Akiyama, M.; Unoki, T.; Shinkai, Y.; Ishii, I.; Ida, T.; Akaike, T.; Yamamoto, M.; Kumagai, Y. Environmental electrophile-mediated toxicity in mice lacking nrf2, cse, or both. *Environ. Health Perspect.* **2019**, *127*, 67002. [CrossRef] [PubMed]
198. Shinkai, Y.; Masuda, A.; Akiyama, M.; Xian, M.; Kumagai, Y. Cadmium-mediated activation of the hsp90/hsf1 pathway regulated by reactive persulfides/polysulfides. *Toxicol. Sci.* **2017**, *156*, 412–421. [PubMed]
199. Akiyama, M.; Shinkai, Y.; Unoki, T.; Shim, I.; Ishii, I.; Kumagai, Y. The capture of cadmium by reactive polysulfides attenuates cadmium-induced adaptive responses and hepatotoxicity. *Chem. Res. Toxicol.* **2017**, *30*, 2209–2217. [CrossRef]
200. Bonventre, J.V. Dedifferentiation and proliferation of surviving epithelial cells in acute renal failure. *J. Am. Soc. Nephrol.* **2003**, *14*, S55–S61. [CrossRef]
201. Padanilam, B.J. Cell death induced by acute renal injury: A perspective on the contributions of apoptosis and necrosis. *Am. J. Physiol. Renal Physiol.* **2003**, *284*, F608–F627. [CrossRef]
202. Chang-Panesso, M.; Humphreys, B.D. Cellular plasticity in kidney injury and repair. *Nat. Rev. Nephrol.* **2017**, *13*, 39–46. [CrossRef]
203. Chang-Panesso, M.; Kadyrov, F.F.; Lalli, M.; Wu, H.; Ikeda, S.; Kefaloyianni, E.; Abdelmageed, M.M.; Herrlich, A.; Kobayashi, A.; Humphreys, B.D. Foxm1 drives proximal tubule proliferation during repair from acute ischemic kidney injury. *J. Clin. Investig.* **2019**, *129*, 5501–5517. [CrossRef]
204. Moulis, J.M. Cellular mechanisms of cadmium toxicity related to the homeostasis of essential metals. *Biometals* **2010**, *23*, 877–896. [CrossRef] [PubMed]
205. Fujiwara, Y.; Lee, J.Y.; Tokumoto, M.; Satoh, M. Cadmium renal toxicity via apoptotic pathways. *Biol. Pharm. Bull.* **2012**, *35*, 1892–1897. [CrossRef] [PubMed]

biomolecules

MDPI

Review

NADH/NAD+ Redox Imbalance and Diabetic Kidney Disease

Liang-Jun Yan

Department of Pharmaceutical Sciences, College of Pharmacy, University of North Texas Health Science Center, Fort Worth, TX 76107, USA; liang-jun.yan@unthsc.edu; Tel.: +1-817-735-2386; Fax: +1-817-735-2603

Abstract: Diabetic kidney disease (DKD) is a common and severe complication of diabetes mellitus. If left untreated, DKD can advance to end stage renal disease that requires either dialysis or kidney replacement. While numerous mechanisms underlie the pathogenesis of DKD, oxidative stress driven by NADH/NAD+ redox imbalance and mitochondrial dysfunction have been thought to be the major pathophysiological mechanism of DKD. In this review, the pathways that increase NADH generation and those that decrease NAD+ levels are overviewed. This is followed by discussion of the consequences of NADH/NAD+ redox imbalance including disruption of mitochondrial homeostasis and function. Approaches that can be applied to counteract DKD are then discussed, which include mitochondria-targeted antioxidants and mimetics of superoxide dismutase, caloric restriction, plant/herbal extracts or their isolated compounds. Finally, the review ends by pointing out that future studies are needed to dissect the role of each pathway involved in NADH-NAD+ metabolism so that novel strategies to restore NADH/NAD+ redox balance in the diabetic kidney could be designed to combat DKD.

Keywords: diabetic kidney disease; caloric restriction; NADH/NAD+; redox imbalance; mitochondrial homeostasis; mitophagy; oxidative stress

Citation: Yan, L.-J. NADH/NAD+ Redox Imbalance and Diabetic Kidney Disease. *Biomolecules* **2021**, *11*, 730. https://doi.org/10.3390/biom11050730

Academic Editor: Theodoros Eleftheriadis

Received: 22 April 2021
Accepted: 12 May 2021
Published: 14 May 2021

1. Introduction

Constituting approximately 0.51% to 1.08% of the body weight, the kidneys are an energy-demanding organ and receive approximately 20–25% of the cardiac output [1–3]. Every 24 min, it filters a volume equal to that of whole plasma volume; and every 6 h, it filters a volume equal to that of total body water [3]. Given this workload, the kidney needs a large amount of ATP produced by mitochondria, which, unfortunately, also generate reactive oxygen species (ROS) as metabolic byproducts [4–6]. Therefore, the kidney is under constant attack from ROS. Such is indeed the case in diabetic kidney disease (DKD) whereby oxidative stress is elevated and mitochondrial dysfunction is aggravated, leading to renal injury [7,8]. DKD, also known as diabetic nephropathy [9–13], is a common complication of diabetic mellitus, including both type 1 and type 2 diabetes. While type 1 diabetes is caused by lack of insulin due to pancreatic β cell destruction [14–16], type 2 diabetes could be caused by insulin resistance or insulin deficiency [17–22]. The hallmark of diabetes is a persistent high blood glucose content (hyperglycemia) that can damage a variety of tissues and cells [15,21–23]. In the kidney, renal microvascular structures are the major targets of high blood glucose [24–27]. Additionally, given the facts that the kidney is the organ where mature or active form of vitamin D is made [28–30] and erythropoiesis erythropoietin is produced [31–33], DKD can also lead to vitamin D deficiency and anemia [34–40]. Therefore, while there is great understanding of the pathophysiology and progression of DKD, novel and effective treatment approaches are still needed as current therapeutic options remain limited.

While many mechanisms underlie the pathogenesis of DKD including protein kinase C pathway [41,42], hexosamine pathway [43,44], formation of advanced glycation end products [45,46] and the polyol pathway [47,48]; at the molecular level, redox imbalance of NADH/NAD+ caused by deranged glucose metabolism [49–51] may stand out as a distinct mechanism of diabetic kidney injury [52–55]. This is because electrons from breakdown of glucose and other nutrients such as fatty acids and amino acids are stored in NADH using NAD^+ as the electron acceptor [56–58]. Therefore, a key feature of diabetes mellitus is oversupply of NADH and under supply of NAD^+ [48,51,59].

2. Sources of Elevated NADH in Diabetes

In addition to the conventional metabolic pathways that extract electrons by breaking the chemical bonds in carbohydrates and fatty acids (Figure 1), other glucose utilization pathways are activated by hyperglycemia [47]. One of these pathways is the polyol pathway [60,61] (Figure 2), which can burn up to 30% of the glucose pool in a diabetic patient [62]. This pathway converts glucose to fructose and also converts NADPH to NADH. There is also an intermediate product known as sorbitol, which could accumulate and impair cellular osmosis in the kidney [63,64]. While the first reaction from glucose to sorbitol is catalyzed by aldose reductase, the second reaction from sorbitol to fructose is catalyzed by sorbitol dehydrogenase. In this pathway, aldose reductase is the rate-limiting enzyme that has a high Km value for glucose [65]. Therefore, numerous studies have focused on aldose reductase as a potential therapeutic target in diabetes [66–71]. In particular, attention has been paid to develop small molecule compounds that can inhibit aldose reductase [72–76] to prevent accumulation of sorbitol and fructose and to prevent build-up of NADH, the elevation of which can perturb NADH/NAD^+ redox balance, initiating reductive stress and oxidative stress. Furthermore, the contribution of the polyol pathway to diabetes development has been demonstrated by the use of aldose reductase animal models whereby lack of aldose reductase prevents the development of diabetes [76]. It should be noted that this NADH/NAD^+ redox imbalance is also termed as pseudohypoxia in diabetes [77,78] because hypoxia and ischemia often leads to NADH accumulation and NAD^+ depletion [79–81]. It should also be noted that endogenous production of fructose via the polyol pathway has been shown to cause increased fructose and fructose-1-phosphate contents in the kidney, leading to aggravation of DKD [82].

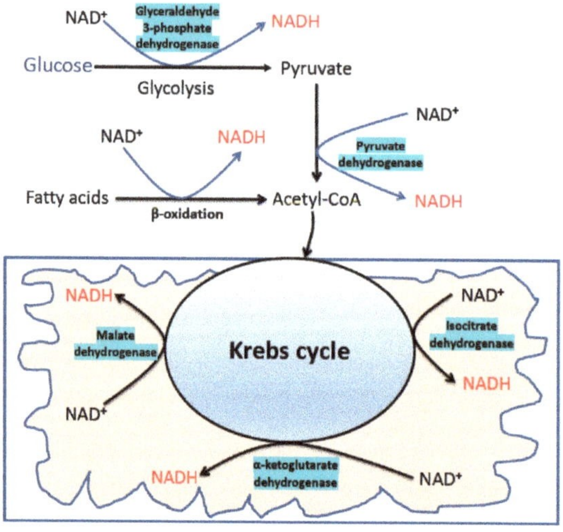

Figure 1. The conventional metabolic pathways that generate NADH from NAD^+. Shown are the glycolytic pathway, fatty acid oxidation, and the Krebs cycle. These are the major pathways that store electrons in NADH by breaking the chemical bonds in dietary components including glucose, fatty acids. Enzymes involved in direct production of NADH are also indicated in the diagram.

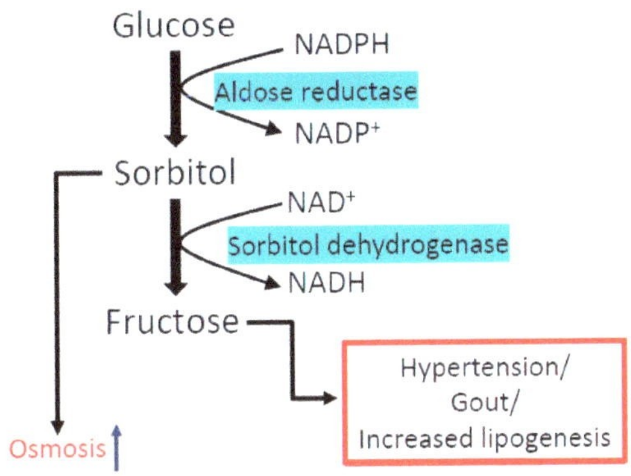

Figure 2. The polyol pathway. This pathway contains two reactions. The first reaction converting glucose to sorbitol is catalyzed by aldose reductase. This enzyme is rate-limiting for the whole pathway. The second reaction converting sorbitol to fructose is catalyzed by sorbitol dehydrogenase. The final products are NADH and fructose, and sorbitol is an intermediate product. Note that NADPH is consumed by aldose reductase in the first reaction. Additionally, accumulation of sorbitol in the kidney could cause osmotic problems for nephrons [63,64].

3. Pathways of NAD^+ Consumption in Diabetes

3.1. The Poly ADP Ribosylase Pathway

While NADH is over-supplied in diabetes, NAD^+ could be depleted in diabetes. One major pathway utilizing NAD^+ is the poly ADP ribosylase catalyzed reaction (Figure 3A), which is activated due to DNA damage by ROS in diabetes and uses NAD^+ as a substrate thereby leading to degradation of NAD^+ [83–85]. The contribution of this pathway to the pathogenesis of diabetes has been confirmed by studies using poly ATP ribosylase deficient mouse, in which lack of the enzyme prevents development of diabetes [86,87], demonstrating the detrimental effects of NAD^+ depletion in diabetes.

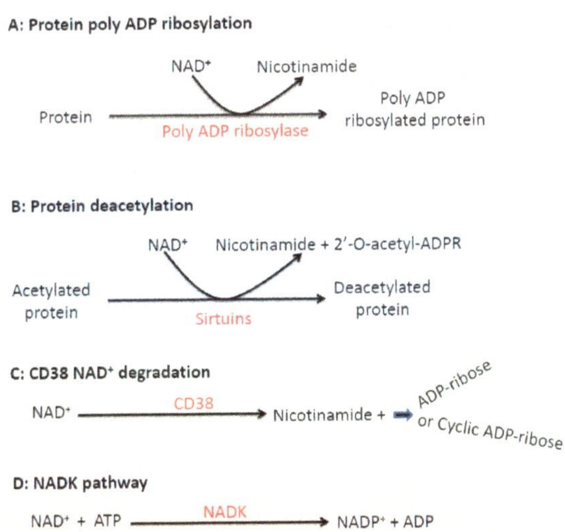

Figure 3. Major pathways that consume NAD^+. Shown are (**A**) the poly ADP ribosylase reaction; (**B**) the sirtuin-catalyzed deacetylation reaction; (**C**) the CD38 NAD^+ degradation pathway; (**D**) the NAD kinase pathway converting NAD^+ to $NADP^+$. All the shown pathways or reactions use NAD^+ as the respective enzyme's substrate.

3.2. *The Sirtuins Pathway*

Another pathway that also consumes NAD^+ is the sirtuin proteins [88,89] (Figure 3B), which remove acetyl groups from acetylated proteins using NAD^+ as a substrate. This pathway may play an important role in lowering NAD^+ levels in early stages of diabetes, but at advanced stages of diabetes, sirtuin protein contents tend to be down regulated [90]. Therefore, it is likely that sirtuin deficiency in advanced stages of diabetes contributes less to NAD^+ depletion in diabetes.

3.3. *The CD38 Pathway*

CD38 is an NADase that catalyzes the degradation of NAD^+ [91–93] (Figure 3C). This enzyme has been shown to be upregulated in a variety of diseases as well as aging [92,94], leading to decreased content of NAD^+ that would impair the function of sirtuins and poly ADP ribosylase [95,96]. CD38-driven NAD^+ deficiency has been shown to be responsible for organ fibrosis and diabetic kidney dysfunction [97]. Conversely, CD38 inhibitors have been shown to mitigate mitochondrial oxidative stress in DKD via restoration of $NADH/NAD^+$ redox balance [98].

3.4. *The NAD Kinase Pathway*

NAD kinase (NADK) exists both in the cytosol and in the mitochondria [99,100]. This protein is the sole enzyme responsible for conversion of NAD^+ to $NADP^+$ [101,102] (Figure 3D). Given the key role of $NADP^+$ in maintaining the levels of cellular antioxidant glutathione [103–105], NADK is an indispensable element in the redox metabolic pathways. Although many studies have been conducted on NADK in a variety of experimental systems, the role of this protein in DKD has yet to be explored. Furthermore, as NADK consumes NAD^+, how it is involved in maintaining or perturbing $NADH/NAD^+$ redox balance in DKD will also need to be investigated. The major pathways causing increase in NADH and decrease in NAD^+ as well as NAD^+ regeneration by mitochondrial complex I and lactate dehydrogenase (under hypoxia) are summarized in Figure 4.

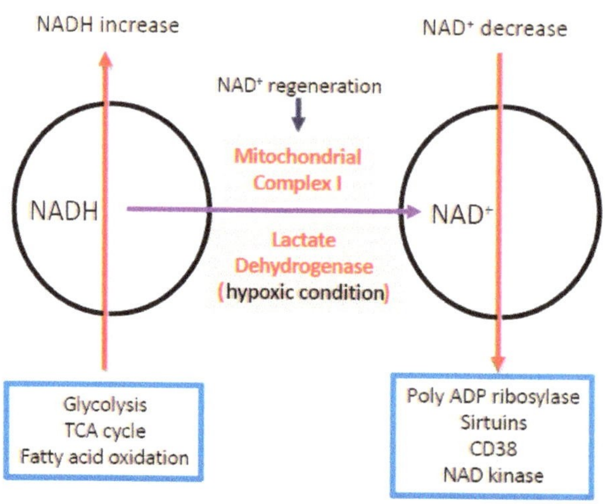

Figure 4. Diagram summarizing the pathways that cause NADH increase and NAD^+ decrease in the diabetic kidneys. Regeneration of NAD^+ from NADH by either mitochondrial complex I or lactate dehydrogenase (under hypoxic conditions) is also shown.

4. Redox Imbalance-linked Mitochondrial Dysfunction in DKD

One of the major consequences of $NADH/NAD^+$ redox imbalance is mitochondrial oxidative stress due to oversupply of NADH to the mitochondrial electron transport chain [47,106,107]. This is caused by electron leakage from the electron transport chain [6,108–112], as it cannot use all of the NADH for ATP production [90,113]. As such, oxygen is partially reduced to form superoxide anion via the electron transport chain, mainly through complexes I, III and IV [114,115]. This mitochondrial superoxide, regardless of the exact site of its generation, is the original source of oxidative stress that can cause oxidative damage to DNA, proteins and lipids [116–119]. Accumulation of these oxidatively damaged macromolecule adducts can eventually lead to cell death and kidney failure [120,121].

While NADH/NAD$^+$ redox imbalance drives the initial event of superoxide production in mitochondria [47,48,122], other abnormalities of mitochondria could also manifest in DKD, culminating in decreased oxygen consumption and ATP production. As mitochondrion is a dynamic organelle, disruption of its fission and fusion processes [123], also knowns as mitochondrial homeostasis [123–125], can also worsen diabetic kidney injury [126]. Indeed, dynamin-related protein 1 (Drp1), well known for its role in regulating mitochondrial fission, has been shown to be upregulated to cause mitochondrial fragmentation in DKD [127–129]. Conversely, mitochondrial fusion regulating proteins such as optic atrophy-1 (opa1) and mitochondrial fusion proteins, in particular, mitochondrial fusion protein 2 (Mfn2), have been shown to be down regulated to impair mitochondrial fusion in DKD [130–132]. This disruption of mitochondrial homeostasis is linked with redox imbalance and oxidative stress accompanied with impairment of mitochondrial membrane potential and release of apoptosis stimulating factors such as cytochrome c and apoptosis inducing factor (AIF) [133–136]. These deranged mitochondrial dynamics, if left unattended or uncorrected, would eventually lead to accumulation of damaged mitochondria, which could overwhelm the mitophagy capacity that is regulated by key proteins such as PINK1 and Parkin [137–140], resulting in cell death and worsened diabetic kidney injury. Therefore, mitochondrial homeostasis and dynamics can also serve as targets for renal therapy in DKD. Figure 5 outlines the potential deleterious mitochondrial consequences of NADH/NAD$^+$ redox imbalance implicated in the pathogenesis of DKD.

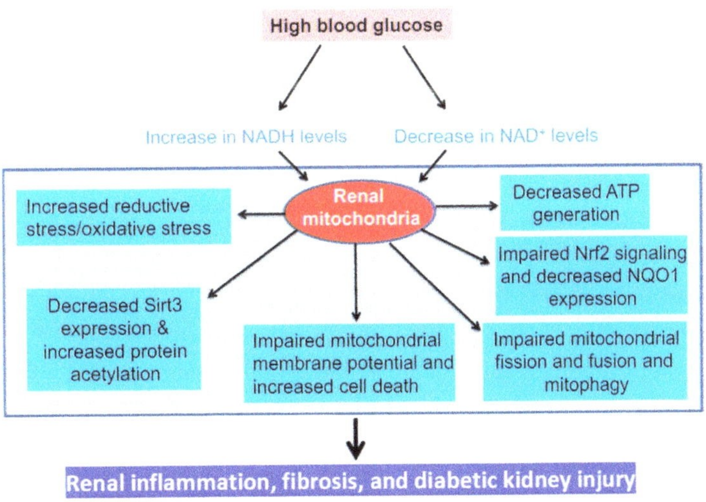

Figure 5. Mitochondrial dysfunction driven by NADH/NAD$^+$ redox imbalance and the potential mitochondrial mechanisms underlying pathophysiology of DKD. These mechanisms include increased mitochondrial oxidative damage, decreased ATP production, perturbed mitochondrial membrane potential and deranged mitochondrial homeostasis and impaired sirt3 pathway as well as Nrf2 signaling pathway. The ultimate manifestation of these mitochondrial dysfunctional mechanisms is renal inflammation, fibrosis and diabetic kidney injury.

It is also worth mentioning that a hallmark of the diabetic kidney is hyperfiltration, so that the energy demands of the proximal tubule are greatly increased [141–143]. This may temporarily ameliorate the NADH/NAD^+ redox imbalance as NADH utilization is increased for ATP production. However, as increased NADH consumption means more oxygen consumption and more electron leakage from mitochondria for superoxide production [114,144,145], tubular cells could exhibit increased oxidative stress, which could eventually lead to hyperfiltration linked diabetic nephropathy [53,146]. Nonetheless, whether there is an increased mitochondrial superoxide production linked to hyperfiltration and increased ATP demands remains to be determined.

5. Therapeutic Approaches to Counteracting DKD

5.1. Superoxide Dismutation and Suppression

There are at least 11 mitochondrial sites that are involved in superoxide generation [147]. Therefore, overall approaches of dismutating superoxide could alleviate DKD [148]. Recently, small molecules that can suppress or inhibit mitochondrial superoxide production have been developed. Typical examples of these small molecules are S1QELs and S3QELs [145,149], which do not interfere with the process of oxidative phosphorylation or ATP production [144]. S1QELs acts at the site I_Q of complex I [145,147] while S3QELs acts at the Q site of complex III [150,151]. The usefulness of these suppressors in combating oxidative stress has been tested in certain experimental systems [151–154]. However, studies of these suppressors in alleviation of DKD have yet to be conducted. It is anticipated that these compounds could attenuate the severity of DKD in diabetic subjects.

5.2. Mitochondria-targeted Antioxidants and Superoxide Dismutase (SOD) Mimetics

Antioxidants that go into mitochondria are a class of compounds that can be used to counteract mitochondrial oxidative stress. These are generally purposely synthesized for targeting mitochondria. One well-known compound is mitoQ that has been investigated in a variety of diseases including kidney disease [155–162]. Using type 1 diabetic Akita mouse model, Chacko et al. [163] demonstrated that mitoQ administration over a 12-week period improved tubular and glomerular function in the Akita diabetic mice and decreased urinary albumin content to the level as observed in healthy controls. Moreover, mitoQ-treated Akita mice yielded mitochondria that functioned similar to those isolated from healthy control animals, resulting in attenuation of interstitial fibrosis and glomerular damage. MitoQ could also ameliorate tubular injury by enhancing mitophagy via the Nrf2/PINK1 pathway [164]. In fact, the efficacy of mitoQ in diabetic renal protection is nearly equal to that of angiotensin converting enzyme inhibition [165]. MitoQ could also decrease mitochondrial fragmentation mediated by the JNK signaling pathway in DKD [166]. All these protective effects of mitoQ on DKD can be attributed to its capacity in destroying ROS [167]. It should be noted that while the protective effects of an SOD mimetic tempol has been investigated in DKD [168–171], the protective effects of other SOD mimetics such as GC4419 [172,173] and EUK189 [174–176] are yet to be evaluated in DKD.

5.3. Plant and Herb Derived Antioxidants

Numerous plant- or herbal extracts or plant/herb-derived natural products have been tested for their capacity in fighting DKD. A major representative of these extracts is polyphenols that can scavenge ROS [177], leading to correction of redox imbalance and enhancement of mitochondrial function [178–180]. Moreover, these plants extracts can also activate the Nrf2 signaling pathway thereby leading to upregulation of the so-called second cellular defense system including antioxidant proteins such as heme oxygenase-1 and NQO1 [181]. As chronic inflammation is implicated in the pathogenesis of DKD, many studies involving plant extracts have also demonstrated their anti-inflammation properties in preclinical DKD [182]. Table 1 shows selected representatives of plant/herb extracts or plant/herb-derived compounds and their redox balanced-related anti-DKD mechanisms.

Table 1. Selected representatives of plant/herbal extracts/components in DKD from the literature. Experimental models and the major underlying renoprotective mechanisms are also given in the table.

Extracts/Components	Experimental Model	Major Mechanisms	Refs.
Azuki bean extract	* STZ-rat	Autophagy stimulation	[183]
Acacia nilotica	STZ-rat	Antioxidant/anti-hyperglycemia	[184]
Anogeissus acuminate leaf	STZ-rat	Antioxidation	[185]
Broccoli	STZ-rat	Mitigating oxidative damage	[186]
Curcumin	STZ-rat	Inhibiting PKC beta	[187]
Coccinia indica	STZ-rat	Increased antioxidant enzymes	[188]
Coffea arabica pulp	HFD/STZ	Antioxidation upregulation	[189]
Ganoderma lucidum	STZ-rat	TGFβ-1, NFkB	[190]
Garlic extract	STZ-rat	Anti-glycation	[191]
Geraniin	* HFD	Inhibiting oxidative stress	[192]
Ginger extract	STZ-rat	Apoptosis attenuation	[193]
Ginkgo biloba EGB761	HFD/STZ mouse	Mitigating ECM * accumulation	[194]
Berberine	db/db mouse	Mitochondrial fission	[195]
Cupuacu extract	STZ-rat	Mitigating nitrosation	[196]
Anchomanes difformis (leaf)	STZ-rat	Nrf2 activation	[197]
Abelmoschus manihot	HFD/STZ mouse	Autophagy activation	[198]
Hibiscus sabdariffa Linnaeus	STZ-rat	Akt regulating	[199]
Mulberry leaf	HFD/STZ rat	Inhibiting TGF-β1	[200]
Liriope spicata var. prolifera	STZ-rat	Suppressing inflammation	[201]
Nelumbo nucifera leaf	HFD/STZ rat	Antioxidative	[202]
Coreopsis tinctoria nutt	High glucose/HFD/STZ	Anti-fibrotic	[203]
Oil palm	STZ-rat	Attenuating oxidative stress	[204]
Armillariella tabescens	STZ-mouse	Anti-inflammation	[205]
Red ginseng	STZ-rat	Autophagy acceleration	[206]
Paederia foetida leaf	Alloxan-rat	Antioxidative effects	[207]
Tiliacora triandra	HFD/STZ rat	Redox imbalance modulation	[208]
Flavonoids (review article)	Numerous models	Miscellaneous mechanisms	[209]
Grape seed	STZ-rat	Reduce apoptosis	[210]
Grape seed/proanthocyanidins	STZ-rat	Mitigating ER stress	[211]
Grape seed procyanidin B2	db/db mouse	Targeting MFG-E8*	[212]
Grape seed polyphenols	Cell culture	Mitigating oxidative stress	[213]
Catlpol	db/db mouse	Improving lipid metabolism	[214]
Cudrania tricuspidata root	Human kidney cells	Preventing inflammation	[215]
Hyperoside	HFD/STZ mouse	Targeting miR-499-5p?APC	[216]
Phyllanthus niruri leaf	STZ/nicotinamide rat	Anti-fibrosis/apoptosis	[217]
Pomegranate peel extract	STZ-mouse	Nrf2 signaling pathway	[218]
Quercetin	STZ-mouse	Anti-apoptosis/oxidative stress	[219]
Resveratrol	STZ-mouse	Sirt1 activation	[220]

* Abbreviations: HFD, high fat diet; STZ, streptozotocin; ECM, extracellular matrix; MFG-E8, milk fat globule EGF-8. Please note that this table is not meant to exhaust the literature on plant/herbal extracts and DKD.

5.4. Caloric Restriction

Caloric restriction (CR) [221–223], sometimes also called energy restriction [224,225], is a well-established approach for extending the lifespan of many species. CR can also prolong the health span of many organs including the kidney [226–229]. As CR has a direct impact on energy supply that involves NADH and NAD^+, it thus is involved in eliciting antioxidative responses in DKD by restoring redox balance and mitigating diabetic kidney injury [230,231]. Such responses include AMPK activation, autophagy, ROS elimination, Nrf2 signaling pathway activation and enhancement of antioxidative capacity in the kidney [231–235]. In certain studies, exercise has been shown to have a synergistic effect on CR [236,237]. Therefore, CR and exercise may be applied simultaneously to enhance kidney function in diabetes [238,239]. Moreover, intermittent fasting, a different version of CR, has also been demonstrated to prevent progression of DKD via NAD^+ dependent sirtuin pathway [230]. Additionally, the restriction of single element in a given diet such as iron can also afford renoprotection in diabetes via attenuation of oxidative stress [240].

6. Magnitude of Redox Imbalance and Progression of DKD

While it is now known that $NADH/NAD^+$ redox imbalance is one of the underlying mechanisms of DKD and this redox imbalance drives reductive stress to oxidative stress [47], culminating in renal dysfunction in DKD, whether the magnitude of $NADH/NAD^+$ redox imbalance can be associated with the indices of DKD progression has not been established. DKD progression can be determined by the ratio of urinary albumin to urinary creatinine [127] and by estimated glomerular flow rate (eGFR) [241,242], but whether $NADH/NAD^+$ ratio would also advance from low to high during DKD progression is unknown at this time and needs to be investigated. It is conceivable that with the progression of DKD quantitated by the above-mentioned parameters, values of the $NADH/NAD^+$ ratio would also increase gradually to reflect the severity of DKD. Conversely, the value of the $NADH/NAD^+$ ratio should go down upon remission of DKD after treatment. Regardless, this would need to be evaluated using proper animal models that can show clearly an association of the value of $NADH/NAD^+$ to progression of DKD until the end stage of renal disease.

7. Conclusions

$NADH/NAD^+$ redox imbalance, driven by persistent hyperglycemia and oversupply of other nutrients, is the initiator of reductive stress and oxidative stress in DKD [47]. More studies would be needed to dissect the role of each and every player in this cascade of redox imbalance biochemistry mechanism. Complete and comprehensive studies not only will shed insights into the mechanisms of DKD but will also facilitate identification of targets that can be explored for DKD therapy. As indicated in a recent review article by Matoba et al. [243], targeting $NADH/NAD^+$ redox imbalance would be a valuable approach for combating DKD. Finally, it should be pointed out that in terms of potential injury caused by redox imbalance, which part of the kidney or what type of cells that sustain the most damage have not been comprehensively evaluated. Therefore, future efforts should be made to assess redox imbalance-induced damage to endothelial cells of the renal vasculature, the podocytes and mesangial cells of the glomerulus and the epithelial cells of the tubule. Additionally, how redox imbalance differs within the tubule should also be measured.

Funding: This research received no external funding.

Institutional Review Board Statement: Not applicable.

Data Availability Statement: Not applicable.

Acknowledgments: The author's work was supported in part by UNTHSC seed grants RI10015 and RI10039.

Conflicts of Interest: The authors declare no conflict of interest.

References

1. Radi, Z.A. Kidney Pathophysiology, Toxicology, and Drug-Induced Injury in Drug Development. *Int. J. Toxicol.* **2019**, *38*, 215–227. [CrossRef]
2. Koeppen, B.M.; Stanton, B.A. *Renal Physiology*, 5th ed.; Elsevier: Philadelphia, PA, USA, 2013.
3. Carroll, R.G. *Elsevier's Integrated Physiology*; Elsevier: Philadelphia, PA, USA, 2007.
4. Zorov, D.B.; Filburn, C.R.; Klotz, L.O.; Zweier, J.L.; Sollott, S.J. Reactive oxygen species (ROS)-induced ROS release: A new phenomenon accompanying induction of the mitochondrial permeability transition in cardiac myocytes. *J. Exp. Med.* **2000**, *192*, 1001–1014. [CrossRef] [PubMed]
5. Kim, J.; Jang, H.S.; Park, K.M. Reactive oxygen species generated by renal ischemia and reperfusion trigger protection against subsequent renal ischemia and reperfusion injury in mice. *Am. J. Physiol. Renal Physiol.* **2010**, *298*, F158–F166. [CrossRef] [PubMed]
6. Rosca, M.G.; Vazquez, E.J.; Chen, Q.; Kerner, J.; Kern, T.S.; Hoppel, C.L. Oxidation of fatty acids is the source of increased mitochondrial reactive oxygen species production in kidney cortical tubules in early diabetes. *Diabetes* **2012**, *61*, 2074–2083. [CrossRef] [PubMed]
7. Wei, P.Z.; Szeto, C.C. Mitochondrial dysfunction in diabetic kidney disease. *Clin. Chim. Acta* **2019**, *496*, 108–116. [CrossRef] [PubMed]
8. Hallan, S.; Sharma, K. The Role of Mitochondria in Diabetic Kidney Disease. *Curr. Diab. Rep.* **2016**, *16*, 61. [CrossRef] [PubMed]
9. Cui, J.; Bai, X.; Chen, X. Autophagy and Diabetic Nephropathy. *Adv. Exp. Med. Biol.* **2020**, *1207*, 487–494. [CrossRef]
10. Ji, J.; Tao, P.; Wang, Q.; Li, L.; Xu, Y. SIRT1: Mechanism and Protective in Diabetic Nephropathy. *Endocr. Metab. Immune Disord. Drug Targets* **2020**. [CrossRef]
11. Zoja, C.; Xinaris, C.; Macconi, D. Diabetic Nephropathy: Novel Molecular Mechanisms and Therapeutic Targets. *Front. Pharmacol.* **2020**, *11*, 586892. [CrossRef]
12. Wang, Y.; Tian, J.; Mi, Y.; Ren, X.; Lian, S.; Kang, J.; Wang, J.; Zang, H.; Wu, Z.; Yang, J.; et al. Experimental study on renoprotective effect of intermedin on diabetic nephropathy. *Mol. Cell. Endocrinol.* **2021**, *528*, 111224. [CrossRef] [PubMed]
13. Lodhi, A.H.; Ahmad, F.U.; Furwa, K.; Madni, A. Role of Oxidative Stress and Reduced Endogenous Hydrogen Sulfide in Diabetic Nephropathy. *Drug Des. Devel. Ther.* **2021**, *15*, 1031–1043. [CrossRef]
14. Powers, A.C. Type 1 diabetes mellitus: Much progress, many opportunities. *J. Clin. Investig.* **2021**, *131*. [CrossRef]
15. Tuch, B.; Dunlop, M.; Proietto, J. *Diabetes Research: A Guide for Postgraduates*; Harwood Academic Publishers: Reading, UK, 2000.
16. Volfson-Sedletsky, V.; Jones, A., IV; Hernandez-Escalante, J.; Dooms, H. Emerging Therapeutic Strategies to Restore Regulatory T Cell Control of Islet Autoimmunity in Type 1 Diabetes. *Front. Immunol.* **2021**, *12*, 635767. [CrossRef]
17. Szkudelski, T. The mechanism of alloxan and streptozotocin action in B cells of the rat pancreas. *Physiol. Res.* **2001**, *50*, 537–546. [PubMed]
18. Szkudelski, T. Streptozotocin-nicotinamide-induced diabetes in the rat. Characteristics of the experimental model. *Exp. Biol. Med.* **2012**, *237*, 481–490. [CrossRef] [PubMed]
19. Kahn, S.E.; Cooper, M.E.; Del Prato, S. Pathophysiology and treatment of type 2 diabetes: Perspectives on the past, present, and future. *Lancet* **2014**, *383*, 1068–1083. [CrossRef]
20. DeFronzo, R.A. Pathogenesis of type 2 diabetes mellitus. *Med. Clin. N. Am.* **2004**, *88*, 787–835, ix. [CrossRef] [PubMed]
21. DeFronzo, R.A. Insulin resistance: A multifaceted syndrome responsible for NIDDM, obesity, hypertension, dyslipidaemia and atherosclerosis. *Neth. J. Med.* **1997**, *50*, 191–197. [CrossRef]
22. Barnett, A.H. *Type 2 Diabetes*, 2nd ed.; Oxford University Press: New York, NY, USA, 2012; p. 162.
23. Abdul-Ghani, M.A.; DeFronzo, R.A. Oxidative stress in type 2 diabetes. In *Oxidative Stress in Aging*; Miwa, S., Beckman, K.B., Muller, F.L., Eds.; Humana Press: Totowa, NJ, USA, 2008; pp. 191–212.
24. Maqbool, M.; Cooper, M.E.; Jandeleit-Dahm, K.A.M. Cardiovascular Disease and Diabetic Kidney Disease. *Semin. Nephrol.* **2018**, *38*, 217–232. [CrossRef]
25. Su, J.; Ye, D.; Gao, C.; Huang, Q.; Gui, D. Mechanism of progression of diabetic kidney disease mediated by podocyte mitochondrial injury. *Mol. Biol. Rep.* **2020**, *47*, 8023–8035. [CrossRef] [PubMed]
26. Sharma, K.; McCue, P.; Dunn, S.R. Diabetic kidney disease in the db/db mouse. *Am. J. Physiol. Renal Physiol.* **2003**, *284*, F1138–F1144. [CrossRef]
27. Tesch, G.H.; Lim, A.K. Recent insights into diabetic renal injury from the db/db mouse model of type 2 diabetic nephropathy. *Am. J. Physiol. Renal Physiol.* **2011**, *300*, F301–F310. [CrossRef]
28. Martin, K.J.; Gonzalez, E.A. Vitamin D and the kidney. *Mo. Med.* **2012**, *109*, 124–126.
29. Dusso, A.S. Kidney disease and vitamin D levels: 25-hydroxyvitamin D, 1,25-dihydroxyvitamin D, and VDR activation. *Kidney Int. Suppl.* **2011**, *1*, 136–141. [CrossRef]
30. Iida, K.; Shinki, T.; Yamaguchi, A.; DeLuca, H.F.; Kurokawa, K.; Suda, T. A possible role of vitamin D receptors in regulating vitamin D activation in the kidney. *Proc. Natl. Acad. Sci. USA* **1995**, *92*, 6112–6116. [CrossRef] [PubMed]
31. Souma, T.; Suzuki, N.; Yamamoto, M. Renal erythropoietin-producing cells in health and disease. *Front. Physiol.* **2015**, *6*, 167. [CrossRef] [PubMed]
32. Suzuki, N.; Yamamoto, M. Roles of renal erythropoietin-producing (REP) cells in the maintenance of systemic oxygen homeostasis. *Pflugers Arch.* **2016**, *468*, 3–12. [CrossRef] [PubMed]
33. Hu, X.; Xie, J.; Chen, N. Hypoxia-Inducible Factor-Proline Hydroxylase Inhibitor in the Treatment of Renal Anemia. *Kidney Dis.* **2021**, *7*, 1–9. [CrossRef] [PubMed]

34. Galuska, D.; Pacal, L.; Kankova, K. Pathophysiological Implication of Vitamin D in Diabetic Kidney Disease. *Kidney Blood Press Res.* **2021**, *46*, 152–161. [CrossRef] [PubMed]
35. Xiao, X.; Wang, Y.; Hou, Y.; Han, F.; Ren, J.; Hu, Z. Vitamin D deficiency and related risk factors in patients with diabetic nephropathy. *J. Int. Med. Res.* **2016**, *44*, 673–684. [CrossRef] [PubMed]
36. Xie, S.; Huang, L.; Cao, W.; Hu, Y.; Sun, H.; Cao, L.; Liu, K.; Liu, C. Association between serum 25-hydroxyvitamin D and diabetic kidney disease in Chinese patients with type 2 diabetes. *PLoS ONE* **2019**, *14*, e0214728. [CrossRef]
37. Tsai, S.F.; Tarng, D.C. Anemia in patients of diabetic kidney disease. *J. Chin. Med. Assoc.* **2019**, *82*, 752–755. [CrossRef]
38. Sonkar, S.K.; Singh, H.P.; Sonkar, G.K.; Pandey, S. Association of Vitamin D and secondary hyperparathyroidism with anemia in diabetic kidney disease. *J. Family Med. Prim. Care* **2018**, *7*, 815–818. [CrossRef]
39. Thomas, M.; Tsalamandris, C.; MacIsaac, R.; Jerums, G. Anaemia in diabetes: An emerging complication of microvascular disease. *Curr. Diabetes Rev.* **2005**, *1*, 107–126. [CrossRef]
40. Ravanan, R.; Spiro, J.R.; Mathieson, P.W.; Smith, R.M. Impact of diabetes on haemoglobin levels in renal disease. *Diabetologia* **2007**, *50*, 26–31. [CrossRef]
41. Feng, B.; Ruiz, M.A.; Chakrabarti, S. Oxidative-stress-induced epigenetic changes in chronic diabetic complications. *Can. J. Physiol. Pharmacol.* **2013**, *91*, 213–220. [CrossRef] [PubMed]
42. Naruse, K.; Rask-Madsen, C.; Takahara, N.; Ha, S.W.; Suzuma, K.; Way, K.J.; Jacobs, J.R.; Clermont, A.C.; Ueki, K.; Ohshiro, Y.; et al. Activation of vascular protein kinase C-beta inhibits Akt-dependent endothelial nitric oxide synthase function in obesity-associated insulin resistance. *Diabetes* **2006**, *55*, 691–698. [CrossRef]
43. Gurel, Z.; Sieg, K.M.; Shallow, K.D.; Sorenson, C.M.; Sheibani, N. Retinal O-linked N-acetylglucosamine protein modifications: Implications for postnatal retinal vascularization and the pathogenesis of diabetic retinopathy. *Mol. Vis.* **2013**, *19*, 1047–1059. [PubMed]
44. McLarty, J.L.; Marsh, S.A.; Chatham, J.C. Post-translational protein modification by O-linked N-acetyl-glucosamine: Its role in mediating the adverse effects of diabetes on the heart. *Life Sci.* **2013**, *92*, 621–627. [CrossRef] [PubMed]
45. Lovestone, S.; Smith, U. Advanced glycation end products, dementia, and diabetes. *Proc. Natl. Acad. Sci. USA* **2014**, *111*, 4743–4744. [CrossRef] [PubMed]
46. Mellor, K.M.; Brimble, M.A.; Delbridge, L.M. Glucose as an agent of post-translational modification in diabetes—New cardiac epigenetic insights. *Life Sci.* **2014**. [CrossRef] [PubMed]
47. Yan, L.J. Pathogenesis of Chronic Hyperglycemia: From Reductive Stress to Oxidative Stress. *J. Diabetes Res.* **2014**, *2014*, 137919. [CrossRef] [PubMed]
48. Yan, L.J. Redox imbalance stress in diabetes mellitus: Role of the polyol pathway. *Animal Model. Exp. Med.* **2018**, *1*, 7–13. [CrossRef]
49. Berg, D.; Youdim, M.B.; Riederer, P. Redox imbalance. *Cell Tissue Res.* **2004**, *318*, 201–213. [CrossRef] [PubMed]
50. Hayden, M.R.; Sowers, J.R. Redox imbalance in diabetes. *Antioxid. Redox Signal.* **2007**, *9*, 865–867. [CrossRef] [PubMed]
51. Wu, J.; Jin, Z.; Zheng, H.; Yan, L.J. Sources and implications of NADH/NAD+ redox imbalance in diabetes and its complications. *Diabetes Metab. Syndr. Obes.* **2016**, *9*, 145–153. [CrossRef] [PubMed]
52. Zhan, M.; Usman, I.M.; Sun, L.; Kanwar, Y.S. Disruption of renal tubular mitochondrial quality control by Myo-inositol oxygenase in diabetic kidney disease. *J. Am. Soc. Nephrol.* **2015**, *26*, 1304–1321. [CrossRef]
53. Miranda-Diaz, A.G.; Pazarin-Villasenor, L.; Yanowsky-Escatell, F.G.; Andrade-Sierra, J. Oxidative Stress in Diabetic Nephropathy with Early Chronic Kidney Disease. *J. Diabetes Res.* **2016**, *2016*, 7047238. [CrossRef] [PubMed]
54. Yanar, K.; Coskun, Z.M.; Beydogan, A.B.; Aydin, S.; Bolkent, S. The effects of delta-9-tetrahydrocannabinol on Kruppel-like factor-4 expression, redox homeostasis, and inflammation in the kidney of diabetic rat. *J. Cell. Biochem.* **2019**, *120*, 16219–16228. [CrossRef] [PubMed]
55. Amorim, R.G.; Guedes, G.D.S.; Vasconcelos, S.M.L.; Santos, J.C.F. Kidney Disease in Diabetes Mellitus: Cross-Linking between Hyperglycemia, Redox Imbalance and Inflammation. *Arq. Bras. Cardiol.* **2019**, *112*, 577–587. [CrossRef] [PubMed]
56. Tilton, R.G.; Baier, L.D.; Harlow, J.E.; Smith, S.R.; Ostrow, E.; Williamson, J.R. Diabetes-induced glomerular dysfunction: Links to a more reduced cytosolic ratio of NADH/NAD+. *Kidney Int.* **1992**, *41*, 778–788. [CrossRef]
57. Luo, X.; Li, R.; Yan, L.J. Roles of Pyruvate, NADH, and Mitochondrial Complex I in Redox Balance and Imbalance in β Cell Function and Dysfunction. *J. Diabetes Res.* **2015**, *2015*, 512618. [CrossRef]
58. Luo, X.; Wu, J.; Jing, S.; Yan, L.J. Hyperglycemic stress and carbon stress in diabetic glucotoxicity. *Aging Dis.* **2016**, *7*, 90–110. [CrossRef]
59. Ido, Y. Pyridine nucleotide redox abnormalities in diabetes. *Antioxid. Redox Signal.* **2007**, *9*, 931–942. [CrossRef] [PubMed]
60. Chung, S.S.; Ho, E.C.; Lam, K.S.; Chung, S.K. Contribution of polyol pathway to diabetes-induced oxidative stress. *J. Am. Soc. Nephrol.* **2003**, *14*, S233–S236. [CrossRef]
61. Dunlop, M. Aldose reductase and the role of the polyol pathway in diabetic nephropathy. *Kidney Int. Suppl.* **2000**, *77*, S3–S12. [CrossRef]
62. Fantus, I.G. The pathogenesis of the chronic complications of the diabetes mellitus. *Endocrinol. Rounds* **2002**, *2*, 1–8.
63. Chatzilias, A.A.; Whiteside, C.I. Cellular mechanisms of glucose-induced myo-inositol transport upregulation in rat mesangial cells. *Am. J. Physiol.* **1994**, *267*, F459–F466. [CrossRef] [PubMed]
64. Yancey, P.H.; Haner, R.G.; Freudenberger, T.H. Effects of an aldose reductase inhibitor on organic osmotic effectors in rat renal medulla. *Am. J. Physiol.* **1990**, *259*, F733–F738. [CrossRef] [PubMed]

65. Brownlee, M. The pathobiology of diabetic complications: A unifying mechanism. *Diabetes* **2005**, *54*, 1615–1625. [CrossRef] [PubMed]
66. Veeresham, C.; Rama Rao, A.; Asres, K. Aldose reductase inhibitors of plant origin. *Phytother. Res.* **2014**, *28*, 317–333. [CrossRef] [PubMed]
67. El Gamal, H.; Eid, A.H.; Munusamy, S. Renoprotective Effects of Aldose Reductase Inhibitor Epalrestat against High Glucose-Induced Cellular Injury. *Biomed. Res. Int.* **2017**, *2017*, 5903105. [CrossRef] [PubMed]
68. Shukla, K.; Pal, P.B.; Sonowal, H.; Srivastava, S.K.; Ramana, K.V. Aldose Reductase Inhibitor Protects against Hyperglycemic Stress by Activating Nrf2-Dependent Antioxidant Proteins. *J. Diabetes Res.* **2017**, *2017*, 6785852. [CrossRef] [PubMed]
69. Artasensi, A.; Pedretti, A.; Vistoli, G.; Fumagalli, L. Type 2 Diabetes Mellitus: A Review of Multi-Target Drugs. *Molecules* **2020**, *25*, 1987. [CrossRef]
70. Sonowal, H.; Ramana, K.V. Development of aldose reductase inhibitors for the treatment of inflammatory disorders and cancer: Current drug design strategies and future directions. *Curr. Med. Chem.* **2020**. [CrossRef]
71. Jannapureddy, S.; Sharma, M.; Yepuri, G.; Schmidt, A.M.; Ramasamy, R. Aldose Reductase: An Emerging Target for Development of Interventions for Diabetic Cardiovascular Complications. *Front. Endocrinol.* **2021**, *12*, 636267. [CrossRef]
72. Zhou, X.; Liu, Z.; Ying, K.; Wang, H.; Liu, P.; Ji, X.; Chi, T.; Zou, L.; Wang, S.; He, Z. WJ-39, an Aldose Reductase Inhibitor, Ameliorates Renal Lesions in Diabetic Nephropathy by Activating Nrf2 Signaling. *Oxid. Med. Cell. Longev.* **2020**, *2020*, 7950457. [CrossRef]
73. Suzen, S.; Buyukbingol, E. Recent studies of aldose reductase enzyme inhibition for diabetic complications. *Curr. Med. Chem.* **2003**, *10*, 1329–1352. [CrossRef] [PubMed]
74. Reddy, A.B.; Ramana, K.V. Aldose reductase inhibition: Emerging drug target for the treatment of cardiovascular complications. *Recent Pat. Cardiovasc. Drug Discov.* **2010**, *5*, 25–32. [CrossRef]
75. Ohmura, C.; Watada, H.; Azuma, K.; Shimizu, T.; Kanazawa, A.; Ikeda, F.; Yoshihara, T.; Fujitani, Y.; Hirose, T.; Tanaka, Y.; et al. Aldose reductase inhibitor, epalrestat, reduces lipid hydroperoxides in type 2 diabetes. *Endocr. J.* **2009**, *56*, 149–156. [CrossRef]
76. Tang, J.; Du, Y.; Petrash, J.M.; Sheibani, N.; Kern, T.S. Deletion of aldose reductase from mice inhibits diabetes-induced retinal capillary degeneration and superoxide generation. *PLoS ONE* **2013**, *8*, e62081. [CrossRef]
77. Williamson, J.R.; Chang, K.; Frangos, M.; Hasan, K.S.; Ido, Y.; Kawamura, T.; Nyengaard, J.R.; van den Enden, M.; Kilo, C.; Tilton, R.G. Hyperglycemic pseudohypoxia and diabetic complications. *Diabetes* **1993**, *42*, 801–813. [CrossRef] [PubMed]
78. Ido, Y.; Williamson, J.R. Hyperglycemic cytosolic reductive stress 'pseudohypoxia': Implications for diabetic retinopathy. *Invest. Ophthalmol. Vis. Sci.* **1997**, *38*, 1467–1470.
79. Yang, L.; Garcia Canaveras, J.C.; Chen, Z.; Wang, L.; Liang, L.; Jang, C.; Mayr, J.A.; Zhang, Z.; Ghergurovich, J.M.; Zhan, L.; et al. Serine Catabolism Feeds NADH when Respiration Is Impaired. *Cell. Metab.* **2020**, *31*, 809–821. [CrossRef]
80. Harden, W.R., III; Barlow, C.H.; Simson, M.B.; Harken, A.H. Temporal relation between onset of cell anoxia and ischemic contractile failure. Myocardial ischemia and left ventricular failure in the isolated, perfused rabbit heart. *Am. J. Cardiol.* **1979**, *44*, 741–746. [CrossRef]
81. He, M.D.; Xu, S.C.; Zhang, X.; Wang, Y.; Xiong, J.C.; Zhang, X.; Lu, Y.H.; Zhang, L.; Yu, Z.P.; Zhou, Z. Disturbance of aerobic metabolism accompanies neurobehavioral changes induced by nickel in mice. *Neurotoxicology* **2013**, *38*, 9–16. [CrossRef]
82. Doke, T.; Ishimoto, T.; Hayasaki, T.; Ikeda, S.; Hasebe, M.; Hirayama, A.; Soga, T.; Kato, N.; Kosugi, T.; Tsuboi, N.; et al. Lacking ketohexokinase-A exacerbates renal injury in streptozotocin-induced diabetic mice. *Metabolism* **2018**, *85*, 161–170. [CrossRef]
83. Zakaria, E.M.; El-Maraghy, N.N.; Ahmed, A.F.; Ali, A.A.; El-Bassossy, H.M. PARP inhibition ameliorates nephropathy in an animal model of type 2 diabetes: Focus on oxidative stress, inflammation, and fibrosis. *Naunyn Schmiedebergs Arch. Pharmacol.* **2017**, *390*, 621–631. [CrossRef] [PubMed]
84. Massudi, H.; Grant, R.; Guillemin, G.J.; Braidy, N. NAD+ metabolism and oxidative stress: The golden nucleotide on a crown of thorns. *Redox Rep.* **2012**, *17*, 28–46. [CrossRef] [PubMed]
85. Nikiforov, A.; Kulikova, V.; Ziegler, M. The human NAD metabolome: Functions, metabolism and compartmentalization. *Crit. Rev. Biochem. Mol. Biol.* **2015**, *50*, 284–297. [CrossRef]
86. Pieper, A.A.; Brat, D.J.; Krug, D.K.; Watkins, C.C.; Gupta, A.; Blackshaw, S.; Verma, A.; Wang, Z.Q.; Snyder, S.H. Poly (ADP-ribose) polymerase-deficient mice are protected from streptozotocin-induced diabetes. *Proc. Natl. Acad. Sci. USA* **1999**, *96*, 3059–3064. [CrossRef]
87. Masutani, M.; Suzuki, H.; Kamada, N.; Watanabe, M.; Ueda, O.; Nozaki, T.; Jishage, K.; Watanabe, T.; Sugimoto, T.; Nakagama, H.; et al. Poly (ADP-ribose) polymerase gene disruption conferred mice resistant to streptozotocin-induced diabetes. *Proc. Natl. Acad. Sci. USA* **1999**, *96*, 2301–2304. [CrossRef]
88. Vedantham, S.; Thiagarajan, D.; Ananthakrishnan, R.; Wang, L.; Rosario, R.; Zou, Y.S.; Goldberg, I.; Yan, S.F.; Schmidt, A.M.; Ramasamy, R. Aldose reductase drives hyperacetylation of Egr-1 in hyperglycemia and consequent upregulation of proinflammatory and prothrombotic signals. *Diabetes* **2014**, *63*, 761–774. [CrossRef]
89. Sauve, A.A. Sirtuin chemical mechanisms. *Biochim. Biophys. Acta* **2010**, *1804*, 1591–1603. [CrossRef]
90. Wu, J.; Jin, Z.; Yan, L.J. Redox imbalance and mitochondrial abnormalities in the diabetic lung. *Redox Biol.* **2017**, *11*, 51–59. [CrossRef] [PubMed]
91. Chini, C.C.S.; Tarrago, M.G.; Chini, E.N. NAD and the aging process: Role in life, death and everything in between. *Mol. Cell. Endocrinol.* **2017**, *455*, 62–74. [CrossRef] [PubMed]

92. Benzi, A.; Sturla, L.; Heine, M.; Fischer, A.W.; Spinelli, S.; Magnone, M.; Sociali, G.; Parodi, A.; Fenoglio, D.; Emionite, L.; et al. CD38 downregulation modulates NAD+ and NADP (H) levels in thermogenic adipose tissues. *Biochim. Biophys. Acta Mol. Cell. Biol. Lipids* **2021**, *1866*, 158819. [CrossRef] [PubMed]
93. Peclat, T.R.; Shi, B.; Varga, J.; Chini, E.N. The NADase enzyme CD38: An emerging pharmacological target for systemic sclerosis, systemic lupus erythematosus and rheumatoid arthritis. *Curr. Opin. Rheumatol.* **2020**, *32*, 488–496. [CrossRef]
94. Covarrubias, A.J.; Kale, A.; Perrone, R.; Lopez-Dominguez, J.A.; Pisco, A.O.; Kasler, H.G.; Schmidt, M.S.; Heckenbach, I.; Kwok, R.; Wiley, C.D.; et al. Senescent cells promote tissue NAD+ decline during ageing via the activation of CD38+ macrophages. *Nat. Metab.* **2020**, *2*, 1265–1283. [CrossRef]
95. Zuo, W.; Liu, N.; Zeng, Y.; Liu, Y.; Li, B.; Wu, K.; Xiao, Y.; Liu, Q. CD38: A Potential Therapeutic Target in Cardiovascular Disease. *Cardiovasc. Drugs Ther.* **2020**, 1–14. [CrossRef]
96. Sun, L.; Adebanjo, O.A.; Moonga, B.S.; Corisdeo, S.; Anandatheerthavarada, H.K.; Biswas, G.; Arakawa, T.; Hakeda, Y.; Koval, A.; Sodam, B.; et al. CD38/ADP-ribosyl cyclase: A new role in the regulation of osteoclastic bone resorption. *J. Cell. Biol.* **1999**, *146*, 1161–1172. [CrossRef] [PubMed]
97. Shi, B.; Wang, W.; Korman, B.; Kai, L.; Wang, Q.; Wei, J.; Bale, S.; Marangoni, R.G.; Bhattacharyya, S.; Miller, S.; et al. Targeting CD38-dependent NAD+ metabolism to mitigate multiple organ fibrosis. *iScience* **2021**, *24*, 101902. [CrossRef] [PubMed]
98. Ogura, Y.; Kitada, M.; Xu, J.; Monno, I.; Koya, D. CD38 inhibition by apigenin ameliorates mitochondrial oxidative stress through restoration of the intracellular NAD+/NADH ratio and Sirt3 activity in renal tubular cells in diabetic rats. *Aging* **2020**, *12*, 11325–11336. [CrossRef] [PubMed]
99. Tedeschi, P.M.; Bansal, N.; Kerrigan, J.E.; Abali, E.E.; Scotto, K.W.; Bertino, J.R. NAD+ Kinase as a Therapeutic Target in Cancer. *Clin. Cancer Res.* **2016**, *22*, 5189–5195. [CrossRef]
100. Zhang, R. MNADK, a Long-Awaited Human Mitochondrion-Localized NAD Kinase. *J. Cell. Physiol.* **2015**, *230*, 1697–1701. [CrossRef]
101. Shi, F.; Li, Y.; Li, Y.; Wang, X. Molecular properties, functions, and potential applications of NAD kinases. *Acta Biochim. Biophys. Sin.* **2009**, *41*, 352–361. [CrossRef] [PubMed]
102. Hoxhaj, G.; Ben-Sahra, I.; Lockwood, S.E.; Timson, R.C.; Byles, V.; Henning, G.T.; Gao, P.; Selfors, L.M.; Asara, J.M.; Manning, B.D. Direct stimulation of NADP+ synthesis through Akt-mediated phosphorylation of NAD kinase. *Science* **2019**, *363*, 1088–1092. [CrossRef] [PubMed]
103. Chen, L.; Zhang, Z.; Hoshino, A.; Zheng, H.D.; Morley, M.; Arany, Z.; Rabinowitz, J.D. NADPH production by the oxidative pentose-phosphate pathway supports folate metabolism. *Nat. Metab.* **2019**, *1*, 404–415. [CrossRef]
104. Spencer, N.Y.; Stanton, R.C. Glucose 6-phosphate dehydrogenase and the kidney. *Curr. Opin. Nephrol. Hypertens.* **2017**, *26*, 43–49. [CrossRef]
105. Yan, L.J.; Christians, E.S.; Liu, L.; Xiao, X.; Sohal, R.S.; Benjamin, I.J. Mouse heat shock transcription factor 1 deficiency alters cardiac redox homeostasis and increases mitochondrial oxidative damage. *EMBO J.* **2002**, *21*, 5164–5172. [CrossRef]
106. Ogura, Y.; Kitada, M.; Monno, I.; Kanasaki, K.; Watanabe, A.; Koya, D. Renal mitochondrial oxidative stress is enhanced by the reduction of Sirt3 activity, in Zucker diabetic fatty rats. *Redox Rep.* **2018**, *23*, 153–159. [CrossRef] [PubMed]
107. Forbes, J.M.; Coughlan, M.T.; Cooper, M.E. Oxidative stress as a major culprit in kidney disease in diabetes. *Diabetes* **2008**, *57*, 1446–1454. [CrossRef] [PubMed]
108. Turrens, J.F. Superoxide production by the mitochondrial respiratory chain. *Biosci. Rep.* **1997**, *17*, 3–8. [CrossRef]
109. Turrens, J.F.; Alexandre, A.; Lehninger, A.L. Ubisemiquinone is the electron donor for superoxide formation by complex III of heart mitochondria. *Arch. Biochem. Biophys.* **1985**, *237*, 408–414. [CrossRef]
110. Vujic, A.; Koo, A.N.M.; Prag, H.A.; Krieg, T. Mitochondrial redox and TCA cycle metabolite signaling in the heart. *Free Radic. Biol. Med.* **2021**, *166*, 287–296. [CrossRef] [PubMed]
111. Lee, S.R.; An, E.J.; Kim, J.; Bae, Y.S. Function of NADPH Oxidases in Diabetic Nephropathy and Development of Nox Inhibitors. *Biomol. Ther.* **2020**, *28*, 25–33. [CrossRef]
112. Granata, S.; Dalla Gassa, A.; Tomei, P.; Lupo, A.; Zaza, G. Mitochondria: A new therapeutic target in chronic kidney disease. *Nutr. Metab.* **2015**, *12*, 49. [CrossRef]
113. Wu, J.; Luo, X.; Thangthaeng, N.; Sumien, N.; Chen, Z.; Rutledge, M.A.; Jing, S.; Forster, M.J.; Yan, L.J. Pancreatic mitochondrial complex I exhibits aberrant hyperactivity in diabetes. *Biochem. Biophys. Rep.* **2017**, *11*, 119–129. [CrossRef]
114. Murphy, M.P. How mitochondria produce reactive oxygen species. *Biochem. J.* **2009**, *417*, 1–13. [CrossRef]
115. Drose, S.; Brandt, U. Molecular mechanisms of superoxide production by the mitochondrial respiratory chain. *Adv. Exp. Med. Biol.* **2012**, *748*, 145–169.
116. Yan, L.J.; Sumien, N.; Thangthaeng, N.; Forster, M.J. Reversible inactivation of dihydrolipoamide dehydrogenase by mitochondrial hydrogen peroxide. *Free Radic. Res.* **2013**, *47*, 123–133. [CrossRef]
117. Zheng, H.; Wu, J.; Jin, Z.; Yan, L.J. Protein Modifications as Manifestations of Hyperglycemic Glucotoxicity in Diabetes and Its Complications. *Biochem. Insights* **2016**, *9*, 1–9. [CrossRef]
118. Yan, L.J.; Lodge, J.K.; Traber, M.G.; Matsugo, S.; Packer, L. Comparison between copper-mediated and hypochlorite-mediated modifications of human low density lipoproteins evaluated by protein carbonyl formation. *J. Lipid Res.* **1997**, *38*, 992–1001. [CrossRef]
119. Ames, B.N.; Shigenaga, M.K. Oxidants are a major contributor to aging. *Ann. N. Y. Acad. Sci.* **1992**, *663*, 85–96. [CrossRef] [PubMed]

120. Stefek, M.; Gajdosik, A.; Tribulova, N.; Navarova, J.; Volkovova, K.; Weismann, P.; Gajdosikova, A.; Drimal, J.; Mihalova, D. The pyridoindole antioxidant stobadine attenuates albuminuria, enzymuria, kidney lipid peroxidation and matrix collagen cross-linking in streptozotocin-induced diabetic rats. *Methods Find. Exp. Clin. Pharmacol.* **2002**, *24*, 565–571.
121. Wang, W.X.; Luo, S.B.; Jiang, P.; Xia, M.M.; Hei, A.L.; Mao, Y.H.; Li, C.B.; Hu, G.X.; Cai, J.P. Increased Oxidative Damage of RNA in Early-Stage Nephropathy in db/db Mice. *Oxid. Med. Cell. Longev.* **2017**, *2017*, 2353729. [CrossRef]
122. Wu, J.; Luo, X.; Jing, S.; Yan, L.J. Two-dimensional gel electrophoretic detection of protein carbonyls derivatized with biotin-hydrazide. *J. Chromatogr. B* **2016**, *1019*, 128–131. [CrossRef]
123. Bhargava, P.; Schnellmann, R.G. Mitochondrial energetics in the kidney. *Nat. Rev. Nephrol.* **2017**, *13*, 629–646. [CrossRef] [PubMed]
124. Higgins, G.C.; Coughlan, M.T. Mitochondrial dysfunction and mitophagy: The beginning and end to diabetic nephropathy? *Br. J. Pharmacol.* **2014**, *171*, 1917–1942. [CrossRef] [PubMed]
125. Yi, H.S.; Chang, J.Y.; Shong, M. The mitochondrial unfolded protein response and mitohormesis: A perspective on metabolic diseases. *J. Mol. Endocrinol.* **2018**, *61*, R91–R105. [CrossRef] [PubMed]
126. Galvan, D.L.; Green, N.H.; Danesh, F.R. The hallmarks of mitochondrial dysfunction in chronic kidney disease. *Kidney Int.* **2017**, *92*, 1051–1057. [CrossRef]
127. Ayanga, B.A.; Badal, S.S.; Wang, Y.; Galvan, D.L.; Chang, B.H.; Schumacker, P.T.; Danesh, F.R. Dynamin-Related Protein 1 Deficiency Improves Mitochondrial Fitness and Protects against Progression of Diabetic Nephropathy. *J. Am. Soc. Nephrol.* **2016**, *27*, 2733–2747. [CrossRef]
128. Galvan, D.L.; Long, J.; Green, N.; Chang, B.H.; Lin, J.S.; Schumacker, P.; Truong, L.D.; Overbeek, P.; Danesh, F.R. Drp1S600 phosphorylation regulates mitochondrial fission and progression of nephropathy in diabetic mice. *J. Clin. Investig.* **2019**, *129*, 2807–2823. [CrossRef]
129. Wang, Y.; Lu, M.; Xiong, L.; Fan, J.; Zhou, Y.; Li, H.; Peng, X.; Zhong, Z.; Wang, Y.; Huang, F.; et al. Drp1-mediated mitochondrial fission promotes renal fibroblast activation and fibrogenesis. *Cell. Death Dis.* **2020**, *11*, 29. [CrossRef] [PubMed]
130. Wang, Y.; Zhang, X.; Wang, P.; Shen, Y.; Yuan, K.; Li, M.; Liang, W.; Que, H. Sirt3 overexpression alleviates hyperglycemia-induced vascular inflammation through regulating redox balance, cell survival, and AMPK-mediated mitochondrial homeostasis. *J. Recept. Signal Transduct.* **2019**, *39*, 341–349. [CrossRef] [PubMed]
131. Wang, Y.; Katayama, A.; Terami, T.; Han, X.; Nunoue, T.; Zhang, D.; Teshigawara, S.; Eguchi, J.; Nakatsuka, A.; Murakami, K.; et al. Translocase of inner mitochondrial membrane 44 alters the mitochondrial fusion and fission dynamics and protects from type 2 diabetes. *Metabolism* **2015**, *64*, 677–688. [CrossRef]
132. Agil, A.; Chayah, M.; Visiedo, L.; Navarro-Alarcon, M.; Rodriguez Ferrer, J.M.; Tassi, M.; Reiter, R.J.; Fernandez-Vazquez, G. Melatonin Improves Mitochondrial Dynamics and Function in the Kidney of Zucker Diabetic Fatty Rats. *J. Clin. Med.* **2020**, *9*, 2916. [CrossRef]
133. Zhan, M.; Usman, I.; Yu, J.; Ruan, L.; Bian, X.; Yang, J.; Yang, S.; Sun, L.; Kanwar, Y.S. Perturbations in mitochondrial dynamics by p66Shc lead to renal tubular oxidative injury in human diabetic nephropathy. *Clin. Sci.* **2018**, *132*, 1297–1314. [CrossRef] [PubMed]
134. Mishra, R.; Emancipator, S.N.; Kern, T.; Simonson, M.S. High glucose evokes an intrinsic proapoptotic signaling pathway in mesangial cells. *Kidney Int.* **2005**, *67*, 82–93. [CrossRef]
135. Coughlan, M.T.; Higgins, G.C.; Nguyen, T.V.; Penfold, S.A.; Thallas-Bonke, V.; Tan, S.M.; Ramm, G.; van Bergen, N.J.; Henstridge, D.C.; Sourris, K.C.; et al. Deficiency in Apoptosis-Inducing Factor Recapitulates Chronic Kidney Disease via Aberrant Mitochondrial Homeostasis. *Diabetes* **2016**, *65*, 1085–1098. [CrossRef]
136. Kostic, S.; Hauke, T.; Ghahramani, N.; Filipovic, N.; Vukojevic, K. Expression pattern of apoptosis-inducing factor in the kidneys of streptozotocin-induced diabetic rats. *Acta Histochem.* **2020**, *122*, 151655. [CrossRef]
137. Zhao, Y.; Guo, Y.; Jiang, Y.; Zhu, X.; Liu, Y.; Zhang, X. Mitophagy regulates macrophage phenotype in diabetic nephropathy rats. *Biochem. Biophys. Res. Commun.* **2017**, *494*, 42–50. [CrossRef] [PubMed]
138. Zuo, Z.; Jing, K.; Wu, H.; Wang, S.; Ye, L.; Li, Z.; Yang, C.; Pan, Q.; Liu, W.J.; Liu, H.F. Mechanisms and Functions of Mitophagy and Potential Roles in Renal Disease. *Front. Physiol.* **2020**, *11*, 935. [CrossRef] [PubMed]
139. Nguyen, T.N.; Padman, B.S.; Lazarou, M. Deciphering the Molecular Signals of PINK1/Parkin Mitophagy. *Trends Cell Biol.* **2016**, *26*, 733–744. [CrossRef]
140. Springer, M.Z.; Macleod, K.F. In Brief: Mitophagy: Mechanisms and role in human disease. *J. Pathol.* **2016**, *240*, 253–255. [CrossRef] [PubMed]
141. Persson, M.F.; Franzen, S.; Catrina, S.B.; Dallner, G.; Hansell, P.; Brismar, K.; Palm, F. Coenzyme Q10 prevents GDP-sensitive mitochondrial uncoupling, glomerular hyperfiltration and proteinuria in kidneys from db/db mice as a model of type 2 diabetes. *Diabetologia* **2012**, *55*, 1535–1543. [CrossRef]
142. Fabris, B.; Candido, R.; Armini, L.; Fischetti, F.; Calci, M.; Bardelli, M.; Fazio, M.; Campanacci, L.; Carretta, R. Control of glomerular hyperfiltration and renal hypertrophy by an angiotensin converting enzyme inhibitor prevents the progression of renal damage in hypertensive diabetic rats. *J. Hypertens.* **1999**, *17*, 1925–1931. [CrossRef] [PubMed]
143. Tan, A.L.; Sourris, K.C.; Harcourt, B.E.; Thallas-Bonke, V.; Penfold, S.; Andrikopoulos, S.; Thomas, M.C.; O'Brien, R.C.; Bierhaus, A.; Cooper, M.E.; et al. Disparate effects on renal and oxidative parameters following RAGE deletion, AGE accumulation inhibition, or dietary AGE control in experimental diabetic nephropathy. *Am. J. Physiol. Renal Physiol.* **2010**, *298*, F763–F770. [CrossRef] [PubMed]

144. Brand, M.D. Riding the tiger—Physiological and pathological effects of superoxide and hydrogen peroxide generated in the mitochondrial matrix. *Crit. Rev. Biochem. Mol. Biol.* **2020**, *55*, 592–661. [CrossRef]
145. Brand, M.D.; Goncalves, R.L.; Orr, A.L.; Vargas, L.; Gerencser, A.A.; Borch Jensen, M.; Wang, Y.T.; Melov, S.; Turk, C.N.; Matzen, J.T.; et al. Suppressors of Superoxide-H_2O_2 Production at Site IQ of Mitochondrial Complex I Protect against Stem Cell Hyperplasia and Ischemia-Reperfusion Injury. *Cell Metab.* **2016**, *24*, 582–592. [CrossRef]
146. Apakkan Aksun, S.; Ozmen, B.; Ozmen, D.; Parildar, Z.; Senol, B.; Habif, S.; Mutaf, I.; Turgan, N.; Bayindir, O. Serum and urinary nitric oxide in Type 2 diabetes with or without microalbuminuria: Relation to glomerular hyperfiltration. *J. Diabetes Complicat.* **2003**, *17*, 343–348. [CrossRef]
147. Fang, J.; Wong, H.S.; Brand, M.D. Production of superoxide and hydrogen peroxide in the mitochondrial matrix is dominated by site IQ of complex I in diverse cell lines. *Redox Biol.* **2020**, *37*, 101722. [CrossRef] [PubMed]
148. Kuo, C.W.; Shen, C.J.; Tung, Y.T.; Chen, H.L.; Chen, Y.H.; Chang, W.H.; Cheng, K.C.; Yang, S.H.; Chen, C.M. Extracellular superoxide dismutase ameliorates streptozotocin-induced rat diabetic nephropathy via inhibiting the ROS/ERK1/2 signaling. *Life Sci.* **2015**, *135*, 77–86. [CrossRef] [PubMed]
149. Watson, M.A.; Wong, H.S.; Brand, M.D. Use of S1QELs and S3QELs to link mitochondrial sites of superoxide and hydrogen peroxide generation to physiological and pathological outcomes. *Biochem. Soc. Trans.* **2019**, *47*, 1461–1469. [CrossRef] [PubMed]
150. Plecita-Hlavata, L.; Engstova, H.; Jezek, J.; Holendova, B.; Tauber, J.; Petraskova, L.; Kren, V.; Jezek, P. Potential of Mitochondria-Targeted Antioxidants to Prevent Oxidative Stress in Pancreatic beta-cells. *Oxid. Med. Cell. Longev.* **2019**, *2019*, 1826303. [CrossRef]
151. Homma, T.; Kobayashi, S.; Sato, H.; Fujii, J. Superoxide produced by mitochondrial complex III plays a pivotal role in the execution of ferroptosis induced by cysteine starvation. *Arch. Biochem. Biophys.* **2021**, *700*, 108775. [CrossRef]
152. Heslop, K.A.; Rovini, A.; Hunt, E.G.; Fang, D.; Morris, M.E.; Christie, C.F.; Gooz, M.B.; DeHart, D.N.; Dang, Y.; Lemasters, J.J.; et al. JNK activation and translocation to mitochondria mediates mitochondrial dysfunction and cell death induced by VDAC opening and sorafenib in hepatocarcinoma cells. *Biochem. Pharmacol.* **2020**, *171*, 113728. [CrossRef] [PubMed]
153. Wong, H.S.; Mezera, V.; Dighe, P.; Melov, S.; Gerencser, A.A.; Sweis, R.F.; Pliushchev, M.; Wang, Z.; Esbenshade, T.; McKibben, B.; et al. Superoxide produced by mitochondrial site IQ inactivates cardiac succinate dehydrogenase and induces hepatic steatosis in Sod2 knockout mice. *Free Radic. Biol. Med.* **2021**, *164*, 223–232. [CrossRef]
154. Hatinguais, R.; Pradhan, A.; Brown, G.D.; Brown, A.J.P.; Warris, A.; Shekhova, E. Mitochondrial Reactive Oxygen Species Regulate Immune Responses of Macrophages to Aspergillus fumigatus. *Front. Immunol.* **2021**, *12*, 641495. [CrossRef]
155. Dare, A.J.; Bolton, E.A.; Pettigrew, G.J.; Bradley, J.A.; Saeb-Parsy, K.; Murphy, M.P. Protection against renal ischemia-reperfusion injury in vivo by the mitochondria targeted antioxidant MitoQ. *Redox Biol.* **2015**, *5*, 163–168. [CrossRef]
156. Dare, A.J.; Logan, A.; Prime, T.A.; Rogatti, S.; Goddard, M.; Bolton, E.M.; Bradley, J.A.; Pettigrew, G.J.; Murphy, M.P.; Saeb-Parsy, K. The mitochondria-targeted anti-oxidant MitoQ decreases ischemia-reperfusion injury in a murine syngeneic heart transplant model. *J. Heart Lung Transplant.* **2015**, *34*, 1471–1480. [CrossRef]
157. James, A.M.; Sharpley, M.S.; Manas, A.R.; Frerman, F.E.; Hirst, J.; Smith, R.A.; Murphy, M.P. Interaction of the mitochondria-targeted antioxidant MitoQ with phospholipid bilayers and ubiquinone oxidoreductases. *J. Biol. Chem.* **2007**, *282*, 14708–14718. [CrossRef] [PubMed]
158. Saretzki, G.; Murphy, M.P.; von Zglinicki, T. MitoQ counteracts telomere shortening and elongates lifespan of fibroblasts under mild oxidative stress. *Aging Cell* **2003**, *2*, 141–143. [CrossRef]
159. Yang, M.Y.; Fan, Z.; Zhang, Z.; Fan, J. MitoQ protects against high glucose-induced brain microvascular endothelial cells injury via the Nrf2/HO-1 pathway. *J. Pharmacol. Sci.* **2021**, *145*, 105–114. [CrossRef]
160. Hou, L.; Zhang, J.; Liu, Y.; Fang, H.; Liao, L.; Wang, Z.; Yuan, J.; Wang, X.; Sun, J.; Tang, B.; et al. MitoQ alleviates LPS-mediated acute lung injury through regulating Nrf2/Drp1 pathway. *Free Radic. Biol. Med.* **2021**, *165*, 219–228. [CrossRef]
161. Cen, M.; Ouyang, W.; Zhang, W.; Yang, L.; Lin, X.; Dai, M.; Hu, H.; Tang, H.; Liu, H.; Xia, J.; et al. MitoQ protects against hyperpermeability of endothelium barrier in acute lung injury via a Nrf2-dependent mechanism. *Redox Biol.* **2021**, *41*, 101936. [CrossRef]
162. Fortner, K.A.; Blanco, L.P.; Buskiewicz, I.; Huang, N.; Gibson, P.C.; Cook, D.L.; Pedersen, H.L.; Yuen, P.S.T.; Murphy, M.P.; Perl, A.; et al. Targeting mitochondrial oxidative stress with MitoQ reduces NET formation and kidney disease in lupus-prone MRL-lpr mice. *Lupus Sci. Med.* **2020**, *7*, e000387. [CrossRef]
163. Chacko, B.K.; Reily, C.; Srivastava, A.; Johnson, M.S.; Ye, Y.; Ulasova, E.; Agarwal, A.; Zinn, K.R.; Murphy, M.P.; Kalyanaraman, B.; et al. Prevention of diabetic nephropathy in Ins2+/- AkitaJ mice by the mitochondria-targeted therapy MitoQ. *Biochem. J.* **2010**, *432*, 9–19. [CrossRef] [PubMed]
164. Xiao, L.; Xu, X.; Zhang, F.; Wang, M.; Xu, Y.; Tang, D.; Wang, J.; Qin, Y.; Liu, Y.; Tang, C.; et al. The mitochondria-targeted antioxidant MitoQ ameliorated tubular injury mediated by mitophagy in diabetic kidney disease via Nrf2/PINK1. *Redox Biol.* **2017**, *11*, 297–311. [CrossRef] [PubMed]
165. Ward, M.S.; Flemming, N.B.; Gallo, L.A.; Fotheringham, A.K.; McCarthy, D.A.; Zhuang, A.; Tang, P.H.; Borg, D.J.; Shaw, H.; Harvie, B.; et al. Targeted mitochondrial therapy using MitoQ shows equivalent renoprotection to angiotensin converting enzyme inhibition but no combined synergy in diabetes. *Sci. Rep.* **2017**, *7*, 15190. [CrossRef] [PubMed]
166. Gao, P.; Yang, M.; Chen, X.; Xiong, S.; Liu, J.; Sun, L. DsbA-L deficiency exacerbates mitochondrial dysfunction of tubular cells in diabetic kidney disease. *Clin. Sci.* **2020**, *134*, 677–694. [CrossRef]

167. Han, Y.; Xu, X.; Tang, C.; Gao, P.; Chen, X.; Xiong, X.; Yang, M.; Yang, S.; Zhu, X.; Yuan, S.; et al. Reactive oxygen species promote tubular injury in diabetic nephropathy: The role of the mitochondrial ros-txnip-nlrp3 biological axis. *Redox Biol.* **2018**, *16*, 32–46. [CrossRef]
168. Fujita, H.; Fujishima, H.; Chida, S.; Takahashi, K.; Qi, Z.; Kanetsuna, Y.; Breyer, M.D.; Harris, R.C.; Yamada, Y.; Takahashi, T. Reduction of renal superoxide dismutase in progressive diabetic nephropathy. *J. Am. Soc. Nephrol.* **2009**, *20*, 1303–1313. [CrossRef]
169. Peixoto, E.B.; Pessoa, B.S.; Biswas, S.K.; Lopes de Faria, J.B. Antioxidant SOD mimetic prevents NADPH oxidase-induced oxidative stress and renal damage in the early stage of experimental diabetes and hypertension. *Am. J. Nephrol.* **2009**, *29*, 309–318. [CrossRef] [PubMed]
170. Asaba, K.; Tojo, A.; Onozato, M.L.; Goto, A.; Fujita, T. Double-edged action of SOD mimetic in diabetic nephropathy. *J. Cardiovasc. Pharmacol.* **2007**, *49*, 13–19. [CrossRef]
171. Rafikova, O.; Salah, E.M.; Tofovic, S.P. Renal and metabolic effects of tempol in obese ZSF1 rats—Distinct role for superoxide and hydrogen peroxide in diabetic renal injury. *Metabolism* **2008**, *57*, 1434–1444. [CrossRef] [PubMed]
172. El-Mahdy, M.A.; Alzarie, Y.A.; Hemann, C.; Badary, O.A.; Nofal, S.; Zweier, J.L. The novel SOD mimetic GC4419 increases cancer cell killing with sensitization to ionizing radiation while protecting normal cells. *Free Radic. Biol. Med.* **2020**, *160*, 630–642. [CrossRef] [PubMed]
173. Anderson, C.M.; Lee, C.M.; Saunders, D.P.; Curtis, A.; Dunlap, N.; Nangia, C.; Lee, A.S.; Gordon, S.M.; Kovoor, P.; Arevalo-Araujo, R.; et al. Phase IIb, Randomized, Double-Blind Trial of GC4419 Versus Placebo to Reduce Severe Oral Mucositis Due to Concurrent Radiotherapy and Cisplatin for Head and Neck Cancer. *J. Clin. Oncol.* **2019**, *37*, 3256–3265. [CrossRef] [PubMed]
174. Langan, A.R.; Khan, M.A.; Yeung, I.W.; van Dyk, J.; Hill, R.P. Partial volume rat lung irradiation: The protective/mitigating effects of Eukarion-189, a superoxide dismutase-catalase mimetic. *Radiother. Oncol.* **2006**, *79*, 231–238. [CrossRef]
175. Garcia-Quintans, N.; Prieto, I.; Sanchez-Ramos, C.; Luque, A.; Arza, E.; Olmos, Y.; Monsalve, M. Regulation of endothelial dynamics by PGC-1alpha relies on ROS control of VEGF-A signaling. *Free Radic. Biol. Med.* **2016**, *93*, 41–51. [CrossRef]
176. Hosakote, Y.M.; Komaravelli, N.; Mautemps, N.; Liu, T.; Garofalo, R.P.; Casola, A. Antioxidant mimetics modulate oxidative stress and cellular signaling in airway epithelial cells infected with respiratory syncytial virus. *Am. J. Physiol. Lung Cell. Mol. Physiol.* **2012**, *303*, L991–L1000. [CrossRef]
177. Torbati, F.A.; Ramezani, M.; Dehghan, R.; Amiri, M.S.; Moghadam, A.T.; Shakour, N.; Elyasi, S.; Sahebkar, A.; Emami, S.A. Ethnobotany, Phytochemistry and Pharmacological Features of Centella asiatica: A Comprehensive Review. *Adv. Exp. Med. Biol.* **2021**, *1308*, 451–499. [CrossRef] [PubMed]
178. Liu, I.M.; Tzeng, T.F.; Liou, S.S.; Chang, C.J. Angelica acutiloba root attenuates insulin resistance induced by high-fructose diet in rats. *Phytother. Res.* **2011**, *25*, 1283–1293. [CrossRef] [PubMed]
179. Wang, H.; Guan, Y.; Widlund, A.L.; Becker, L.B.; Baur, J.A.; Reilly, P.M.; Sims, C.A. Resveratrol ameliorates mitochondrial dysfunction but increases the risk of hypoglycemia following hemorrhagic shock. *J. Trauma Acute Care Surg.* **2014**, *77*, 926–933. [CrossRef]
180. Bao, L.; Cai, X.; Zhang, Z.; Li, Y. Grape seed procyanidin B2 ameliorates mitochondrial dysfunction and inhibits apoptosis via the AMP-activated protein kinase-silent mating type information regulation 2 homologue 1-PPARgamma co-activator-1alpha axis in rat mesangial cells under high-dose glucosamine. *Br. J. Nutr.* **2015**, *113*, 35–44. [CrossRef]
181. Guerrero-Hue, M.; Rayego-Mateos, S.; Vazquez-Carballo, C.; Palomino-Antolin, A.; Garcia-Caballero, C.; Opazo-Rios, L.; Morgado-Pascual, J.L.; Herencia, C.; Mas, S.; Ortiz, A.; et al. Protective Role of Nrf2 in Renal Disease. *Antioxidants* **2020**, *10*, 39. [CrossRef]
182. Menini, S.; Iacobini, C.; Vitale, M.; Pugliese, G. The Inflammasome in Chronic Complications of Diabetes and Related Metabolic Disorders. *Cells* **2020**, *9*, 1812. [CrossRef]
183. Sato, S.; Kataoka, S.; Kimura, A.; Mukai, Y. Azuki bean (Vigna angularis) extract reduces oxidative stress and stimulates autophagy in the kidneys of streptozotocin-induced early diabetic rats. *Can. J. Physiol. Pharmacol.* **2016**, *94*, 1298–1303. [CrossRef]
184. Omara, E.A.; Nada, S.A.; Farrag, A.R.; Sharaf, W.M.; El-Toumy, S.A. Therapeutic effect of Acacia nilotica pods extract on streptozotocin induced diabetic nephropathy in rat. *Phytomedicine* **2012**, *19*, 1059–1067. [CrossRef]
185. Navale, A.M.; Paranjape, A. Antidiabetic and renoprotective effect of Anogeissus acuminata leaf extract on experimentally induced diabetic nephropathy. *J. Basic Clin. Physiol. Pharmacol.* **2018**, *29*, 359–364. [CrossRef] [PubMed]
186. Cho, E.J.; Lee, Y.A.; Yoo, H.H.; Yokozawa, T. Protective effects of broccoli (Brassica oleracea) against oxidative damage in vitro and in vivo. *J. Nutr. Sci. Vitaminol.* **2006**, *52*, 437–444. [CrossRef]
187. ALTamimi, J.Z.; AlFaris, N.A.; Al-Farga, A.M.; Alshammari, G.M.; BinMowyna, M.N.; Yahya, M.A. Curcumin reverses diabetic nephropathy in streptozotocin-induced diabetes in rats by inhibition of PKCβ/p66Shc axis and activation of FOXO-3a. *J. Nutr. Biochem.* **2021**, *87*, 108515. [CrossRef]
188. Gurukar, M.S.; Mahadevamma, S.; Chilkunda, N.D. Renoprotective effect of Coccinia indica fruits and leaves in experimentally induced diabetic rats. *J. Med. Food* **2013**, *16*, 839–846. [CrossRef]
189. Boonphang, O.; Ontawong, A.; Pasachan, T.; Phatsara, M.; Duangjai, A.; Amornlerdpison, D.; Jinakote, M.; Srimaroeng, C. Antidiabetic and Renoprotective Effects of Coffea arabica Pulp Aqueous Extract through Preserving Organic Cation Transport System Mediated Oxidative Stress Pathway in Experimental Type 2 Diabetic Rats. *Molecules* **2021**, *26*, 1907. [CrossRef]
190. Hassan, H.M.; Mahran, Y.F.; Ghanim, A.M.H. Ganoderma lucidum ameliorates the diabetic nephropathy via down-regulatory effect on TGFβ-1 and TLR-4/NFκB signalling pathways. *J. Pharm. Pharmacol.* **2021**. [CrossRef]

191. Shiju, T.M.; Rajesh, N.G.; Viswanathan, P. Renoprotective effect of aged garlic extract in streptozotocin-induced diabetic rats. *Indian J. Pharmacol.* **2013**, *45*, 18–23. [CrossRef]
192. Chung, A.; Gurtu, S.; Chakravarthi, S.; Moorthy, M.; Palanisamy, U.D. Geraniin Protects High-Fat Diet-Induced Oxidative Stress in Sprague Dawley Rats. *Front. Nutr.* **2018**, *5*, 17. [CrossRef]
193. Al Hroob, A.M.; Abukhalil, M.H.; Alghonmeen, R.D.; Mahmoud, A.M. Ginger alleviates hyperglycemia-induced oxidative stress, inflammation and apoptosis and protects rats against diabetic nephropathy. *Biomed. Pharmacother.* **2018**, *106*, 381–389. [CrossRef]
194. Han, J.; Pang, X.; Shi, X.; Zhang, Y.; Peng, Z.; Xing, Y. Ginkgo Biloba Extract EGB761 Ameliorates the Extracellular Matrix Accumulation and Mesenchymal Transformation of Renal Tubules in Diabetic Kidney Disease by Inhibiting Endoplasmic Reticulum Stress. *Biomed. Res. Int.* **2021**, *2021*, 6657206. [CrossRef]
195. Qin, X.; Zhao, Y.; Gong, J.; Huang, W.; Su, H.; Yuan, F.; Fang, K.; Wang, D.; Li, J.; Zou, X.; et al. Berberine Protects Glomerular Podocytes via Inhibiting Drp1-Mediated Mitochondrial Fission and Dysfunction. *Theranostics* **2019**, *9*, 1698–1713. [CrossRef]
196. Punaro, G.R.; Lima, D.Y.; Rodrigues, A.M.; Pugliero, S.; Mouro, M.G.; Rogero, M.M.; Higa, E.M.S. Cupuacu extract reduces nitrosative stress and modulates inflammatory mediators in the kidneys of experimental diabetes. *Clin. Nutr.* **2019**, *38*, 364–371. [CrossRef]
197. Alabi, T.D.; Brooks, N.L.; Oguntibeju, O.O. Leaf Extracts of Anchomanes difformis Ameliorated Kidney and Pancreatic Damage in Type 2 Diabetes. *Plants* **2021**, *10*, 300. [CrossRef] [PubMed]
198. Kim, H.; Dusabimana, T.; Kim, S.R.; Je, J.; Jeong, K.; Kang, M.C.; Cho, K.M.; Kim, H.J.; Park, S.W. Supplementation of Abelmoschus manihot Ameliorates Diabetic Nephropathy and Hepatic Steatosis by Activating Autophagy in Mice. *Nutrients* **2018**, *10*, 1703. [CrossRef]
199. Wang, S.C.; Lee, S.F.; Wang, C.J.; Lee, C.H.; Lee, W.C.; Lee, H.J. Aqueous Extract from Hibiscus sabdariffa Linnaeus Ameliorate Diabetic Nephropathy via Regulating Oxidative Status and Akt/Bad/14-3-3γ in an Experimental Animal Model. *Evid. Based Complement. Alternat. Med.* **2011**, *2011*, 938126. [CrossRef]
200. Zhang, Q.; Lu, Y.; Ma, Z.; Li, Y.; Guo, J.; Meng, Q.; Bian, H. A novel formula from mulberry leaf ameliorates diabetic nephropathy in rats via inhibiting the TGF-β1 pathway. *Food Funct.* **2015**, *6*, 3307–3315. [CrossRef]
201. Lu, H.J.; Tzeng, T.F.; Hsu, J.C.; Kuo, S.H.; Chang, C.H.; Huang, S.Y.; Chang, F.Y.; Wu, M.C.; Liu, I.M. An Aqueous-Ethanol Extract of Liriope spicata var. prolifera Ameliorates Diabetic Nephropathy through Suppression of Renal Inflammation. *Evid. Based Complement. Alternat. Med.* **2013**, *2013*, 201643. [CrossRef]
202. Chen, H.W.; Yang, M.Y.; Hung, T.W.; Chang, Y.C.; Wang, C.J. Nelumbo nucifera leaves extract attenuate the pathological progression of diabetic nephropathy in high-fat diet-fed and streptozotocin-induced diabetic rats. *J. Food Drug Anal.* **2019**, *27*, 736–748. [CrossRef]
203. Yao, L.; Li, L.; Li, X.; Li, H.; Zhang, Y.; Zhang, R.; Wang, J.; Mao, X. The anti-inflammatory and antifibrotic effects of Coreopsis tinctoria Nutt on high-glucose-fat diet and streptozotocin-induced diabetic renal damage in rats. *BMC Complement. Altern. Med.* **2015**, *15*, 314. [CrossRef]
204. Rajavel, V.; Abdul Sattar, M.Z.; Abdulla, M.A.; Kassim, N.M.; Abdullah, N.A. Chronic Administration of Oil Palm (Elaeis guineensis) Leaves Extract Attenuates Hyperglycaemic-Induced Oxidative Stress and Improves Renal Histopathology and Function in Experimental Diabetes. *Evid. Based Complement. Alternat. Med.* **2012**, *2012*, 195367. [CrossRef] [PubMed]
205. Yang, R.; Li, Y.; Mehmood, S.; Yan, C.; Huang, Y.; Cai, J.; Ji, J.; Pan, W.; Zhang, W.; Chen, Y. Polysaccharides from Armillariella tabescens mycelia ameliorate renal damage in type 2 diabetic mice. *Int. J. Biol. Macromol.* **2020**, *162*, 1682–1691. [CrossRef] [PubMed]
206. Karunasagara, S.; Hong, G.L.; Park, S.R.; Lee, N.H.; Jung, D.Y.; Kim, T.W.; Jung, J.Y. Korean red ginseng attenuates hyperglycemia-induced renal inflammation and fibrosis via accelerated autophagy and protects against diabetic kidney disease. *J. Ethnopharmacol.* **2020**, *254*, 112693. [CrossRef] [PubMed]
207. Borgohain, M.P.; Chowdhury, L.; Ahmed, S.; Bolshette, N.; Devasani, K.; Das, T.J.; Mohapatra, A.; Lahkar, M. Renoprotective and antioxidative effects of methanolic Paederia foetida leaf extract on experimental diabetic nephropathy in rats. *J. Ethnopharmacol.* **2017**, *198*, 451–459. [CrossRef]
208. Song, P.; Sun, C.; Li, J.; Long, T.; Yan, Y.; Qin, H.; Makinde, E.A.; Famurewa, A.C.; Jaisi, A.; Nie, Y.; et al. Tiliacora triandra extract and its major constituent attenuates diabetic kidney and testicular impairment by modulating redox imbalance and pro-inflammatory responses in rats. *J. Sci. Food Agric.* **2021**, *101*, 1598–1608. [CrossRef]
209. Vargas, F.; Romecin, P.; Garcia-Guillen, A.I.; Wangesteen, R.; Vargas-Tendero, P.; Paredes, M.D.; Atucha, N.M.; Garcia-Estan, J. Flavonoids in Kidney Health and Disease. *Front. Physiol.* **2018**, *9*, 394. [CrossRef]
210. Guclu, A.; Yonguc, N.; Dodurga, Y.; Gundogdu, G.; Guclu, Z.; Yonguc, T.; Adiguzel, E.; Turkmen, K. The effects of grape seed on apoptosis-related gene expression and oxidative stress in streptozotocin-induced diabetic rats. *Ren. Fail.* **2015**, *37*, 192–197. [CrossRef] [PubMed]
211. Gao, Z.; Liu, G.; Hu, Z.; Shi, W.; Chen, B.; Zou, P.; Li, X. Grape seed proanthocyanidins protect against streptozotocininduced diabetic nephropathy by attenuating endoplasmic reticulum stressinduced apoptosis. *Mol. Med. Rep.* **2018**, *18*, 1447–1454. [CrossRef] [PubMed]
212. Zhang, Z.; Li, B.Y.; Li, X.L.; Cheng, M.; Yu, F.; Lu, W.D.; Cai, Q.; Wang, J.F.; Zhou, R.H.; Gao, H.Q.; et al. Proteomic analysis of kidney and protective effects of grape seed procyanidin B2 in db/db mice indicate MFG-E8 as a key molecule in the development of diabetic nephropathy. *Biochim. Biophys. Acta* **2013**, *1832*, 805–816. [CrossRef]

213. Fujii, H.; Yokozawa, T.; Kim, Y.A.; Tohda, C.; Nonaka, G. Protective effect of grape seed polyphenols against high glucose-induced oxidative stress. *Biosci. Biotechnol. Biochem.* **2006**, *70*, 2104–2111. [CrossRef]
214. Jiang, P.; Xiang, L.; Chen, Z.; Lu, H.; Zhou, L.; Yang, L.; Ji, Y.; Liu, Y.; Sun, X.; Deng, Y.; et al. Catalpol alleviates renal damage by improving lipid metabolism in diabetic db/db mice. *Am. J. Transl. Res.* **2018**, *10*, 1750–1761.
215. Kim, D.; Cheon, J.; Yoon, H.; Jun, H.S. Cudrania tricuspidata Root Extract Prevents Methylglyoxal-Induced Inflammation and Oxidative Stress via Regulation of the PKC-NOX4 Pathway in Human Kidney Cells. *Oxid. Med. Cell. Longev.* **2021**, *2021*, 5511881. [CrossRef]
216. Zhou, J.; Zhang, S.; Sun, X.; Lou, Y.; Bao, J.; Yu, J. Hyperoside ameliorates diabetic nephropathy induced by STZ via targeting the miR-499-5p/APC axis. *J. Pharmacol. Sci.* **2021**, *146*, 10–20. [CrossRef]
217. Giribabu, N.; Karim, K.; Kilari, E.K.; Salleh, N. Phyllanthus niruri leaves aqueous extract improves kidney functions, ameliorates kidney oxidative stress, inflammation, fibrosis and apoptosis and enhances kidney cell proliferation in adult male rats with diabetes mellitus. *J. Ethnopharmacol.* **2017**, *205*, 123–137. [CrossRef] [PubMed]
218. Manna, K.; Mishra, S.; Saha, M.; Mahapatra, S.; Saha, C.; Yenge, G.; Gaikwad, N.; Pal, R.; Oulkar, D.; Banerjee, K.; et al. Amelioration of diabetic nephropathy using pomegranate peel extract-stabilized gold nanoparticles: Assessment of NF-κB and Nrf2 signaling system. *Int. J. Nanomed.* **2019**, *14*, 1753–1777. [CrossRef] [PubMed]
219. Gomes, I.B.; Porto, M.L.; Santos, M.C.; Campagnaro, B.P.; Pereira, T.M.; Meyrelles, S.S.; Vasquez, E.C. Renoprotective, anti-oxidative and anti-apoptotic effects of oral low-dose quercetin in the C57BL/6J model of diabetic nephropathy. *Lipids Health Dis.* **2014**, *13*, 184. [CrossRef] [PubMed]
220. Zhang, T.; Chi, Y.; Kang, Y.; Lu, H.; Niu, H.; Liu, W.; Li, Y. Resveratrol ameliorates podocyte damage in diabetic mice via SIRT1/PGC-1alpha mediated attenuation of mitochondrial oxidative stress. *J. Cell. Physiol.* **2019**, *234*, 5033–5043. [CrossRef]
221. Sohal, R.S.; Forster, M.J. Caloric restriction and the aging process: A critique. *Free Radic. Biol. Med.* **2014**, *73*, 366–382. [CrossRef]
222. Sohal, R.S.; Weindruch, R. Oxidative stress, caloric restriction, and aging. *Science* **1996**, *273*, 59–63. [CrossRef]
223. Diaz-Ruiz, A.; Di Francesco, A.; Carboneau, B.A.; Levan, S.R.; Pearson, K.J.; Price, N.L.; Ward, T.M.; Bernier, M.; de Cabo, R.; Mercken, E.M. Benefits of Caloric Restriction in Longevity and Chemical-Induced Tumorigenesis Are Transmitted Independent of NQO1. *J. Gerontol. A* **2019**, *74*, 155–162. [CrossRef]
224. Wang, S.Y.; Cai, G.Y.; Chen, X.M. Energy restriction in renal protection. *Br. J. Nutr.* **2018**, *120*, 1149–1158. [CrossRef]
225. Gray, K.L.; Clifton, P.M.; Keogh, J.B. The effect of intermittent energy restriction on weight loss and diabetes risk markers in women with a history of gestational diabetes: A 12-month randomized control trial. *Am. J. Clin. Nutr.* **2021**. [CrossRef]
226. Estrela, G.R.; Wasinski, F.; Batista, R.O.; Hiyane, M.I.; Felizardo, R.J.; Cunha, F.; de Almeida, D.C.; Malheiros, D.M.; Camara, N.O.; Barros, C.C.; et al. Caloric Restriction Is More Efficient than Physical Exercise to Protect from Cisplatin Nephrotoxicity via PPAR-Alpha Activation. *Front. Physiol.* **2017**, *8*, 116. [CrossRef] [PubMed]
227. Ning, Y.C.; Cai, G.Y.; Zhuo, L.; Gao, J.J.; Dong, D.; Cui, S.Y.; Shi, S.Z.; Feng, Z.; Zhang, L.; Sun, X.F.; et al. Beneficial effects of short-term calorie restriction against cisplatin-induced acute renal injury in aged rats. *Nephron Exp. Nephrol.* **2013**, *124*, 19–27. [CrossRef]
228. Stern, J.S.; Gades, M.D.; Wheeldon, C.M.; Borchers, A.T. Calorie restriction in obesity: Prevention of kidney disease in rodents. *J. Nutr.* **2001**, *131*, 913S–917S. [CrossRef]
229. Singh, G.; Krishan, P. Dietary restriction regimens for fighting kidney disease: Insights from rodent studies. *Exp. Gerontol.* **2019**, *128*, 110738. [CrossRef]
230. Tikoo, K.; Lodea, S.; Karpe, P.A.; Kumar, S. Calorie restriction mimicking effects of roflumilast prevents diabetic nephropathy. *Biochem. Biophys. Res. Commun.* **2014**, *450*, 1581–1586. [CrossRef] [PubMed]
231. Pall, M.L.; Levine, S. Nrf2, a master regulator of detoxification and also antioxidant, anti-inflammatory and other cytoprotective mechanisms, is raised by health promoting factors. *Sheng Li Xue Bao* **2015**, *67*, 1–18.
232. Juszczak, F.; Caron, N.; Mathew, A.V.; Decleves, A.E. Critical Role for AMPK in Metabolic Disease-Induced Chronic Kidney Disease. *Int. J. Mol. Sci.* **2020**, *21*, 7994. [CrossRef] [PubMed]
233. Nowak, K.L.; Hopp, K. Metabolic Reprogramming in Autosomal Dominant Polycystic Kidney Disease: Evidence and Therapeutic Potential. *Clin. J. Am. Soc. Nephrol.* **2020**, *15*, 577–584. [CrossRef]
234. Kume, S.; Koya, D. Autophagy: A Novel Therapeutic Target for Diabetic Nephropathy. *Diabetes Metab. J.* **2015**, *39*, 451–460. [CrossRef]
235. Sharma, K. Mitochondrial hormesis and diabetic complications. *Diabetes* **2015**, *64*, 663–672. [CrossRef]
236. Aydemir, N.; Pike, M.M.; Alsouqi, A.; Headley, S.A.E.; Tuttle, K.; Evans, E.E.; Milch, C.M.; Moody, K.A.; Germain, M.; Lipworth, L.; et al. Effects of diet and exercise on adipocytokine levels in patients with moderate to severe chronic kidney disease. *Nutr. Metab. Cardiovasc. Dis.* **2020**, *30*, 1375–1381. [CrossRef] [PubMed]
237. Malin, S.K.; Navaneethan, S.D.; Fealy, C.E.; Scelsi, A.; Huang, H.; Rocco, M.; Kirwan, J.P. Exercise plus caloric restriction lowers soluble RAGE in adults with chronic kidney disease. *Obes. Sci. Pract.* **2020**, *6*, 307–312. [CrossRef] [PubMed]
238. Doshi, S.M.; Friedman, A.N. Diagnosis and Management of Type 2 Diabetic Kidney Disease. *Clin. J. Am. Soc. Nephrol.* **2017**, *12*, 1366–1373. [CrossRef]
239. Ikizler, T.A.; Robinson-Cohen, C.; Ellis, C.; Headley, S.A.E.; Tuttle, K.; Wood, R.J.; Evans, E.E.; Milch, C.M.; Moody, K.A.; Germain, M.; et al. Metabolic Effects of Diet and Exercise in Patients with Moderate to Severe CKD: A Randomized Clinical Trial. *J. Am. Soc. Nephrol.* **2018**, *29*, 250–259. [CrossRef]

240. Ikeda, Y.; Enomoto, H.; Tajima, S.; Izawa-Ishizawa, Y.; Kihira, Y.; Ishizawa, K.; Tomita, S.; Tsuchiya, K.; Tamaki, T. Dietary iron restriction inhibits progression of diabetic nephropathy in db/db mice. *Am. J. Physiol. Renal Physiol.* **2013**, *304*, F1028–F1036. [CrossRef]
241. Lin, C.H.; Chang, Y.C.; Chuang, L.M. Early detection of diabetic kidney disease: Present limitations and future perspectives. *World J. Diabetes* **2016**, *7*, 290–301. [CrossRef] [PubMed]
242. Ibarra-Gonzalez, I.; Cruz-Bautista, I.; Bello-Chavolla, O.Y.; Vela-Amieva, M.; Pallares-Mendez, R.; Ruiz de Santiago, Y.N.D.; Salas-Tapia, M.F.; Rosas-Flota, X.; Gonzalez-Acevedo, M.; Palacios-Penaloza, A.; et al. Optimization of kidney dysfunction prediction in diabetic kidney disease using targeted metabolomics. *Acta Diabetol.* **2018**, *55*, 1151–1161. [CrossRef]
243. Matoba, K.; Takeda, Y.; Nagai, Y.; Yokota, T.; Utsunomiya, K.; Nishimura, R. Targeting Redox Imbalance as an Approach for Diabetic Kidney Disease. *Biomedicines* **2020**, *8*, 40. [CrossRef]

biomolecules

MDPI

Review

Mitochondrial Oxidative Stress and Cell Death in Podocytopathies

Yu-Ting Zhu [1], Cheng Wan [1], Ji-Hong Lin [2], Hans-Peter Hammes [2] and Chun Zhang [1,*]

1 Department of Nephrology, Union Hospital, Tongji Medical College, Huazhong University of Science and Technology, Wuhan 430022, China; drzhuyuting@hust.edu.cn (Y.-T.Z.); stellarwane@126.com (C.W.)
2 5th Medical Department, Medical Faculty Mannheim, University of Heidelberg, D-68167 Mannheim, Germany; jihong.lin@medma.uni-heidelberg.de (J.-H.L.); hans-peter.hammes@medma.uni-heidelberg.de (H.-P.H.)
* Correspondence: drzhangchun@hust.edu.cn; Tel.: +86-02785726712

Abstract: Podocytopathies are kidney diseases that are driven by podocyte injury with proteinuria and proteinuria-related symptoms as the main clinical presentations. Albeit podocytopathies are the major contributors to end-stage kidney disease, the underlying molecular mechanisms of podocyte injury remain to be elucidated. Mitochondrial oxidative stress is associated with kidney diseases, and increasing evidence suggests that oxidative stress plays a vital role in the pathogenesis of podocytopathies. Accumulating evidence has placed mitochondrial oxidative stress in the focus of cell death research. Excessive generated reactive oxygen species over antioxidant defense under pathological conditions lead to oxidative damage to cellular components and regulate cell death in the podocyte. Conversely, exogenous antioxidants can protect podocyte from cell death. This review provides an overview of the role of mitochondrial oxidative stress in podocytopathies and discusses its role in the cell death of the podocyte, aiming to identify the novel targets to improve the treatment of patients with podocytopathies.

Keywords: podocytopathies; mitochondrial oxidative stress; reactive oxygen species (ROS); antioxidant defense; cell death

Citation: Zhu, Y.-T.; Wan, C.; Lin, J.-H.; Hammes, H.-P.; Zhang, C. Mitochondrial Oxidative Stress and Cell Death in Podocytopathies. *Biomolecules* **2022**, *12*, 403. https://doi.org/10.3390/biom12030403

Academic Editor: Liang-Jun Yan

Received: 14 December 2021
Accepted: 1 March 2022
Published: 4 March 2022

1. Introduction

Podocytopathies are defined as kidney diseases that are driven by podocyte injury with proteinuria and proteinuria-related symptoms as the main clinical features [1]. The incidence of podocytopathies seems to be gradually rising and they are the leading cause of end-stage kidney disease around the world [1,2]. However, efficient therapies for podocytopathies are lacking and current treatment can only retard the progression of diseases. Podocytes are highly specialized epithelial cells that are located in the glomerulus and constitute the filtration barrier with a glomerular basement membrane (GBM) and endothelial cells [2,3]. The interdigitated foot processes and slit diaphragm of podocytes are elementary structures for the selective filtration function of the glomerulus [1]. Excessive stress and harmful stimuli are likely to cause podocyte injury, possibly even death, which is clinically characterized by proteinuria and pathologically characterized by podocyte foot process effacement (FPE), detachment, and loss [1,2]. Considering the poor proliferation capacity of the podocyte, excessive podocyte loss progressively aggravates podocyte damage and eventually leads to global glomerulosclerosis [2]. Understanding how such detrimental stress and stimuli cause podocyte injury can help us to advance our acknowledgement of the mechanisms underlying the occurrence and progression of podocytopathies.

Mitochondrial oxidative stress refers to disrupted redox homeostasis by the elevated generation of reactive oxygen species (ROS) and (or) declined antioxidant defense capacity [4,5]. Increasing evidence suggests that oxidative stress plays a vital role in the pathogenesis of podocytopathies [6–9]. The excessive accumulation of ROS causes damage to intracellular components and impairs the normal structure and function of podocytes [7,8,10–12].

Accumulating evidence has placed mitochondrial oxidative stress in the focus of cell death research [4,13]. Classic cell death includes apoptosis, necrosis, necroptosis, pyroptosis, and ferroptosis [14,15]. Under pathological conditions, when redundant ROS-induced damage is beyond the compensatory capacity of podocytes, cell death occurs [8,16–18]. Conversely, the application of exogenous antioxidants can protect podocytes from cell death and improve kidney function [8,19]. Hence, this review provides an overview of the role of mitochondrial oxidative stress in podocytopathies and discusses its role in the cell death of the podocyte, aiming to identify novel targets to improve the treatment of patients with podocytopathies.

2. Spectrum of Podocytopathies

Genetic factors and non-genetic factors, such as immune, infectious, metabolic, and hemodynamic factors, can cause damage to the podocyte [2]. Therefore, podocytopathies can be further divided into genetic and non-genetic podocytopathies on the basis of the causes.

Numerous genetic researches identified many susceptibility genes relevant to podocytopathies, which can be divided into podocyte genes and syndromal non-specific genes in terms of the types of cells that experience genetic variation [1]. For example, genetic variants in *PLCE1* and *WT1*, two podocyte-expressed genes, result in the arrested development of glomeruli and the onset of diffuse mesangial sclerosis [20,21]. *APOL1* podocytopathy is the best studied podocytopathy that is associated with genetic variants of susceptibility genes [1]. In Africans carrying a high frequency of *APOL1* alleles, the prevalence of chronic kidney disease (CKD) is up to 16% [22].

Non-genetic podocytopathies consist of kidney diseases of many distinct causes. Immune injury to the podocyte can induce the development of podocytopathies, such as IgA nephropathy (IgAN), lupus nephritis (LN), and membranous nephropathy (MN) [2]. Metabolic and hemodynamic abnormalities damage the podocyte as well. Long-term poor glucose control and the hemodynamic changes in diabetics contribute to diabetic nephropathy (DN) [1]. Elevated blood pressure and the accompanying hyperfiltration can also induce podocyte injury, which plays an important role in the pathogenesis of hypertensive nephropathy [23]. Podocytopathies caused by infections and nephrotoxic substances are not negligible, for instance, HIV-associated nephropathy and collapsing glomerulopathy induced by pamidronate [24,25].

3. Oxidative Stress in Podocytopathies

Under physiological conditions, a homeostasis between the production of ROS and the antioxidant defense system exists in the podocyte. ROS are a collection of chemical substances originated from incomplete reduced oxygen, which mainly consist of superoxide anion, hydrogen peroxide (H_2O_2), singlet oxygen, and hydroxyl radical [13,26]. In the podocyte, ROS mainly come from the mitochondrial respiration chain and NADPH oxidase (NOX) [27–29]. The mitochondrial respiration chain is mainly composed of NADH dehydrogenase (complex I), succinate dehydrogenase (complex II), ubiquinol-cytochrome c reductase (complex III), cytochrome c oxidase (complex IV), cytochrome c (Cyt C), and quinone [30,31]. Mitochondrial respiration chain dysfunction contributes to excessive ROS generation. Antioxidant defense systems are developed in the organism to eliminate ROS generated from various sources. Enzymatic defense systems include superoxide dismutase (SOD), glutathione peroxidase (GPX), catalase (CAT), and thioredoxin reductase (TrxR) [4,32]. Additionally, antioxidants are comprised of ascorbic acid (Vitamin C), α-tocopherol (Vitamin E), glutathione (GSH), thioredoxin, peroxiredoxin (Prdx), and carotenoids [4,32]. When the redox homeostasis is disrupted due to external stimuli under pathological conditions or inherent defects of podocyte, oxidative stress occurs and leads to podocyte injury and renal damage (Figure 1). Stimuli that are harmful to the podocyte, such as puromycin aminonucleoside (PA), high glucose (HG), and angiotensin II (Ang II), can all contribute to the intracellular accumulation of ROS [28,33,34].

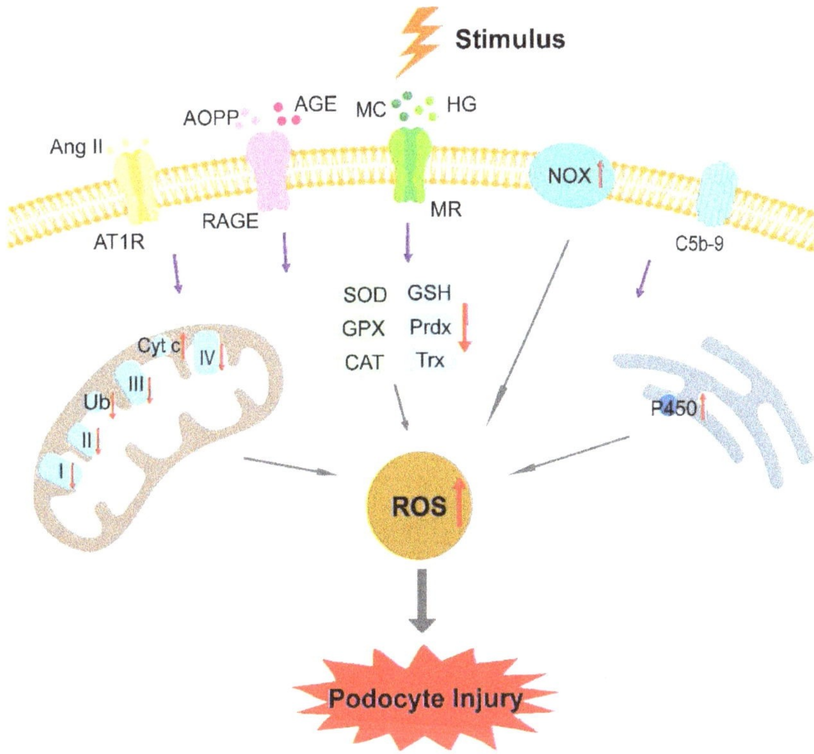

Figure 1. Oxidative stress plays a significant role in the pathogenesis of podocytopathies. Various harmful stimuli, such as puromycin aminonucleoside, immune complexes, HG, and Ang II, upregulate NOX, Cyt C, and P450, and downregulate mitochondrial respiration chain complexes (complex I, II, III, and IV), Ub, and antioxidant defense systems, including SOD, GPX, CAT, GSH, Prdx, and Trx. Excessive ROS accumulation in the podocyte causes damage to DNA, lipids, and proteins, and activates downstream signaling pathways, leading to podocyte foot process effacement, loss, and detachment, with clinical presentations of proteinuria and proteinuria-related symptoms. HG, high glucose; Ang II, angiotensin II; AT1R, angiotensin II type 1 receptor; AOPPs, advanced oxidation protein products; AGEs, advanced glycation end-products; RAGE, receptor of advanced glycation end-products; MC, mineralocorticoid; MR, mineralocorticoid receptor; C5b-9, C5b-9 membrane attack complex; NOX, NADPH oxidase; I, complex I; II, complex II; III, complex III; IV, complex IV; Ub, quinone; Cyt c, cytochrome c; SOD, superoxide dismutase; GPX, glutathione peroxidase; CAT, catalase; GSH, glutathione; Prdx, peroxiredoxin; Trx, thioredoxin; P450, cytochrome P450; and ROS, reactive oxygen species.

In this section, we discuss and summarize the mechanisms by which ROS production is elevated in podocytopathies.

3.1. Elevated ROS Production

In focal segmental glomerulosclerosis (FSGS), the level of NOX4, a subunit of NOX complex, is higher in patients with steroid-resistant nephrotic syndrome (SRNS) than those with steroid-sensitive nephrotic syndrome, and the level of ROS is also higher in the glomeruli isolated from patients with SRNS [35]. The induction of NOX4 is observed in the PA-treated podocyte as well [27]. In the Imai rats, a spontaneous FSGS model, the activation of the angiotensin II type 1 receptor (AT1R) by Ang II leads to oxidative stress via upregulating the expression of NOX and downregulating the expression of Nrf2 [36].

In the FSGS phase of puromycin aminonucleoside nephrosis, cytochrome c oxidase subunit 1 (COX1) in the glomerulus is decreased [6]. Mice that lack the cytochrome c oxidase assembly cofactor heme A, farnesyltransferase (COX10), develop severe FSGS at an early age [37]. In the PA-treated podocyte, a decrease in COX1, 2, and 4, and a reduction in complexes I and IV are observed [27]. A mutation in *ADCK4*, which participates in the biosynthesis of coenzyme Q10 (CoQ10), causes the development of FSGS [38]. Whole-exome sequencing and Sanger sequencing suggests that a coenzyme Q10 mono-oxygenase 6 (COQ6) mutation might increase the production of ROS and associate with the occurrence of FSGS [39]. In PA-induced minimal change disease (MCD), CYP2B1, a cytochrome P450 isozyme, is a source of ROS and catalytic iron formation in the podocyte, with CYP2B1 inhibitors cimetidine and piperine alleviating PA-induced proteinuria and attenuating the increase in H_2O_2 and catalytic iron [33,40]. Adriamycin can induce the downregulation of complex I subunits both in vitro and in vivo [41]. In patients with a congenital nephrotic syndrome of the Finnish type, the kidney cortex shows the downregulation of complexes II and IV, and the mRNA levels of COX1 and COX2 are reduced [9,42,43].

For patients with IgAN, the copy numbers of COX3 and nicotinamide adenine dinucleotide dehydrogenase subunit 1 (ND1) in the urine are higher than the healthy controls [44]. After treatment, the changes in urinary COX3 and ND1 levels are positively associated with the changes in proteinuria and negatively associated with the changes in eGFR [44]. In the plasma of IgAN patients, the levels of advanced oxidation protein products (AOPPs) and malonaldehyde (MDA) are elevated, which can be attenuated by angiotensin-converting enzyme inhibitors (ACEIs) [7]. In the model of MN, complement activation in the podocyte reinforces the assembly of the C5b-9 membrane attack complex to induce the generation of ROS, which is mainly mediated via the upregulation of NOX [45–47]. An overexpression of CYP2B1 markedly elevates the generation of ROS and damages the cytoskeleton of the podocyte in MN, while silencing CYP2B1 ameliorates podocyte injury [48]. In podocytes treated with LN plasma, the production of ROS by mitochondria increases and the level of mitophagy decreases [49]. IgG from LN patients can elicit the generation of ROS as well [50].

HG-induced mitochondrial ROS generation is regarded as the main mechanism of DN [28,51]. Extracellular HG induces the formation of intracellular ROS in the podocyte via NOX and mitochondrial respiration [8,52,53]. In the kidney of Zucker obese rats, the expressions of NOX2, NOX4, and AT1R are upregulated [54]. Silencing NOX4 or pretreatment with NOX inhibitor GKT137831, diminishes high glucose-elicited ROS generation and the upregulation of profibrotic markers of the podocyte [55]. NOX5 is also upregulated in human diabetic kidney biopsies and type I diabetic mice [56,57]. Transgenic mice expressing podocyte-specific NOX5 develop albuminuria, podocyte FPE, and the elevation of systolic BP through the generation of ROS, which are further aggravated under streptozotocin (STZ)-induced diabetes [56]. Hyperglycemia can also activate the mineralocorticoid receptor (MR) to induce podocyte injury and proteinuria via the generation of ROS from NOX [58]. CYP4A, a member of the cytochrome P450 family, is upregulated and 20-hydroxyeicosatetraenoic acid is increased, followed by NOX activation [53]. In db/db mice and the HG-cultured podocyte, the activity of complexes I and III is reduced, while the functional expression of complex I in the cultured podocyte prevents the increase in mitochondrial ROS caused by HG [28,59]. In diabetic mice, the enhanced expression of p66Shc, a redox-regulating protein associated with Cyt C, promotes the generation of ROS from mitochondria [60,61].

A high-salt diet and Ang II both increase NADPH-dependent ROS formation in the kidney cortex [34,62]. Ang II induces an ROS-dependent rearrangement of the cytoskeleton to facilitate podocyte migration and eventual podocyte depletion, which can be inhibited by an NOX4 knockdown [12,63]. MR activation is involved in the podocyte damage in hypertensive kidney damage and metabolic syndrome as well [64,65]. Sprague–Dawley rats administrated with aldosterone and high salt display podocyte injury and proteinuria, resulting from elevated NOX activity and ROS production in the podocyte, while selective aldosterone blocker eplerenone and antioxidant tempol can protect against aldosterone-induced injury and proteinuria [64,65].

Homocysteine-induced podocyte injury and glomerulosclerosis involve NOX activation and endogenously generated superoxide anion and H_2O_2 that mediate the NLR family pyrin domain containing 3 (NLRP3) inflammasome formation [66]. In spontaneously hypertensive rats, a high level of homocysteine (Hcy) upregulates the mRNA levels of NOX2 and NOX4 [29]. AOPP has been shown to be closely associated with the severity of CKD and proteinuria [67]. It can bind to the receptor of advanced glycation end-products (RAGE) to induce NOX-mediated ROS generation, which further activates the Wnt/β-catenin pathway, resulting in FPE, matrix accumulation, and proteinuria [67–69]. Transforming growth factor β (TGFβ) has been shown to accumulate in renal diseases, and it can induce the upregulation of NOX4 via the Smad-dependent pathway and following ERK1/2-mTOR activation [70,71].

3.2. Deficient Antioxidant Defense Systems

Patients diagnosed as FSGS show markedly lower plasma and urinary GPX levels compared with MCD and healthy controls [72]. The glomerular GPX immunostaining score in FSGS is also lower than MCD and the normal controls both in patients and rats [72]. CoQ10 acts not only as an electron carrier, but also as an important antioxidant. In patients with idiopathic FSGS, CoQ10 partial deficiency might disturb podocyte biological function and induce renal lesions [73]. Proteinuria and podocyte FPE, resulting from podocyte lesions caused by a superoxide anion formed upon PA via xanthine oxidase, can be attenuated by intravenous SOD treatment [74]. Mitotempo, a SOD mimetic, reduces urinary protein excretion and lipid peroxidation via inhibiting oxidative stress and tissue damage in the MCD model [75]. In adriamycin-induced nephropathy, a model of FSGS, decreased activities of CAT, GPX, and SOD have a vital role in the occurrence of tubulointerstitial injury and glomerulosclerosis [27,76]. Mice lacking CAT are more vulnerable to the renal toxicity of adriamycin than wild mice, and manifest severe proteinuria and pathological lesions [77]. The supplementation of vitamin E in the diet of rats ameliorates adriamycin-induced renal injury via decreasing the activity of SOD1, SOD2, CAT, and GPX, both in the cortex and glomeruli [78].

In MN, anti-SOD2 IgG4 is detected in the plasma of patients, and it is co-localized with C5b-9 in the immune deposits, suggesting that SOD2 has an important role in the pathogenesis of MN [79]. A lower level of SOD in the plasma of IgAN patients is observed as well, which can be significantly reversed by treatment with ACEI [7]. IgAN patients treated with vitamin E show significantly lower proteinuria [80]. In the kidney of pristane-induced LN, the expressions of SOD1 and CAT are significantly reduced [81].

In the DN murine model and HG-treated podocyte, the expression levels of SOD2 and total SOD are significantly lower than controls [59,82,83]. Additionally, HG-induced oxidative stress-dependent podocyte loss and proteinuria can be attenuated by mitotempol [84]. The activities of CAT and GPX are lower at an early age of Zucker obese rats compared with controls, resulting in the accumulation of H_2O_2 [85]. HG treatment decreases the expression of Prdx6, an isoform of the Prdx family of antioxidant enzymes, which catalyze the reduction in H_2O_2 and ROS, but increases the expression of the thioredoxin-interacting protein (TxNIP) to inhibit the antioxidative function of Prdx and thioredoxin, leading to the accumulation of ROS and oxidative stress [17,86,87]. However, the overexpression of Prdx6 in the mouse podocyte restrains HG-induced ROS and MDA generation and recovers the activities of SOD and GSH [17]. In streptozotocin-induced TxNIP KO mice, albuminuria, serum creatinine, podocyte FPE and loss, GBM thickening, mesangial matrix expansion, and glomerulosclerosis are effectively attenuated, as well as oxidative stress and inflammation [86].

A high-salt diet and Ang II both downregulate the expression level of SOD in the kidney cortex [34]. Ang II stimulates the downregulation of Prdx2 in the podocyte, which results in elevated ROS release and protein peroxidation [88]. In spontaneously hypertensive rats administrated with high Hcy, a significant decrease in SOD and increase in MDA are observed [29].

Intracellular ROS excessive accumulation can induce damage to cellular macromolecules, such as DNA, proteins, and lipids, ultimately leading to podocyte injury, renal lesions, and the occurrence and progression of podocytopathies [7,10,11,32]. ROS-mediated nuclear and mitochondrial DNA damage and the upregulation of cell-cycle checkpoint proteins result in a poor proliferation of podocytes, which can be mitigated by radical scavenger 1,3-dimethyl-2-thiourea (DMTU) [10,89]. The generation and deposition of lipoperoxides, such as MDA and 4-hydroxynonenal (4-HNE) caused by ROS from damaged mitochondrion, are observed [7,43]. Additionally, elevated ROS also disturb the balance among key regulators of actin cytoskeletons, including RhoA and Rac1, resulting in actin cytoskeleton reorganization and subsequent podocyte FPE, depletion, and glomerulosclerosis [12,47,48]. Deglycosilation of alpha-dystroglycan in the glomeruli mediated by ROS leads to podocyte detachment from the GBM [11]. Moreover, ROS mediate the upregulation of profibrotic markers in the podocyte and promotes the inflammatory responses resulting in podocyte injury [55]. Pathological and ultrastructural features of podocytopathies, such as podocyte FPE, detachment, and loss, GBM thickening, mesangial matrix expansion, tubulointerstitial fibrosis, and glomerulosclerosis, as well as renal dysfunction indicators, such as urinary protein excretion and eGFR, are closely related to oxidative stress in the podocyte [8,44,56,67–69,90].

4. Oxidative Stress and Cell Death in Podocytopathies

Cell death is a highly conserved process that not only participates in the morphogenesis and development of organisms, but also the pathophysiological process. Different patterns of cell death manifest with discriminating morphological alterations and mediate the pathogenesis of various diseases [14]. Apoptosis is perhaps the most widely recognized pattern of cell death, which involves the activation of caspases (CASPs), especially CASP3, with morphological manifestations of cytoplasmic shrinkage, chromatin condensation, nuclear fragmentation, and ultimate apoptotic body formations [4,14,91]. Necrosis is another form of cell death characterized by the swelling of organelles and the whole cell coupling with negative morphological features of apoptosis and autophagy, and triggered by cyclophilin D-dependent permeability transition pore complex formation [14,92]. Necroptosis is generally manifested as a necrotic morphology and depends on the sequential activation of receptor interacting with serine/threonine kinase 3 (RIPK3) and a mixed lineage kinase domain, such as pseudokinase (MLKL) [14]. Pyroptosis is an inflammatory response depending on the activation of CASP1 and the formation of the plasma membrane pore by gasdermin-D (GSDMD), and manifests chromatin condensation that is different from apoptosis, cellular swelling, and plasma membrane permeabilization [14,91]. Ferroptosis is a newly discovered iron-dependent cell death that has gained significant attention, and is initiated by lipid peroxidation and manifests as necrotic morphology [14,93].

The death of the podocyte contributes to the progressive loss of the podocyte and subsequent glomerular filtration barrier destruction, and podocyte loss is the primary feature for the progression of podocytopathies [18,90,94,95]. In patients with DN, the number of podocytes in the kidney is decreased, and this is the strongest predictor of the progression of DN [96]. A large amount of research has confirmed that ROS play a significant role in regulating cell death [4,13,14]. Thus, in this section, we summarize the mechanisms by which the accumulation of intracellular ROS can regulate podocyte death.

Apoptosis has been the focus of extensive researches and is widely confirmed to be mediated mainly by two primary signaling pathways: the intrinsic apoptosis pathway and the extrinsic apoptosis pathway [14,97]. The intrinsic apoptosis pathway, or mitochondria-mediated pathway, can be activated by DNA damage, endoplasmic reticulum (ER) stress, oxidative stress, and other stimuli [14,97]. Proapoptotic members of the B-cell lymphoma-2 (BCL2) protein family, such as the BCL2-interacting mediator of cell death (BIM), BH3 interacting domain death agonist (BID), and BCL2 binding component 3 (BBC3), are activated transcriptionally or post-transcriptionally to induce the oligomerization of BCL2 associated X (BAX) and BCL2 antagonist/killer 1 (BAK) and the formation of mitochondrial outer-membrane permeabilization (MOMP) [14,97]. However, pro-survival members of the BCL2 protein family, such as BCL2 and BCL2-like 1 (BCL2L1), can antagonize MOMP to inhibit the occurrence of apoptosis [14,98]. MOMP facilitates the release of Cyt C, which can bind apoptotic peptidase activating factor 1 (APAF1) to mediate the activation of pro-CASP9 [14,97]. The extrinsic apoptosis pathway, or death receptor-mediated pathway, can be triggered by the Fas ligand (FasL), TNF-α, and the TNF-related apoptosis-inducing ligand (TRAIL) [97]. The death receptor executes the assembly of the death-inducing signaling complex (DISC), which regulates the activation of pro-CASP8 [97,99]. CASP9 from the intrinsic apoptosis pathway and CASP8 from the extrinsic apoptosis pathway can both cleave pro-CASP3 to activate CASP3, which cleaves cellular proteins leading to morphological damage and cell death [14,97].

Plenty of studies have shown that excessive ROS accumulation triggers podocyte apoptosis both in vivo or in vitro [8,68,100–102]. In type 1 and type 2 diabetic models, podocyte apoptosis is correlated with urinary albumin excretion and precedes the loss of the podocyte [8]. The molecular mechanisms for oxidative stress-triggered podocyte apoptosis are summarized in Figure 2.

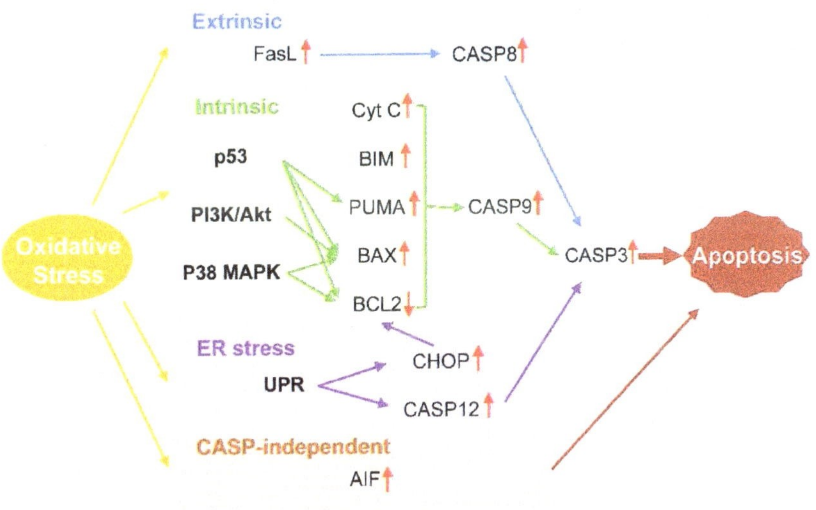

Figure 2. Molecular mechanisms for oxidative stress-triggered podocyte apoptosis. Oxidative stress upregulates FasL to activate the CASP8-mediated extrinsic apoptosis pathway. As for intrinsic apoptosis, p53, PI3K/Akt, and P38 MAPK pathways are activated by oxidative stress, and the expression levels of BAX, PUMA, BIM, and Cyt C are elevated, whereas anti-apoptotic BCL2 is reduced. Whereafter, CASP9 is activated to cleave pro-CASP3 to activate CASP3, which ultimately cleaves cellular components and leads to apoptosis. ER stress is also activated, UPR triggers CHOP- and CASP12-dependent apoptosis, and CHOP activation downregulates BCL2, indicating that intrinsic apoptosis is involved. Moreover, oxidative stress activates CASP-independent apoptosis via upregulating AIF. FasL, Fas ligand; CASP, caspase; Cyt C, cytochrome c; BCL2, B-cell lymphoma-2; BIM, BCL2-interacting mediator of cell death; PUMA, p53-upregulated modulator of apoptosis; BAX, BCL2-associated X; ER, endoplasmic reticulum; UPR, unfolded protein response; CHOP, CCAAT/enhancer-binding protein-homologous protein; and AIF, apoptosis-inducing factor.

The ROS-mediated intrinsic apoptosis pathway has been shown to participate in palmitic acid and HG-induced podocyte apoptosis, which can be markedly alleviated by mitotempo and NOX inhibitor apocynin [18,27,84]. Smad3 is activated to upregulate

NOX4 after PA treatment [27]. However, extrinsic apoptosis pathway-associated indicators, including the Fas-associated protein with the death domain (FADD) and CASP8, are not changed by an intervention of palmitic acid [18]. Meanwhile, decreased BCL2 expression, increased BAX expression, and the translocation to mitochondria are observed [18,84]. Additionally, more Cyt C is released from the mitochondria and the levels of CASP9 and CASP3 are increased [18,27,84,103].

The ROS-triggered extrinsic apoptosis pathway also has a vital role in the pathogenesis of podocytopathies. AOPP increases the protein level of FOXO3 in the cultured podocyte in a ROS/mTOR-dependent pathway to induce proapoptotic FasL and BIM expression and the sequential upregulation of CASP3, leading to death receptor mediated apoptosis [104].

Mitogen-activated protein kinases (MAPKs) belong to the serine/threonine kinase superfamily that can regulate the activities of organisms, including growth, differentiation, and apoptosis [13]. Among MAPKs, extracellular signal-regulated kinases 1 or 2 (ERK1/2), Jun N-terminal kinase or stress-activated kinases (JNKs/SAPKs), and p38 MAPKs have been widely investigated [13,105]. HG and PA-mediated oxidative stress induces podocyte apoptosis through the activation of the p38 MAPK pathway, which can be attenuated by NOX inhibitor apocynin and antioxidant N-acetylcysteine (NAC) [8,19,106]. In type 1 diabetes and HG treated podocytes, protein kinase A is excited via dopamine 1 receptor activation to upregulate the expression of NOX5, which contributes to oxidative stress and induces the phosphorylation of p38 MAPK and downstream signaling cascades [57]. Moreover, HG also upregulates the expression of NOX4 to induce the phosphorylation of p38 MAPK [83]. The expression of BCL2 is lower in HG-treated podocytes, while BAX, cleaved caspases 3 and 9, are higher when compared with controls, indicating that the mitochondria-mediated apoptosis pathway is activated [57,83,106]. Advanced glycation end-products (AGEs) binding to RAGE can trigger ROS generation to induce podocyte apoptosis through the activation of Forkhead transcription factor 4 (FOXO4) and the p38 MAPK pathway, whereas the inhibition of p38 MAPK or siRNA for FOXO4 abolishes AGE-induced podocyte apoptosis [101].

The phosphatidylinositol 3-kinase (PI3K)/Akt signaling pathway can be activated by various cellular stimuli and regulate cellular functions, such as growth, proliferation, cell cycle, and survival [107]. Previous studies demonstrated that HG increases the level of ROS in the podocyte and induces oxidative stress [8,28]. Meanwhile, ROS downregulate the expression of PI3K and suppress the phosphorylation of Akt [101,108]. Downregulated Akt phosphorylation activates BAX and CASP3 to trigger podocyte apoptosis [108].

p53 is a tumor suppressor transcriptive factor, which responds to various stresses to maintain homeostasis via triggering cell cycle arrest, apoptosis, senescence, and DNA repair [109]. The HG stimulus downregulates AMP-activated protein kinase (AMPK) to upregulate NOX4 expression and activity, which further activates p53 [110,111]. AOPP can also activate NOX to induce the expression of p53 [69,112]. p53 can regulate downstream proapoptotic genes, including the p53-upregulated modulator of apoptosis (PUMA) and BAX [69,110–112]. Downregulated BCL2, upregulated CASP3 and CASP9, and DNA fragmentation are consistent with intrinsic apoptosis pathway activation [110–112].

ER stress is caused by misfolded or unfolded protein accumulation and an intracellular calcium imbalance, with the unfolded protein response (UPR) determining the cell fate under ER stress [113]. Palmitic acid and AOPP trigger ROS-mediated ER stress to induce podocyte apoptosis [68,100]. Three UPR pathways, including protein kinase-like ER kinase (PERK), activating transcription factor 6 (ATF6), and inositol requiring 1 (IRE1) pathways, are activated to trigger CCAAT/enhancer-binding protein-homologous protein (CHOP)- and CASP12-dependent podocyte apoptosis with BCL2 downregulation involved in CHOP-dependent apoptosis [68,100].

Moreover, the cytosolic apoptosis-inducing factor (AIF) is also increased in the HG-intervened podocyte, which can activate caspase-independent apoptosis [103,114].

Other forms of cell death also take part in ROS-mediated podocyte damage. Pyroptosis has been shown to participate in HIV-associated nephropathy [16]. HIV-transgenic mice exhibit increased mRNA levels and the expression of the NLR family pyrin domain containing 3 (NLRP3), CARD domain containing adaptor protein (ASC), and CASP1 and IL-1β in the kidney, which can be partially inhibited by tempo, suggesting that ROS are crucial to HIV-induced podocyte pyroptosis [16]. Ferroptosis is involved in ROS-mediated podocytopathies as well. Solute carrier family 7 member 11 (SLC7A11) is a member of Xc-system, which transports cysteine and glutamate, and GPX4 is the main endogenous inhibitor of ferroptosis, which reduces the production of phospholipid hydroperoxide [115]. In the HG-stimulated podocyte, the expressions of SLC7A11 and GPX4 are reduced, which can be eliminated by Prdx6 overexpression [17]. Additionally, the protective effect of Prdx6 overexpression on the podocyte can be suppressed by the ferroptosis-inducing agent erastin [17].

Both necrosis and necroptosis are found to participate in the pathogenesis of podocytopathies, such as LN and DN [116–118]. However, the evidence for the relationship between oxidative stress, necrosis, and necroptosis in podocytopathies is still lacking, possibly owing to technical restrictions in detecting and validating necrosis and necroptosis in vivo [15].

Since the major researches focus on oxidative stress-mediated apoptosis in DN, further research focusing on the underlying mechanisms of ROS mediating other types of cell death in podocytes exposed to different stimuli under other pathophysiologic conditions is desirable. In conclusion, oxidative stress can trigger podocyte death through a variety of molecular mechanisms. Targeting oxidative stress-mediated podocyte death may be a potential novel therapy for patients with podocytopathies.

5. Therapeutic Implications

Since ROS-mediated cell death is of great importance for the pathogenesis of podocytopathies, antioxidants that can protect the podocyte from death are promising for the treatment of patients diagnosed with podocytopathies.

Mitoquinone (MitoQ) is a mitochondria-targeted antioxidant that consists of CoQ10 and triphenyl phosphate. It has been shown to mitigate oxidative stress in various diseases, such as hypoxia-induced pulmonary hypertension and renal ischemia-reperfusion injury [119,120]. In Ang II-treated podocytes, MitoQ suppresses Ang II-induced mitochondrial dysfunction and oxidative stress and reduces podocyte apoptosis [121].

Taurine is an amino sulfonic acid that has been used as an oral supplement for the treatment of disorders, such as congestive heart failure, hypertension, and diabetes mellitus [122–124]. Antioxidant activity plays an essential role in the protective effect of taurine [125]. In DN and PA-induced podocytopathies, taurine supply alleviates urinary protein excretion and histopathological lesions [126,127]. Mitochondria ROS generation and podocyte apoptosis induced by HG are attenuated by taurine [127].

Melatonin is a hormone secreted by the brain to maintain the circadian rhythm. Some studies revealed that melatonin can mitigate DN via the inhibition of oxidative stress [128,129]. In Ang II-induced podocyte injury, melatonin reduces podocyte apoptosis and enhances the proliferation capacity of the podocyte [130]. Additionally, the protective effect is associated with inhibited oxidative stress and recovered mitochondrial function [130].

Folic acid is a B vitamin that can reduce plasma Hcy and antagonize the harmful effects generated by it [29,131]. In a spontaneously hypertensive model of rats, a high level of Hcy induces mitochondrial oxidative stress and glomerular damage, whereas even a low level of folic acid could attenuate the glomerular damage caused by Hcy via inhibiting oxidative stress and podocyte apoptosis [29]. Additionally, the antioxidant effect of folic acid may be independent of Hcy [29].

Herbal medicine has been applied in the clinic for thousands of years and is regarded as an alternative treatment for a lot of diseases. A great deal of herbs and herb-derived components are found to have an antioxidative capacity, and have been investigated for the therapeutic effectiveness in podocytopathies. Since oxidative stress-regulated cell death has a significant role in the pathogenesis of podocytopathies, those antioxidative herbs and herb-derived components might improve renal injury via the suppression of the death of the podocyte. Herbs and herb-derived components decrease intracellular ROS via downregulating pro-oxidative enzymes, such as NOX, and elevating mitochondrial respiration chain complex activities [59,83]. The levels of antioxidative enzymes, such as CAT and SOD, and antioxidants, such as GSH, are augmented as well [81,132]. ROS-mediated apoptotic pathways, including PI3K/Akt, p38 MAPK, and p53 pathways are found to be regulated by herbs and herb-derived components [83,108,111]. Except for anti-apoptotic ability, geniposide, an ingredient of Gardenia jasminoides Ellis, can alleviate podocyte injury and inhibit the development of DN via the suppression of pyroptosis [133]. In Table 1, we summarize the antioxidative herbs and herb-derived components that were verified to alleviate podocytopathies via the suppression of cell death.

Table 1. A summary of antioxidative herbs and herb-derived components that can protect the podocyte from death.

Herbs/Components	Models of Podocytopathies	Anti-Death Mechanisms	Refs.
Salvia przewalskii	PA-induced rats	Anti-apoptosis	[89]
Berberine	Palmitic acid-cultured podocytes	Anti-apoptosis	[100]
Apigenin	Adriamycin-induced mice	Anti-apoptosis	[132]
Citral	Adriamycin-induced rats	Anti-apoptosis	[134]
Epigallocatechin-3-Gallate	Adriamycin-induced mice	Anti-apoptosis	[135]
Osthole	Adriamycin-induced mice	Anti-apoptosis	[136]
Leonurine	Adriamycin-induced mice	Anti-apoptosis	[137]
Paclitaxel	Palmitate-cultured podocytes	Anti-apoptosis	[138]
Quercetin	Pristane-induced mice	Anti-apoptosis	[81]
Resveratrol	db/db mice STZ-induced mice	Anti-apoptosis	[59,139,140]
Baoshenfang	KK-Ay mice STZ-induced rats	Anti-apoptosis	[83,108]
Icariin	HG-cultured podocytes	Anti-apoptosis	[103]
Tongxinluo	STZ-induced rats	Anti-apoptosis	[106]
Huangqi	STZ-induced mice	Anti-apoptosis	[111]
Geniposide	HFD/STZ-induced mice	Anti-pyroptosis	[133]
Huidouba	Unilateral nephrectomy/STZ-induced rats	Anti-apoptosis	[141]
Salidroside	HG-cultured podocytes	Anti-apoptosis	[142]
Catalpol	KK-Ay/HFD induced mice HG-cultured podocytes AGEs-cultured podocytes	Anti-apoptosis	[143,144]
Loganin	KK-Ay/HFD mice AGEs-cultured podocytes	Anti-apoptosis	[144]
Astragaloside IV	STZ-induced rats	Anti-apoptosis	[145]
Luteolin	HG-cultured podocytes	Anti-apoptosis	[146]
Chrysin	db/db mice	Anti-apoptosis	[147]
Carnosine	STZ-induced rats HG-cultured podocytes	Anti-apoptosis	[148,149]

Table 1. Cont.

Herbs/Components	Models of Podocytopathies	Anti-Death Mechanisms	Refs.
Forsythoside A	HG-cultured podocytes	Anti-apoptosis	[150]
Ginsenoside Rb1	STZ-induced mice	Anti-apoptosis	[151]
Grape seed procyanidin B2	HG-treated podocytes	Anti-apoptosis	[152]
Green tea polyphenols	Diabetic patients' plasma-cultured podocytes	Anti-apoptosis	[153]
Hydroxysafflor yellow A	STZ-induced rats	Anti-apoptosis	[154]
Huaiqihuang	HG-treated podocytes	Anti-apoptosis	[155]
Naringin	STZ-induced rats	Anti-apoptosis	[156]
Mogroside IIIE	HG-treated podocytes	Anti-apoptosis	[157]
Rhizoma Polygonum cuspidatum	STZ-induced rats	Anti-apoptosis	[158]
Tetrahydroxy stilbene glucoside	HG-treated podocytes	Anti-apoptosis	[159]

Abbreviations: PA, puromycin aminonucleoside; STZ, streptozotocin; HG, high glucose; HFD, high-fat diet; and AGEs, advanced glycation end-products.

The therapeutic targets of the antioxidants described above are summarized in Table 2.

Table 2. A summary of the therapeutic targets of antioxidants that can protect the podocyte from death.

Therapeutic Targets		Antioxidants
Mitochondrial respiration chain	Complex I	Resveratrol [59,139,140]
	Complex III	Resveratrol [59,139,140]
	Complex IV	Chrysin [147]; Forsythoside A [150]
	Cyt C	Resveratrol [59,139,140]; Carnosine [149]
	Quinone	MitoQ [121]
NADPH oxidase		Folic Acid [29]; Resveratrol [59,139,140]; Baoshenfang [83,108]; Huangqi [111]; Epigallocatechin-3-Gallate [135]; Paclitaxel [138]; Tongxinluo [106]; Huidouba [141]; Catalpol [143,144]; Loganin [144]; Forsythoside A [150]; Ginsenoside Rb1 [151]; Naringin [156]
Antioxidant defense systems	SOD	Melatonin [129]; Folic Acid [29]; Resveratrol [59,139,140]; Quercetin [81]; Apigenin [132]; Leonurine [137]; Paclitaxel [138]; Geniposide [133]; Catalpol [143,144]; Loganin [144]; Astragaloside IV [145]; Forsythoside A [150]; Grape Seed Procyanidin B2 [152]; Hydroxysafflor Yellow A [154]; Mogroside IIIE [157]
	CAT	Quercetin [81]; Astragaloside IV [145]; Forsythoside A [150]; Mogroside IIIE [157]
	GPX	Osthole [136]; Geniposide [133]; Grape Seed Procyanidin B2 [152]; Naringin [156]
	GSH	Apigenin [132]

Abbreviations: Complex I, NADH dehydrogenase; Complex III, ubiquinol-cytochrome c reductase; Complex IV, cytochrome c oxidase; Cyt C, cytochrome c; SOD, superoxide dismutase; CAT, catalase; GPX, glutathione peroxidase; and GSH, glutathione.

Considering that oxidative stress plays a detrimental role in podocytopathies, drugs targeting oxidative stress may have potential value for clinical application in patients with podocytopathies. Hence, we summarize those oxidative stress-targeted drugs in clinical trials from ClinicalTrials.gov in Table 3 [160].

Table 3. A summary of drugs targeting oxidative stress in clinical trials.

Therapeutic Targets		Drugs in Clinical Trials
Mitochondrial respiration chain	Complex IV	S-equol (NCT02142777), Resveratrol (NCT02123121), etc.
	Quinone	MitoQ (NCT02364648), Ubiquinol (NCT02847585), Coenzyme Q10 (NCT01163500), etc.
NADPH oxidase		GKT137831 (NCT02010242, NCT03865927), Pioglitazone (NCT03060772), Januvia (NCT00659711), ImmunAge (NCT02332993), Threalose plus polyphenols (NCT04061070), Apocynin (NCT03680404), etc.
Antioxidant defense systems	SOD	GC4711 (NCT03762031), PC-SOD (NCT03995732), APN201 (NCT01513278), rhSOD (NCT00264186), Glisodin (NCT03878433), Melatonin (NCT02463318), etc.
	CAT	Melatonin (NCT02463318), Oligopin (NCT03260803), etc.
	GPX	N-acetylcysteine (NCT00493727), Curcumin (NCT03475017), Rutin (NCT04955145), Melatonin (NCT01858909), etc.
	GSH	Glutathione (NCT02948673, NCT02948673), etc.
	Other antioxidants	Vitamin E (NCT00384618), Vitamin C (NCT04210453), Vitamin D3 (NCT03931889), L-carnitine (NCT01819701), etc.

Abbreviations: Complex IV, cytochrome c oxidase; SOD, superoxide dismutase; CAT, catalase; GPX, glutathione peroxidase; and GSH, glutathione.

6. Conclusions

The evidence supports the determination that mitochondrial oxidative stress, driven by excessive stress and harmful stimuli to the podocyte, plays an essential role in the pathogenesis of podocytopathies. Plenty of studies revealed that oxidative stress can mediate the cell death of the podocyte, especially podocyte apoptosis, via many signaling pathways. Remarkably, oxidative stress can also contribute to the development of podocytopathies via other mechanisms, such as cell cycle arrest. Further researches focusing on the relationship between oxidative stress and other types of cell death in the podocyte are needed. Many potential antioxidants are under investigation for the effectiveness in treating podocytopathies. Considering that podocyte injury is the initiation factor of podocytopathies, and oxidative stress and oxidative stress-mediated cell death are consistent mechanisms among podocyte injury under different pathophysiologic conditions, targeted podocyte oxidative stress therapy that can ameliorate cell death might be promising to slow and even halt the progression of podocytopathies.

Author Contributions: Y.-T.Z.: writing—review and editing, visualization; C.W.: writing, funding acquisition; J.-H.L.: writing—review and editing; H.-P.H.: visualization; C.Z.: conceptualization, funding acquisition. All authors have read and agreed to the published version of the manuscript.

Funding: This work was financially supported by the National Natural Science Foundation of China (81961138007, 81974096, 81770711, 81873602), the National Key Research and Development Program (2018YFC1314000), and the Program for HUST Academic Frontier Youth Team (2017QYTD20).

Conflicts of Interest: The authors declare no conflict of interest.

References

1. Kopp, J.B.; Anders, H.J.; Susztak, K.; Podesta, M.A.; Remuzzi, G.; Hildebrandt, F.; Romagnani, P. Podocytopathies. *Nat. Rev. Dis. Primers* **2020**, *6*, 68. [CrossRef] [PubMed]

2. Wiggins, R.C. The Spectrum of Podocytopathies: A Unifying View of Glomerular Diseases. *Kidney Int.* **2007**, *71*, 1205–1214. [CrossRef] [PubMed]
3. Griffin, S.V.; Petermann, A.T.; Durvasula, R.V.; Shankland, S.J. Podocyte Proliferation and Differentiation in Glomerular Disease: Role of Cell-Cycle Regulatory Proteins. *Nephrol. Dial. Transplant.* **2003**, *18* (Suppl. 6), vi8–vi13. [CrossRef] [PubMed]
4. Orrenius, S.; Gogvadze, V.; Zhivotovsky, B. Mitochondrial Oxidative Stress: Implications for Cell Death. *Annu. Rev. Pharmacol. Toxicol.* **2007**, *47*, 143–183. [CrossRef] [PubMed]
5. Sies, H. Oxidative Stress: A Concept in Redox Biology and Medicine. *Redox Biol.* **2015**, *4*, 180–183. [CrossRef]
6. Hagiwara, M.; Yamagata, K.; Capaldi, R.A.; Koyama, A. Mitochondrial Dysfunction in Focal Segmental Glomerulosclerosis of Puromycin Aminonucleoside Nephrosis. *Kidney Int.* **2006**, *69*, 1146–1152. [CrossRef]
7. Pei, Y.; Xu, Y.; Ruan, J.; Rong, L.; Jiang, M.; Mo, Y.; Jiang, X. Plasma Oxidative Stress Level of IgA Nephropathy in Children and the Effect of Early Intervention with Angiotensin-Converting Enzyme Inhibitors. *J. Renin Angiotensin Aldosterone Syst.* **2016**, *17*, 1470320316647240. [CrossRef]
8. Susztak, K.; Raff, A.C.; Schiffer, M.; Bottinger, E.P. Glucose-Induced Reactive Oxygen Species Cause Apoptosis of Podocytes and Podocyte Depletion at the Onset of Diabetic Nephropathy. *Diabetes* **2006**, *55*, 225–233. [CrossRef]
9. Solin, M.L.; Pitkanen, S.; Taanman, J.W.; Holthofer, H. Mitochondrial Dysfunction in Congenital Nephrotic Syndrome. *Lab Investig.* **2000**, *80*, 1227–1232. [CrossRef]
10. Marshall, C.B.; Pippin, J.W.; Krofft, R.D.; Shankland, S.J. Puromycin Aminonucleoside Induces Oxidant-Dependent DNA Damage in Podocytes in Vitro and in Vivo. *Kidney Int.* **2006**, *70*, 1962–1973. [CrossRef]
11. Vogtlander, N.P.; Tamboer, W.P.; Bakker, M.A.; Campbell, K.P.; van der Vlag, J.; Berden, J.H. Reactive Oxygen Species Deglycosilate Glomerular Alpha-Dystroglycan. *Kidney Int.* **2006**, *69*, 1526–1534. [CrossRef] [PubMed]
12. Hsu, H.H.; Hoffmann, S.; Endlich, N.; Velic, A.; Schwab, A.; Weide, T.; Schlatter, E.; Pavenstadt, H. Mechanisms of Angiotensin II Signaling on Cytoskeleton of Podocytes. *J. Mol. Med.* **2008**, *86*, 1379–1394. [CrossRef]
13. Ryter, S.W.; Kim, H.P.; Hoetzel, A.; Park, J.W.; Nakahira, K.; Wang, X.; Choi, A.M. Mechanisms of Cell Death in Oxidative Stress. *Antioxid. Redox Signal.* **2007**, *9*, 49–89. [CrossRef] [PubMed]
14. Galluzzi, L.; Vitale, I.; Aaronson, S.A.; Abrams, J.M.; Adam, D.; Agostinis, P.; Alnemri, E.S.; Altucci, L.; Amelio, I.; Andrews, D.W.; et al. Molecular Mechanisms of Cell Death: Recommendations of the Nomenclature Committee on Cell Death 2018. *Cell Death Differ.* **2018**, *25*, 486–541. [CrossRef] [PubMed]
15. Braun, F.; Becker, J.U.; Brinkkoetter, P.T. Live or Let Die: Is There Any Cell Death in Podocytes? *Semin. Nephrol.* **2016**, *36*, 208–219. [CrossRef] [PubMed]
16. Haque, S.; Lan, X.; Wen, H.; Lederman, R.; Chawla, A.; Attia, M.; Bongu, R.P.; Husain, M.; Mikulak, J.; Saleem, M.A.; et al. HIV Promotes NLRP3 Inflammasome Complex Activation in Murine HIV-Associated Nephropathy. *Am. J. Pathol.* **2016**, *186*, 347–358. [CrossRef] [PubMed]
17. Zhang, Q.; Hu, Y.; Hu, J.E.; Ding, Y.; Shen, Y.; Xu, H.; Chen, H.; Wu, N. Sp1-Mediated Upregulation of Prdx6 Expression Prevents Podocyte Injury in Diabetic Nephropathy via Mitigation of Oxidative Stress and Ferroptosis. *Life Sci.* **2021**, *278*, 119529. [CrossRef]
18. Liu, T.; Chen, X.M.; Sun, J.Y.; Jiang, X.S.; Wu, Y.; Yang, S.; Huang, H.Z.; Ruan, X.Z.; Du, X.G. Palmitic Acid-Induced Podocyte Apoptosis via the Reactive Oxygen Species-Dependent Mitochondrial Pathway. *Kidney Blood Press Res.* **2018**, *43*, 206–219. [CrossRef]
19. Chen, Z.; Wan, X.; Hou, Q.; Shi, S.; Wang, L.; Chen, P.; Zhu, X.; Zeng, C.; Qin, W.; Zhou, W.; et al. GADD45B Mediates Podocyte Injury in Zebrafish by Activating the ROS-GADD45B-P38 Pathway. *Cell Death Dis.* **2016**, *7*, e2068. [CrossRef]
20. Hinkes, B.; Wiggins, R.C.; Gbadegesin, R.; Vlangos, C.N.; Seelow, D.; Nürnberg, G.; Garg, P.; Verma, R.; Chaib, H.; Hoskins, B.E.; et al. Positional Cloning Uncovers Mutations in PLCE1 Responsible for A Nephrotic Syndrome Variant that May Be Reversible. *Nat. Genet.* **2006**, *38*, 1397–1405. [CrossRef]
21. Yang, Y.; Jeanpierre, C.; Dressler, G.R.; Lacoste, M.; Niaudet, P.; Gubler, M.C. WT1 and PAX-2 Podocyte Expression in Denys-Drash Syndrome and Isolated Diffuse Mesangial Sclerosis. *Am. J. Pathol.* **1999**, *154*, 181–192. [CrossRef]
22. Abd ElHafeez, S.; Bolignano, D.; D'Arrigo, G.; Dounousi, E.; Tripepi, G.; Zoccali, C. Prevalence and Burden of Chronic Kidney Disease among the General Population and High-Risk Groups in Africa: A Systematic Review. *BMJ Open* **2018**, *8*, e015069. [CrossRef] [PubMed]
23. Udani, S.; Lazich, I.; Bakris, G.L. Epidemiology of Hypertensive Kidney Disease. *Nat. Rev. Nephrol.* **2011**, *7*, 11–21. [CrossRef] [PubMed]
24. Yang, Y.; Gubler, M.C.; Beaufils, H. Dysregulation of Podocyte Phenotype in Idiopathic Collapsing Glomerulopathy and HIV-Associated Nephropathy. *Nephron* **2002**, *91*, 416–423. [CrossRef]
25. Markowitz, G.S.; Appel, G.B.; Fine, P.L.; Fenves, A.Z.; Loon, N.R.; Jagannath, S.; Kuhn, J.A.; Dratch, A.D.; D'Agati, V.D. Collapsing Focal Segmental Glomerulosclerosis Following Treatment with High-Dose Pamidronate. *J. Am. Soc. Nephrol.* **2001**, *12*, 1164–1172. [CrossRef]
26. Yang, B.; Chen, Y.; Shi, J. Reactive Oxygen Species (ROS)-Based Nanomedicine. *Chem. Rev.* **2019**, *119*, 4881–4985. [CrossRef]
27. Yu, L.; Liu, Y.; Wu, Y.; Liu, Q.; Feng, J.; Gu, X.; Xiong, Y.; Fan, Q.; Ye, J. Smad3/Nox4-Mediated Mitochondrial Dysfunction Plays A Crucial Role in Puromycin Aminonucleoside-Induced Podocyte Damage. *Cell Signal.* **2014**, *26*, 2979–2991. [CrossRef]

28. Galvan, D.L.; Badal, S.S.; Long, J.; Chang, B.H.; Schumacker, P.T.; Overbeek, P.A.; Danesh, F.R. Real-Time in Vivo Mitochondrial Redox Assessment Confirms Enhanced Mitochondrial Reactive Oxygen Species in Diabetic Nephropathy. *Kidney Int.* **2017**, *92*, 1282–1287. [CrossRef]
29. Gao, N.; Zhang, Y.; Lei, L.; Li, L.; Cao, P.; Zhao, X.; Lin, L.; Xu, R. Low Doses of Folic Acid Can Reduce Hyperhomocysteinemia-Induced Glomerular Injury in Spontaneously Hypertensive Rats. *Hypertens. Res.* **2020**, *43*, 1182–1191. [CrossRef]
30. Turrens, J.F. Mitochondrial Formation of Reactive Oxygen Species. *J. Physiol.* **2003**, *552*, 335–344. [CrossRef]
31. Sarewicz, M.; Osyczka, A. Electronic Connection Between the Quinone and Cytochrome C Redox Pools and Its Role in Regulation of Mitochondrial Electron Transport and Redox Signaling. *Physiol. Rev.* **2015**, *95*, 219–243. [CrossRef] [PubMed]
32. Valko, M.; Leibfritz, D.; Moncol, J.; Cronin, M.T.; Mazur, M.; Telser, J. Free Radicals and Antioxidants in Normal Physiological Functions and Human Disease. *Int. J. Biochem. Cell Biol.* **2007**, *39*, 44–84. [CrossRef] [PubMed]
33. Liu, H.; Bigler, S.A.; Henegar, J.R.; Baliga, R. Cytochrome P450 2B1 Mediates Oxidant Injury in Puromycin-Induced Nephrotic Syndrome. *Kidney Int.* **2002**, *62*, 868–876. [CrossRef] [PubMed]
34. Kitiyakara, C.; Chabrashvili, T.; Chen, Y.; Blau, J.; Karber, A.; Aslam, S.; Welch, W.J.; Wilcox, C.S. Salt Intake, Oxidative Stress, and Renal Expression of NADPH Oxidase and Superoxide Dismutase. *J. Am. Soc. Nephrol.* **2003**, *14*, 2775–2782. [CrossRef]
35. Tong, J.; Jin, Y.; Weng, Q.; Yu, S.; Hussain, H.M.J.; Ren, H.; Xu, J.; Zhang, W.; Li, X.; Wang, W.; et al. Glomerular Transcriptome Profiles in Focal Glomerulosclerosis: New Genes and Pathways for Steroid Resistance. *Am. J. Nephrol.* **2020**, *51*, 442–452. [CrossRef] [PubMed]
36. Kim, H.J.; Sato, T.; Rodriguez-Iturbe, B.; Vaziri, N.D. Role of Intrarenal Angiotensin System Activation, Oxidative Stress, Inflammation, and Impaired Nuclear Factor-Erythroid-2-Related Factor 2 Activity in The Progression of Focal Glomerulosclerosis. *J. Pharmacol. Exp. Ther.* **2011**, *337*, 583–590. [CrossRef] [PubMed]
37. Baek, J.H.; Gomez, I.G.; Wada, Y.; Roach, A.; Mahad, D.; Duffield, J.S. Deletion of The Mitochondrial Complex-IV Cofactor Heme A:Farnesyltransferase Causes Focal Segmental Glomerulosclerosis and Interferon Response. *Am. J. Pathol.* **2018**, *188*, 2745–2762. [CrossRef]
38. Ashraf, S.; Gee, H.Y.; Woerner, S.; Xie, L.X.; Vega-Warner, V.; Lovric, S.; Fang, H.; Song, X.; Cattran, D.C.; Avila-Casado, C.; et al. ADCK4 Mutations Promote Steroid-Resistant Nephrotic Syndrome through CoQ10 Biosynthesis Disruption. *J. Clin. Investig.* **2013**, *123*, 5179–5189. [CrossRef]
39. Song, C.C.; Hong, Q.; Geng, X.D.; Wang, X.; Wang, S.Q.; Cui, S.Y.; Guo, M.D.; Li, O.; Cai, G.Y.; Chen, X.M.; et al. New Mutation of Coenzyme Q10 Monooxygenase 6 Causing Podocyte Injury in A Focal Segmental Glomerulosclerosis Patient. *Chin. Med. J.* **2018**, *131*, 2666–2675. [CrossRef]
40. Liu, H.; Baliga, M.; Bigler, S.A.; Baliga, R. Role of Cytochrome P450 2B1 in Puromycin Aminonucleoside-Induced Cytotoxicity to Glomerular Epithelial Cells. *Nephron Exp. Nephrol.* **2003**, *94*, e17–e24. [CrossRef]
41. Xie, K.; Zhu, M.; Xiang, P.; Chen, X.; Kasimumali, A.; Lu, R.; Wang, Q.; Mou, S.; Ni, Z.; Gu, L.; et al. Protein Kinase A/Creb Signaling Prevents Adriamycin-Induced Podocyte Apoptosis via Upregulation of Mitochondrial Respiratory Chain Complexes. *Mol. Cell Biol.* **2018**, *38*, e00181-17. [CrossRef] [PubMed]
42. Holthofer, H.; Kretzler, M.; Haltia, A.; Solin, M.L.; Taanman, J.W.; Schagger, H.; Kriz, W.; Kerjaschki, D.; Schlondorff, D. Altered Gene Expression and Functions of Mitochondria in Human Nephrotic Syndrome. *FASEB J.* **1999**, *13*, 523–532. [CrossRef] [PubMed]
43. Solin, M.L.; Ahola, H.; Haltia, A.; Ursini, F.; Montine, T.; Roveri, A.; Kerjaschki, D.; Holthofer, H. Lipid Peroxidation in Human Proteinuric Disease. *Kidney Int.* **2001**, *59*, 481–487. [CrossRef]
44. Yu, B.C.; Cho, N.-J.; Park, S.; Kim, H.; Choi, S.J.; Kim, J.K.; Hwang, S.D.; Gil, H.-W.; Lee, E.Y.; Jeon, J.S.; et al. IgA Nephropathy Is Associated with Elevated Urinary Mitochondrial DNA Copy Numbers. *Sci. Rep.* **2019**, *9*, 16068. [CrossRef] [PubMed]
45. Takano, T.; Elimam, H.; Cybulsky, A.V. Complement-Mediated Cellular Injury. *Semin. Nephrol.* **2013**, *33*, 586–601. [CrossRef] [PubMed]
46. Nangaku, M.; Shankland, S.J.; Couser, W.G. Cellular Response to Injury in Membranous Nephropathy. *J. Am. Soc. Nephrol.* **2005**, *16*, 1195–1204. [CrossRef] [PubMed]
47. Chen, Z.H.; Qin, W.S.; Zeng, C.H.; Zheng, C.X.; Hong, Y.M.; Lu, Y.Z.; Li, L.S.; Liu, Z.H. Triptolide Reduces Proteinuria in Experimental Membranous Nephropathy and Protects Against C5b-9-Induced Podocyte Injury In Vitro. *Kidney Int.* **2010**, *77*, 974–988. [CrossRef]
48. Liu, H.; Tian, N.; Arany, I.; Bigler, S.A.; Waxman, D.J.; Shah, S.V.; Baliga, R. Cytochrome P450 2B1 Mediates Complement-Dependent Sublytic Injury in A Model of Membranous Nephropathy. *J. Biol. Chem.* **2010**, *285*, 40901–40910. [CrossRef]
49. Tian, Y.; Guo, H.; Miao, X.; Xu, J.; Yang, R.; Zhao, L.; Liu, J.; Yang, L.; Gao, F.; Zhang, W.; et al. Nestin Protects Podocyte from Injury in Lupus Nephritis by Mitophagy and Oxidative Stress. *Cell Death Dis.* **2020**, *11*, 319. [CrossRef]
50. Qi, Y.Y.; Zhou, X.J.; Cheng, F.J.; Hou, P.; Ren, Y.L.; Wang, S.X.; Zhao, M.H.; Yang, L.; Martinez, J.; Zhang, H. Increased Autophagy Is Cytoprotective Against Podocyte Injury Induced by Antibody and Interferon-Alpha in Lupus Nephritis. *Ann. Rheum. Dis.* **2018**, *77*, 1799–1809. [CrossRef]
51. Brownlee, M. Biochemistry and Molecular Cell Biology of Diabetic Complications. *Nature* **2001**, *414*, 813–820. [CrossRef] [PubMed]

52. Eid, S.; Boutary, S.; Braych, K.; Sabra, R.; Massaad, C.; Hamdy, A.; Rashid, A.; Moodad, S.; Block, K.; Gorin, Y.; et al. mTORC2 Signaling Regulates Nox4-Induced Podocyte Depletion in Diabetes. *Antioxid. Redox Signal.* **2016**, *25*, 703–719. [CrossRef] [PubMed]
53. Eid, A.A.; Gorin, Y.; Fagg, B.M.; Maalouf, R.; Barnes, J.L.; Block, K.; Abboud, H.E. Mechanisms of Podocyte Injury in Diabetes: Role of Cytochrome P450 and NADPH Oxidases. *Diabetes* **2009**, *58*, 1201–1211. [CrossRef]
54. Habibi, J.; Aroor, A.R.; Das, N.A.; Manrique-Acevedo, C.M.; Johnson, M.S.; Hayden, M.R.; Nistala, R.; Wiedmeyer, C.; Chandrasekar, B.; DeMarco, V.G. The Combination of A Neprilysin Inhibitor (Sacubitril) and Angiotensin-II Receptor Blocker (Valsartan) Attenuates Glomerular and Tubular Injury in the Zucker Obese Rat. *Cardiovasc Diabetol.* **2019**, *18*, 40. [CrossRef] [PubMed]
55. Jha, J.C.; Gray, S.P.; Barit, D.; Okabe, J.; El-Osta, A.; Namikoshi, T.; Thallas-Bonke, V.; Wingler, K.; Szyndralewiez, C.; Heitz, F.; et al. Genetic Targeting or Pharmacologic Inhibition of NADPH Oxidase Nox4 Provides Renoprotection in Long-Term Diabetic Nephropathy. *J. Am. Soc. Nephrol.* **2014**, *25*, 1237–1254. [CrossRef]
56. Holterman, C.E.; Thibodeau, J.F.; Towaij, C.; Gutsol, A.; Montezano, A.C.; Parks, R.J.; Cooper, M.E.; Touyz, R.M.; Kennedy, C.R. Nephropathy and Elevated BP in Mice with Podocyte-Specific NADPH Oxidase 5 Expression. *J. Am. Soc. Nephrol.* **2014**, *25*, 784–797. [CrossRef]
57. Shao, X.; Zhang, X.; Hu, J.; Gao, T.; Chen, J.; Xu, C.; Wei, C. Dopamine 1 Receptor Activation Protects Mouse Diabetic Podocytes Injury via Regulating the PKA/NOX-5/P38 MAPK Axis. *Exp. Cell Res.* **2020**, *388*, 111849. [CrossRef]
58. Toyonaga, J.; Tsuruya, K.; Ikeda, H.; Noguchi, H.; Yotsueda, H.; Fujisaki, K.; Hirakawa, M.; Taniguchi, M.; Masutani, K.; Iida, M. Spironolactone Inhibits Hyperglycemia-Induced Podocyte Injury by Attenuating ROS Production. *Nephrol. Dial. Transplant.* **2011**, *26*, 2475–2484. [CrossRef]
59. Zhang, T.; Chi, Y.; Kang, Y.; Lu, H.; Niu, H.; Liu, W.; Li, Y. Resveratrol Ameliorates Podocyte Damage in Diabetic Mice via SIRT1/PGC-1alpha Mediated Attenuation of Mitochondrial Oxidative Stress. *J. Cell Physiol.* **2019**, *234*, 5033–5043. [CrossRef]
60. Bock, F.; Shahzad, K.; Wang, H.; Stoyanov, S.; Wolter, J.; Dong, W.; Pelicci, P.G.; Kashif, M.; Ranjan, S.; Schmidt, S.; et al. Activated Protein C Ameliorates Diabetic Nephropathy by Epigenetically Inhibiting the Redox Enzyme P66shc. *Proc. Natl. Acad. Sci. USA* **2013**, *110*, 648–653. [CrossRef]
61. Zheng, D.; Tao, M.; Liang, X.; Li, Y.; Jin, J.; He, Q. P66shc Regulates Podocyte Autophagy in High Glucose Environment Through the Notch-PTEN-PI3k/Akt/mTOR Pathway. *Histol. Histopathol.* **2020**, *35*, 405–415. [CrossRef]
62. Lambeth, J.D. Nox Enzymes and the Biology of Reactive Oxygen. *Nat. Rev. Immunol.* **2004**, *4*, 181–189. [CrossRef] [PubMed]
63. Che, G.; Gao, H.; Hu, Q.; Xie, H.; Zhang, Y. Angiotensin II Promotes Podocyte Injury by Activating Arf6-Erk1/2-Nox4 Signaling Pathway. *PLoS ONE* **2020**, *15*, e0229747. [CrossRef] [PubMed]
64. Shibata, S.; Nagase, M.; Yoshida, S.; Kawachi, H.; Fujita, T. Podocyte as the Target for Aldosterone: Roles of Oxidative Stress and Sgk1. *Hypertension* **2007**, *49*, 355–364. [CrossRef] [PubMed]
65. Nagase, M.; Fujita, T. Aldosterone and Glomerular Podocyte Injury. *Clin. Exp. Nephrol.* **2008**, *12*, 233–242. [CrossRef] [PubMed]
66. Abais, J.M.; Xia, M.; Li, G.; Gehr, T.W.; Boini, K.M.; Li, P.L. Contribution of Endogenously Produced Reactive Oxygen Species to the Activation of Podocyte NLRP3 Inflammasomes in Hyperhomocysteinemia. *Free Radic. Biol. Med.* **2014**, *67*, 211–220. [CrossRef]
67. Zhou, L.; Chen, X.; Lu, M.; Wu, Q.; Yuan, Q.; Hu, C.; Miao, J.; Zhang, Y.; Li, H.; Hou, F.F.; et al. Wnt/Beta-Catenin Links Oxidative Stress to Podocyte Injury and Proteinuria. *Kidney Int.* **2019**, *95*, 830–845. [CrossRef]
68. Rong, G.; Tang, X.; Guo, T.; Duan, N.; Wang, Y.; Yang, L.; Zhang, J.; Liang, X. Advanced Oxidation Protein Products Induce Apoptosis in Podocytes Through Induction of Endoplasmic Reticulum Stress. *J. Physiol. Biochem* **2015**, *71*, 455–470. [CrossRef]
69. Zhou, L.L.; Hou, F.F.; Wang, G.B.; Yang, F.; Xie, D.; Wang, Y.P.; Tian, J.W. Accumulation of Advanced Oxidation Protein Products Induces Podocyte Apoptosis and Deletion Through Nadph-Dependent Mechanisms. *Kidney Int.* **2009**, *76*, 1148–1160. [CrossRef]
70. Das, R.; Xu, S.; Nguyen, T.T.; Quan, X.; Choi, S.K.; Kim, S.J.; Lee, E.Y.; Cha, S.K.; Park, K.S. Transforming Growth Factor Beta1-Induced Apoptosis in Podocytes via the Extracellular Signal-Regulated Kinase-Mammalian Target of Rapamycin Complex 1-NADPH Oxidase 4 Axis. *J. Biol. Chem.* **2015**, *290*, 30830–30842. [CrossRef]
71. Das, R.; Xu, S.; Quan, X.; Nguyen, T.T.; Kong, I.D.; Chung, C.H.; Lee, E.Y.; Cha, S.K.; Park, K.S. Upregulation of Mitochondrial Nox4 Mediates TGF-Beta-Induced Apoptosis in Cultured Mouse Podocytes. *Am. J. Physiol. Renal Physiol.* **2014**, *306*, F155–F167. [CrossRef] [PubMed]
72. Chen, H.C.; Guh, J.Y.; Lai, Y.H. Alterations of Glomerular and Extracellular Glutathione Peroxidase Levels in Patients and Rats with Focal Segmental Glomerulosclerosis. *J. Lab Clin. Med.* **2001**, *137*, 279–283. [CrossRef] [PubMed]
73. Gasser, D.L.; Winkler, C.A.; Peng, M.; An, P.; McKenzie, L.M.; Kirk, G.D.; Shi, Y.; Xie, L.X.; Marbois, B.N.; Clarke, C.F.; et al. Focal Segmental Glomerulosclerosis Is Associated with A PDSS2 Haplotype and, Independently, with A Decreased Content of Coenzyme Q10. *Am. J. Physiol. Renal Physiol.* **2013**, *305*, F1228–F1238. [CrossRef] [PubMed]
74. Diamond, J.R.; Bonventre, J.V.; Karnovsky, M.J. A Role for Oxygen Free Radicals in Aminonucleoside Nephrosis. *Kidney Int.* **1986**, *29*, 478–483. [CrossRef]
75. Fujii, Y.; Matsumura, H.; Yamazaki, S.; Shirasu, A.; Nakakura, H.; Ogihara, T.; Ashida, A. Efficacy of A Mitochondrion-Targeting Agent for Reducing the Level of Urinary Protein in Rats with Puromycin Aminonucleoside-Induced Minimal-Change Nephrotic Syndrome. *PLoS ONE* **2020**, *15*, e0227414. [CrossRef]
76. Deman, A.; Ceyssens, B.; Pauwels, M.; Zhang, J.; Houte, K.V.; Verbeelen, D.; Van den Branden, C. Altered Antioxidant Defence in A Mouse Adriamycin Model of Glomerulosclerosis. *Nephrol. Dial. Transplant.* **2001**, *16*, 147–150. [CrossRef]

77. Takiue, K.; Sugiyama, H.; Inoue, T.; Morinaga, H.; Kikumoto, Y.; Kitagawa, M.; Kitamura, S.; Maeshima, Y.; Wang, D.-H.; Masuoka, N.; et al. Acatalasemic Mice Are Mildly Susceptible to Adriamycin Nephropathy and Exhibit Increased Albuminuria and Glomerulosclerosis. *BMC Nephrol.* **2012**, *13*, 14. [CrossRef]
78. Van den Branden, C.; Deman, A.; Ceyssens, B.; Pauwels, M.; Empsen, C.; Verbeelen, D. Vitamin E Protects Renal Antioxidant Enzymes and Attenuates Glomerulosclerosis in Adriamycin-Treated Rats. *Nephron* **2002**, *91*, 129–133. [CrossRef]
79. Prunotto, M.; Carnevali, M.L.; Candiano, G.; Murtas, C.; Bruschi, M.; Corradini, E.; Trivelli, A.; Magnasco, A.; Petretto, A.; Santucci, L.; et al. Autoimmunity in Membranous Nephropathy Targets Aldose Reductase and SOD2. *J. Am. Soc. Nephrol.* **2010**, *21*, 507–519. [CrossRef]
80. Chan, J.C.; Mahan, J.D.; Trachtman, H.; Scheinman, J.; Flynn, J.T.; Alon, U.S.; Lande, M.B.; Weiss, R.A.; Norkus, E.P. Vitamin E Therapy in IgA Nephropathy: A Double-Blind, Placebo-Controlled Study. *Pediatr. Nephrol.* **2003**, *18*, 1015–1019. [CrossRef]
81. Dos Santos, M.; Poletti, P.T.; Favero, G.; Stacchiotti, A.; Bonomini, F.; Montanari, C.C.; Bona, S.R.; Marroni, N.P.; Rezzani, R.; Veronese, F.V. Protective Effects of Quercetin Treatment in A Pristane-Induced Mouse Model of Lupus Nephritis. *Autoimmunity* **2018**, *51*, 69–80. [CrossRef] [PubMed]
82. Morigi, M.; Perico, L.; Corna, D.; Locatelli, M.; Cassis, P.; Carminati, C.E.; Bolognini, S.; Zoja, C.; Remuzzi, G.; Benigni, A.; et al. C3a Receptor Blockade Protects Podocytes from Injury in Diabetic Nephropathy. *JCI Insight* **2020**, *5*, e131849. [CrossRef] [PubMed]
83. Cui, F.Q.; Tang, L.; Gao, Y.B.; Wang, Y.F.; Meng, Y.; Shen, C.; Shen, Z.L.; Liu, Z.Q.; Zhao, W.J.; Liu, W.J. Effect of Baoshenfang Formula on Podocyte Injury via Inhibiting the NOX-4/ROS/P38 Pathway in Diabetic Nephropathy. *J. Diabetes Res.* **2019**, *2019*, 2981705. [CrossRef]
84. Chen, J.; Chen, J.K.; Harris, R.C. EGR Receptor Deletion in Podocytes Attenuates Diabetic Nephropathy. *J. Am. Soc. Nephrol.* **2015**, *26*, 1115–1125. [CrossRef] [PubMed]
85. Poirier, B.; Lannaud-Bournoville, M.; Conti, M.; Bazin, R.; Michel, O.; Bariety, J.; Chevalier, J.; Myara, I. Oxidative Stress Occurs in Absence of Hyperglycaemia and Inflammation in the Onset of Kidney Lesions in Normotensive Obese Rats. *Nephrol. Dial. Transplant.* **2000**, *15*, 467–476. [CrossRef]
86. Shah, A.; Xia, L.; Masson, E.A.; Gui, C.; Momen, A.; Shikatani, E.A.; Husain, M.; Quaggin, S.; John, R.; Fantus, I.G. Thioredoxin-Interacting Protein Deficiency Protects Against Diabetic Nephropathy. *J. Am. Soc. Nephrol.* **2015**, *26*, 2963–2977. [CrossRef]
87. Wood, Z.A.; Poole, L.B.; Karplus, P.A. Peroxiredoxin Evolution and the Regulation of Hydrogen Peroxide Signaling. *Science* **2003**, *300*, 650–653. [CrossRef]
88. Hsu, H.H.; Hoffmann, S.; Di Marco, G.S.; Endlich, N.; Peter-Katalinic, J.; Weide, T.; Pavenstadt, H. Downregulation of the Antioxidant Protein Peroxiredoxin 2 Contributes to Angiotensin II-Mediated Podocyte Apoptosis. *Kidney Int.* **2011**, *80*, 959–969. [CrossRef]
89. Liu, X.; Liu, Y.; Yang, Y.; Xu, J.; Dai, D.; Yan, C.; Li, X.; Tang, R.; Yu, C.; Ren, H. Antioxidative Stress Effects of Salvia Przewalskii Extract in Experimentally Injured Podocytes. *Nephron* **2016**, *134*, 253–271. [CrossRef]
90. Zheng, S.; Carlson, E.C.; Yang, L.; Kralik, P.M.; Huang, Y.; Epstein, P.N. Podocyte-Specific Overexpression of the Antioxidant Metallothionein Reduces Diabetic Nephropathy. *J. Am. Soc. Nephrol.* **2008**, *19*, 2077–2085. [CrossRef]
91. Bergsbaken, T.; Fink, S.L.; Cookson, B.T. Pyroptosis: Host Cell Death and Inflammation. *Nat. Rev. Microbiol.* **2009**, *7*, 99–109. [CrossRef] [PubMed]
92. Orrenius, S.; Zhivotovsky, B.; Nicotera, P. Regulation of Cell Death: The Calcium-Apoptosis Link. *Nat. Rev. Mol. Cell Biol.* **2003**, *4*, 552–565. [CrossRef] [PubMed]
93. Dixon, S.J.; Lemberg, K.M.; Lamprecht, M.R.; Skouta, R.; Zaitsev, E.M.; Gleason, C.E.; Patel, D.N.; Bauer, A.J.; Cantley, A.M.; Yang, W.S.; et al. Ferroptosis: An Iron-Dependent Form of Nonapoptotic Cell Death. *Cell* **2012**, *149*, 1060–1072. [CrossRef] [PubMed]
94. Tharaux, P.L.; Huber, T.B. How Many Ways Can A Podocyte Die? *Semin. Nephrol.* **2012**, *32*, 394–404. [CrossRef]
95. Kim, S.Y.; Park, S.; Lee, S.W.; Lee, J.H.; Lee, E.S.; Kim, M.; Kim, Y.; Kang, J.S.; Chung, C.H.; Moon, J.S.; et al. RIPK3 Contributes to Lyso-Gb3-Induced Podocyte Death. *Cells* **2021**, *10*, 245. [CrossRef]
96. Meyer, T.W.; Bennett, P.H.; Nelson, R.G. Podocyte Number Predicts Long-Term Urinary Albumin Excretion in Pima Indians with Type II Diabetes and Microalbuminuria. *Diabetologia* **1999**, *42*, 1341–1344. [CrossRef]
97. Nagata, S. Apoptosis and Clearance of Apoptotic Cells. *Annu. Rev. Immunol.* **2018**, *36*, 489–517. [CrossRef]
98. Czabotar, P.E.; Lessene, G.; Strasser, A.; Adams, J.M. Control of Apoptosis by the Bcl-2 Protein Family: Implications for Physiology and Therapy. *Nat. Rev. Mol. Cell Biol.* **2014**, *15*, 49–63. [CrossRef]
99. Dickens, L.S.; Powley, I.R.; Hughes, M.A.; MacFarlane, M. The 'Complexities' of Life and Death: Death Receptor Signalling Platforms. *Experimental. Cell Res.* **2012**, *318*, 1269–1277. [CrossRef]
100. Xiang, X.Y.; Liu, T.; Wu, Y.; Jiang, X.S.; He, J.L.; Chen, X.M.; Du, X.G. Berberine Alleviates Palmitic Acid-Induced Podocyte Apoptosis by Reducing Reactive Oxygen Species-Mediated Endoplasmic Reticulum Stress. *Mol. Med. Rep.* **2021**, *23*. [CrossRef]
101. Chuang, P.Y.; Yu, Q.; Fang, W.; Uribarri, J.; He, J.C. Advanced Glycation Endproducts Induce Podocyte Apoptosis by Activation of the FOXO4 Transcription Factor. *Kidney Int.* **2007**, *72*, 965–976. [CrossRef] [PubMed]
102. Liu, Y.; Hitomi, H.; Diah, S.; Deguchi, K.; Mori, H.; Masaki, T.; Nakano, D.; Kobori, H.; Nishiyama, A. Roles of Na(+)/H(+) Exchanger Type 1 and Intracellular pH in Angiotensin II-Induced Reactive Oxygen Species Generation and Podocyte Apoptosis. *J. Pharmacol. Sci.* **2013**, *122*, 176–183. [CrossRef] [PubMed]

103. Qiao, C.; Ye, W.; Li, S.; Wang, H.; Ding, X. Icariin Modulates Mitochondrial Function and Apoptosis in High Glucose-Induced Glomerular Podocytes Through G Protein-Coupled Estrogen Receptors. *Mol. Cell Endocrinol.* **2018**, *473*, 146–155. [CrossRef] [PubMed]
104. Chen, X.; Liu, W.; Xiao, J.; Zhang, Y.; Chen, Y.; Luo, C.; Huang, Q.; Peng, F.; Gong, W.; Li, S.; et al. FOXO3a Accumulation and Activation Accelerate Oxidative Stress-Induced Podocyte Injury. *FASEB J.* **2020**, *34*, 13300–13316. [CrossRef] [PubMed]
105. Arthur, J.S.; Ley, S.C. Mitogen-Activated Protein Kinases in Innate Immunity. *Nat. Rev. Immunol.* **2013**, *13*, 679–692. [CrossRef] [PubMed]
106. Cui, F.; Gao, Y.; Zhao, W.; Zou, D.; Zhu, Z.; Wu, X.; Tian, N.; Wang, X.; Liu, J.; Tong, Y. Effect of Tongxinluo on Podocyte Apoptosis via Inhibition of Oxidative Stress and P38 Pathway in Diabetic Rats. *Evid. Based Complement. Altern. Med.* **2016**, *2016*, 5957423. [CrossRef]
107. Cantley, L.C. The Phosphoinositide 3-Kinase Pathway. *Science* **2002**, *296*, 1655–1657. [CrossRef]
108. Cui, F.Q.; Wang, Y.F.; Gao, Y.B.; Meng, Y.; Cai, Z.; Shen, C.; Liu, Z.Q.; Jiang, X.C.; Zhao, W.J. Effects of BSF on Podocyte Apoptosis via Regulating the Ros-Mediated PI3K/AKT Pathway in DN. *J. Diabetes Res.* **2019**, *2019*, 9512406. [CrossRef]
109. Vousden, K.H.; Prives, C. Blinded by the Light: The Growing Complexity of P53. *Cell* **2009**, *137*, 413–431. [CrossRef]
110. Eid, A.A.; Ford, B.M.; Block, K.; Kasinath, B.S.; Gorin, Y.; Ghosh-Choudhury, G.; Barnes, J.L.; Abboud, H.E. AMP-Activated Protein Kinase (AMPK) Negatively Regulates Nox4-Dependent Activation of P53 and Epithelial Cell Apoptosis in Diabetes. *J. Biol. Chem.* **2010**, *285*, 37503–37512. [CrossRef]
111. Li, Z.; Deng, W.; Cao, A.; Zang, Y.; Wang, Y.; Wang, H.; Wang, L.; Peng, W. Huangqi Decoction Inhibits Hyperglycemia-Induced Podocyte Apoptosis by Down-Regulated Nox4/P53/Bax Signaling In Vitro and In Vivo. *Am. J. Transl. Res.* **2019**, *11*, 3195–3212. [PubMed]
112. Zhou, L.L.; Cao, W.; Xie, C.; Tian, J.; Zhou, Z.; Zhou, Q.; Zhu, P.; Li, A.; Liu, Y.; Miyata, T.; et al. The Receptor of Advanced Glycation End Products Plays A Central Role in Advanced Oxidation Protein Products-Induced Podocyte Apoptosis. *Kidney Int.* **2012**, *82*, 759–770. [CrossRef] [PubMed]
113. Oakes, S.A.; Papa, F.R. The Role of Endoplasmic Reticulum Stress in Human Pathology. *Annu. Rev. Pathol.* **2015**, *10*, 173–194. [CrossRef] [PubMed]
114. Susin, S.A.; Lorenzo, H.K.; Zamzami, N.; Marzo, I.; Snow, B.E.; Brothers, G.M.; Mangion, J.; Jacotot, E.; Costantini, P.; Loeffler, M.; et al. Molecular Characterization of Mitochondrial Apoptosis-Inducing Factor. *Nature* **1999**, *397*, 441–446. [CrossRef] [PubMed]
115. Tang, D.; Chen, X.; Kang, R.; Kroemer, G. Ferroptosis: Molecular Mechanisms and Health Implications. *Cell Res.* **2020**, *32*, 107–125. [CrossRef]
116. Guo, C.; Fu, R.; Zhou, M.; Wang, S.; Huang, Y.; Hu, H.; Zhao, J.; Gaskin, F.; Yang, N.; Fu, S.M. Pathogenesis of Lupus Nephritis: RIP3 Dependent Necroptosis and NLRP3 Inflammasome Activation. *J. Autoimmun.* **2019**, *103*, 102286. [CrossRef]
117. Xu, Y.; Gao, H.; Hu, Y.; Fang, Y.; Qi, C.; Huang, J.; Cai, X.; Wu, H.; Ding, X.; Zhang, Z. High Glucose-Induced Apoptosis and Necroptosis in Podocytes Is Regulated by UCHL1 via RIPK1/RIPK3 Pathway. *Exp. Cell Res.* **2019**, *382*, 111463. [CrossRef]
118. Lan, X.; Jhaveri, A.; Cheng, K.; Wen, H.; Saleem, M.A.; Mathieson, P.W.; Mikulak, J.; Aviram, S.; Malhotra, A.; Skorecki, K.; et al. APOL1 Risk Variants Enhance Podocyte Necrosis through Compromising Lysosomal Membrane Permeability. *Am. J. Physiol. Renal. Physiol.* **2014**, *307*, F326–F336. [CrossRef]
119. Pak, O.; Scheibe, S.; Esfandiary, A.; Gierhardt, M.; Sydykov, A.; Logan, A.; Fysikopoulos, A.; Veit, F.; Hecker, M.; Kroschel, F.; et al. Impact of the Mitochondria-Targeted Antioxidant MitoQ on Hypoxia-Induced Pulmonary Hypertension. *Eur. Respir. J.* **2018**, *51*, 1701024. [CrossRef]
120. Dare, A.J.; Bolton, E.A.; Pettigrew, G.J.; Bradley, J.A.; Saeb-Parsy, K.; Murphy, M.P. Protection Against Renal Ischemia-Reperfusion Injury in Vivo by the Mitochondria Targeted Antioxidant MitoQ. *Redox. Biol.* **2015**, *5*, 163–168. [CrossRef]
121. Zhu, Z.; Liang, W.; Chen, Z.; Hu, J.; Feng, J.; Cao, Y.; Ma, Y.; Ding, G. Mitoquinone Protects Podocytes from Angiotensin II-Induced Mitochondrial Dysfunction and Injury via the Keap1-Nrf2 Signaling Pathway. *Oxid. Med. Cell Longev.* **2021**, *2021*, 1394486. [CrossRef] [PubMed]
122. Azuma, J.; Sawamura, A.; Awata, N. Usefulness of Taurine in Chronic Congestive Heart Failure and Its Prospective Application. *Jpn. Circ. J.* **1992**, *56*, 95–99. [CrossRef] [PubMed]
123. Katakawa, M.; Fukuda, N.; Tsunemi, A.; Mori, M.; Maruyama, T.; Matsumoto, T.; Abe, M.; Yamori, Y. Taurine and Magnesium Supplementation Enhances the Function of Endothelial Progenitor Cells Through Antioxidation in Healthy Men and Spontaneously Hypertensive Rats. *Hypertens. Res.* **2016**, *39*, 848–856. [CrossRef] [PubMed]
124. Nakaya, Y.; Minami, A.; Harada, N.; Sakamoto, S.; Niwa, Y.; Ohnaka, M. Taurine Improves Insulin Sensitivity in the Otsuka Long-Evans Tokushima Fatty Rat, A Model of Spontaneous Type 2 Diabetes. *Am. J. Clin. Nutr.* **2000**, *71*, 54–58. [CrossRef] [PubMed]
125. Schaffer, S.; Kim, H.W. Effects and Mechanisms of Taurine as A Therapeutic Agent. *Biomol. Ther.* **2018**, *26*, 225–241. [CrossRef]
126. Stacchiotti, A.; Favero, G.; Lavazza, A.; Monsalve, M.; Rodella, L.F.; Rezzani, R. Taurine Supplementation Alleviates Puromycin Aminonucleoside Damage by Modulating Endoplasmic Reticulum Stress and Mitochondrial-Related Apoptosis in Rat Kidney. *Nutrients* **2018**, *10*, 689. [CrossRef]

127. Zhang, R.; Wang, X.; Gao, Q.; Jiang, H.; Zhang, S.; Lu, M.; Liu, F.; Xue, X. Taurine Supplementation Reverses Diabetes-Induced Podocytes Injury via Modulation of the Cse/Trpc6 Axis and Improvement of Mitochondrial Function. *Nephron* **2020**, *144*, 84–95. [CrossRef]
128. Ha, H.; Yu, M.R.; Kim, K.H. Melatonin and Taurine Reduce Early Glomerulopathy in Diabetic Rats. *Free Radic. Biol. Med.* **1999**, *26*, 944–950. [CrossRef]
129. Oktem, F.; Ozguner, F.; Yilmaz, H.R.; Uz, E.; Dundar, B. Melatonin Reduces Urinary Excretion of N-Acetyl-Beta-D-Glucosaminidase, Albumin and Renal Oxidative Markers in Diabetic Rats. *Clin. Exp. Pharmacol. Physiol.* **2006**, *33*, 95–101. [CrossRef]
130. Ji, Z.Z.; Xu, Y.C. Melatonin Protects Podocytes from Angiotensin II-Induced Injury in An in Vitro Diabetic Nephropathy Model. *Mol. Med. Rep.* **2016**, *14*, 920–926. [CrossRef]
131. Romecin, P.; Atucha, N.M.; Navarro, E.G.; Ortiz, M.C.; Iyu, D.; Rosado, J.A.; Garcia-Estan, J. Role of Homocysteine and Folic Acid on the Altered Calcium Homeostasis of Platelets from Rats with Biliary Cirrhosis. *Platelets* **2017**, *28*, 698–705. [CrossRef] [PubMed]
132. Wu, Q.; Li, W.; Zhao, J.; Sun, W.; Yang, Q.; Chen, C.; Xia, P.; Zhu, J.; Zhou, Y.; Huang, G.; et al. Apigenin Ameliorates Doxorubicin-Induced Renal Injury via Inhibition of Oxidative Stress and Inflammation. *Biomed Pharm.* **2021**, *137*, 111308. [CrossRef] [PubMed]
133. Li, F.; Chen, Y.; Li, Y.; Huang, M.; Zhao, W. Geniposide Alleviates Diabetic Nephropathy of Mice through AMPK/SIRT1/NF-kappaB Pathway. *Eur. J. Pharmacol.* **2020**, *886*, 173449. [CrossRef]
134. Yang, S.M.; Hua, K.F.; Lin, Y.C.; Chen, A.; Chang, J.M.; Kuoping Chao, L.; Ho, C.L.; Ka, S.M. Citral Is Renoprotective for Focal Segmental Glomerulosclerosis by Inhibiting Oxidative Stress and Apoptosis and Activating Nrf2 Pathway in Mice. *PLoS ONE* **2013**, *8*, e74871. [CrossRef] [PubMed]
135. Liu, G.; He, L. Epigallocatechin-3-Gallate Attenuates Adriamycin-Induced Focal Segmental Glomerulosclerosis via Suppression of Oxidant Stress and Apoptosis by Targeting Hypoxia-Inducible Factor-1alpha/ Angiopoietin-Like 4 Pathway. *Pharmacology* **2019**, *103*, 303–314. [CrossRef]
136. Yang, S.-M.; Chan, Y.-L.; Hua, K.-F.; Chang, J.-M.; Chen, H.-L.; Tsai, Y.-J.; Hsu, Y.-J.; Chao, L.K.; Feng-Ling, Y.; Tsai, Y.-L.; et al. Osthole Improves An Accelerated Focal Segmental Glomerulosclerosis Model in the Early Stage by Activating the Nrf2 Antioxidant Pathway and Subsequently Inhibiting NF-kappaB-Mediated COX-2 Expression and Apoptosis. *Free Radic. Biol. Med.* **2014**, *73*, 260–269. [CrossRef]
137. Liu, X.; Cao, W.; Qi, J.; Li, Q.; Zhao, M.; Chen, Z.; Zhu, J.; Huang, Z.; Wu, L.; Zhang, B.; et al. Leonurine Ameliorates Adriamycin-Induced Podocyte Injury via Suppression of Oxidative Stress. *Free Radic Res.* **2018**, *52*, 952–960. [CrossRef] [PubMed]
138. Son, S.S.; Kang, J.S.; Lee, E.Y. Paclitaxel Ameliorates Palmitate-Induced Injury in Mouse Podocytes. *Med. Sci. Monit. Basic Res.* **2020**, *26*, e928265. [CrossRef]
139. Wang, F.; Li, R.; Zhao, L.; Ma, S.; Qin, G. Resveratrol Ameliorates Renal Damage by Inhibiting Oxidative Stress-Mediated Apoptosis of Podocytes in Diabetic Nephropathy. *Eur. J. Pharmacol.* **2020**, *885*, 173387. [CrossRef]
140. Zhang, T.; Chi, Y.; Ren, Y.; Du, C.; Shi, Y.; Li, Y. Resveratrol Reduces Oxidative Stress and Apoptosis in Podocytes via Sir2-Related Enzymes, Sirtuins1 (SIRT1)/Peroxisome Proliferator-Activated Receptor Gamma Co-Activator 1alpha (PGC-1alpha) Axis. *Med. Sci. Monit.* **2019**, *25*, 1220–1231. [CrossRef]
141. Yang, K.; Bai, Y.; Yu, N.; Lu, B.; Han, G.; Yin, C.; Pang, Z. Huidouba Improved Podocyte Injury by Down-Regulating Nox4 Expression in Rats with Diabetic Nephropathy. *Front. Pharmacol.* **2020**, *11*, 587995. [CrossRef] [PubMed]
142. Lu, H.; Li, Y.; Zhang, T.; Liu, M.; Chi, Y.; Liu, S.; Shi, Y. Salidroside Reduces High-Glucose-Induced Podocyte Apoptosis and Oxidative Stress via Upregulating Heme Oxygenase-1 (HO-1) Expression. *Med. Sci. Monit.* **2017**, *23*, 4067–4076. [CrossRef] [PubMed]
143. Chen, Y.; Liu, Q.; Shan, Z.; Zhao, Y.; Li, M.; Wang, B.; Zheng, X.; Feng, W. The Protective Effect and Mechanism of Catalpol on High Glucose-Induced Podocyte Injury. *BMC Complement. Altern. Med.* **2019**, *19*, 244. [CrossRef] [PubMed]
144. Chen, Y.; Chen, J.; Jiang, M.; Fu, Y.; Zhu, Y.; Jiao, N.; Liu, L.; Du, Q.; Wu, H.; Xu, H.; et al. Loganin and Catalpol Exert Cooperative Ameliorating Effects on Podocyte Apoptosis upon Diabetic Nephropathy by Targeting Ages-Rage Signaling. *Life Sci.* **2020**, *252*, 117653. [CrossRef]
145. Gui, D.; Guo, Y.; Wang, F.; Liu, W.; Chen, J.; Chen, Y.; Huang, J.; Wang, N. Astragaloside IV, A Novel Antioxidant, Prevents Glucose-Induced Podocyte Apoptosis in Vitro and In Vivo. *PLoS ONE* **2012**, *7*, e39824. [CrossRef]
146. Yu, Q.; Zhang, M.; Qian, L.; Wen, D.; Wu, G. Luteolin Attenuates High Glucose-Induced Podocyte Injury via Suppressing NLRP3 Inflammasome Pathway. *Life Sci.* **2019**, *225*, 1–7. [CrossRef] [PubMed]
147. Kang, M.K.; Park, S.H.; Kim, Y.H.; Lee, E.J.; Antika, L.D.; Kim, D.Y.; Choi, Y.J.; Kang, Y.H. Chrysin Ameliorates Podocyte Injury and Slit Diaphragm Protein Loss via Inhibition of the PERK-eIF2alpha-ATF-CHOP Pathway in Diabetic Mice. *Acta Pharmacol. Sin.* **2017**, *38*, 1129–1140. [CrossRef]
148. Zhao, K.; Li, Y.; Wang, Z.; Han, N.; Wang, Y. Carnosine Protects Mouse Podocytes from High Glucose Induced Apoptosis Through PI3K/AKT and Nrf2 Pathways. *Biomed. Res. Int.* **2019**, *2019*, 4348973. [CrossRef]
149. Riedl, E.; Pfister, F.; Braunagel, M.; Brinkkötter, P.; Sternik, P.; Deinzer, M.; Bakker, S.J.; Henning, R.H.; Born, J.V.D.; Krämer, B.K.; et al. Carnosine Prevents Apoptosis of Glomerular Cells and Podocyte Loss in STZ Diabetic Rats. *Cell Physiol. Biochem.* **2011**, *28*, 279–288. [CrossRef]

150. Quan, X.; Liu, H.; Ye, D.; Ding, X.; Su, X. Forsythoside A Alleviates High Glucose-Induced Oxidative Stress and Inflammation in Podocytes by Inactivating MAPK Signaling via MMP12 Inhibition. *Diabetes Metab. Syndr. Obes.* **2021**, *14*, 1885–1895. [CrossRef]
151. He, J.Y.; Hong, Q.; Chen, B.X.; Cui, S.Y.; Liu, R.; Cai, G.Y.; Guo, J.; Chen, X.M. Ginsenoside Rb1 Alleviates Diabetic Kidney Podocyte Injury by Inhibiting Aldose Reductase Activity. *Acta Pharmacol. Sin.* **2022**, *43*, 342–353. [CrossRef] [PubMed]
152. Cai, X.; Bao, L.; Ren, J.; Li, Y.; Zhang, Z. Grape Seed Procyanidin B2 Protects Podocytes from High Glucose-Induced Mitochondrial Dysfunction and Apoptosis via the AMPK-SIRT1-PGC-1alpha Axis In Vitro. *Food Funct.* **2016**, *7*, 805–815. [CrossRef] [PubMed]
153. Borges, C.M.; Papadimitriou, A.; Duarte, D.A.; Lopes de Faria, J.M.; Lopes de Faria, J.B. The Use of Green Tea Polyphenols for Treating Residual Albuminuria in Diabetic Nephropathy: A Double-Blind Randomised Clinical Trial. *Sci. Rep.* **2016**, *6*, 28282. [CrossRef] [PubMed]
154. Lee, M.; Zhao, H.; Liu, X.; Liu, D.; Chen, J.; Li, Z.; Chu, S.; Kou, X.; Liao, S.; Deng, Y.; et al. Protective Effect of Hydroxysafflor Yellow A on Nephropathy by Attenuating Oxidative Stress and Inhibiting Apoptosis in Induced Type 2 Diabetes in Rat. *Oxid. Med. Cell Longev.* **2020**, *2020*, 7805393. [CrossRef] [PubMed]
155. Li, T.X.; Mao, J.H.; Huang, L.; Fu, H.D.; Chen, S.H.; Liu, A.M.; Liang, Y.Q. Beneficial Effects of Huaiqihuang on Hyperglycemia-Induced MPC5 Podocyte Dysfunction through the Suppression of Mitochondrial Dysfunction and Endoplasmic Reticulum Stress. *Mol. Med. Rep.* **2017**, *16*, 1465–1471. [CrossRef] [PubMed]
156. Zhang, J.; Yang, S.; Li, H.; Chen, F.; Shi, J. Naringin Ameliorates Diabetic Nephropathy by Inhibiting NADPH Oxidase 4. *Eur. J. Pharmacol.* **2017**, *804*, 1–6. [CrossRef]
157. Xue, W.; Mao, J.; Chen, Q.; Ling, W.; Sun, Y. Mogroside IIIE Alleviates High Glucose-Induced Inflammation, Oxidative Stress and Apoptosis of Podocytes by the Activation of AMPK/SIRT1 Signaling Pathway. *Diabetes Metab. Syndr. Obes.* **2020**, *13*, 3821–3830. [CrossRef]
158. Sohn, E.; Kim, J.; Kim, C.S.; Jo, K.; Kim, J.S. Extract of Rhizoma Polygonum Cuspidatum Reduces Early Renal Podocyte Injury in Streptozotocin-Induced Diabetic Rats and Its Active Compound Emodin Inhibits Methylglyoxalmediated Glycation of Proteins. *Mol. Med. Rep.* **2015**, *12*, 5837–5845. [CrossRef]
159. Li, J.; Wang, B.; Zhou, G.; Yan, X.; Zhang, Y. Tetrahydroxy Stilbene Glucoside Alleviates High Glucose-Induced MPC5 Podocytes Injury through Suppression of NLRP3 Inflammasome. *Am. J. Med. Sci.* **2018**, *355*, 588–596. [CrossRef]
160. ClinicalTrials.gov. Available online: https://clinicaltrials.gov/ (accessed on 23 February 2022).

biomolecules

MDPI

Review

Glucose- and Non-Glucose-Induced Mitochondrial Dysfunction in Diabetic Kidney Disease

Marie Ito, Margaret Zvido Gurumani, Sandra Merscher * and Alessia Fornoni *

Department of Medicine, Katz Family Division of Nephrology and Hypertension, Peggy and Harold Katz Family Drug Discovery Center, Miller School of Medicine, University of Miami, Miami, FL 33136, USA; mxi288@miami.edu (M.I.); mzg13@miami.edu (M.Z.G.)

* Correspondence: smerscher@med.miami.edu (S.M.); afornoni@med.miami.edu (A.F.); Tel.: +1-305-243-6567 (S.M.); +1-305-243-7745 (A.F.)

Abstract: Mitochondrial dysfunction plays an important role in the pathogenesis and progression of diabetic kidney disease (DKD). In this review, we will discuss mitochondrial dysfunction observed in preclinical models of DKD as well as in clinical DKD with a focus on oxidative phosphorylation (OXPHOS), mitochondrial reactive oxygen species (mtROS), biogenesis, fission and fusion, mitophagy and urinary mitochondrial biomarkers. Both glucose- and non-glucose-induced mitochondrial dysfunction will be discussed. In terms of glucose-induced mitochondrial dysfunction, the energetic shift from OXPHOS to aerobic glycolysis, called the Warburg effect, occurs and the resulting toxic intermediates of glucose metabolism contribute to DKD-induced injury. In terms of non-glucose-induced mitochondrial dysfunction, we will review the roles of lipotoxicity, hypoxia and vasoactive pathways, including endothelin-1 (Edn1)/Edn1 receptor type A signaling pathways. Although the relative contribution of each of these pathways to DKD remains unclear, the goal of this review is to highlight the complexity of mitochondrial dysfunction in DKD and to discuss how markers of mitochondrial dysfunction could help us stratify patients at risk for DKD.

Keywords: diabetic kidney disease; mitochondrial dysfunction; mitochondrial reactive oxygen species; Warburg effect

Citation: Ito, M.; Gurumani, M.Z.; Merscher, S.; Fornoni, A. Glucose- and Non-Glucose-Induced Mitochondrial Dysfunction in Diabetic Kidney Disease. *Biomolecules* **2022**, *12*, 351. https://doi.org/10.3390/biom12030351

Academic Editor: Liang-Jun Yan

Received: 25 January 2022
Accepted: 21 February 2022
Published: 23 February 2022

1. Introduction

The number of patients with diabetes mellitus is expanding, and diabetic kidney disease (DKD) is an important cause of diabetic microvascular complications that constitutes an independent risk factor of mortality and cardiovascular events [1].

The kidney is one of the most energy-demanding organs and, after the heart, has the second highest expression of proteins involved in mitochondrial function and oxygen consumption [2,3]. The kidney requires energy mainly for solute reabsorption, among other tasks including waste removal, maintenance of electrolyte and fluid balance and acid–base homeostasis [4]. The generation of an ion gradient across the plasma membrane by Na^+/K^+-ATPase is essential for solute reabsorption. Therefore, mitochondrial dysfunction is postulated to play a central role in the pathogenesis and progression of kidney diseases including DKD [5–7].

Since each component of the nephron has a distinct role and different energy requirement, the causes and phenotypes of mitochondrial dysfunction may differ among cell types in the kidney. In this review, we will discuss glucose-related and unrelated pathways of mitochondrial dysfunction contributing to DKD in terms of the categories as well as the causes.

2. Mitochondrial Dysfunction in DKD

Due to the critical role of mitochondria as the powerhouse of cells, mitochondrial dysfunction traditionally referred to an alteration in the production of adenosine triphosphate (ATP) by oxidative phosphorylation (OXPHOS). However, as our understanding of the various roles played by mitochondria has expanded, mitochondrial dysfunction now includes any abnormal biological process in mitochondria [7]. In this section, we will discuss different categories of mitochondrial dysfunction occurring in DKD and discuss their potential causative role (Figure 1).

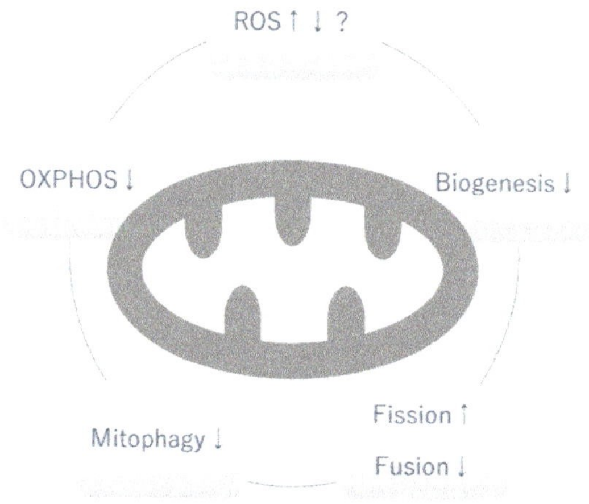

Figure 1. Mitochondrial dysfunction in DKD. Whether the level of mitochondrial ROS is increased or decreased is controversial and can vary depending on the stage of DKD. OXPHOS, mitophagy and biogenesis are generally decreased. Increased fission and decreased fusion causes fragmentation of mitochondria. OXPHOS: oxidative phosphorylation, ROS: reactive oxygen species, ↑: increased, ↓: decreased.

2.1. *Mitochondrial Oxidative Phosphorylation (OXPHOS)*

As the powerhouse of cells, the central role of mitochondria is the production of adenosine triphosphate (ATP). Metabolites from glucose, lipids and amino acids are transported into the mitochondrial matrix, serving as substrates of the tricarboxylic acid (TCA) cycle. NADH and $FADH_2$ are generated along with the reaction feed electrons into Complexes I and II of the electron transport chain (ETC). As electrons are transported through the ETC, H^+ ions are pumped into the intermembrane space. Complex V or ATP synthase uses this proton gradient to generate ATP (Figure 2). The indices of OXPHOS activity and fitness include oxygen consumption rate (OCR), ATP production, membrane potential and the evaluation of each complex (activity, formation). Generally, it has been observed that OCR in the kidney cortex is increased in early DKD, followed by a decrease as DKD progresses, whereas in glomeruli and podocytes the OCR is decreased in both in early and late phases of the disease [5]. Although some discrepancy exists between studies, ATP production and complex activity have been demonstrated to be decreased at least in the late stage of DKD [8,9]. The contribution of decreased activation of OXPHOS to DKD can be inferred from the observation that some genetic mutations in OXPHOS, such as single-nucleotide polymorphisms (SNPs) in coenzyme Q5 (*COQ5*) and cytochrome *c* oxidase (*COX6A1*), are linked to DKD in humans [10]. *COQ5* encodes methyltransferase located in the mitochondrial matrix and *COX6A1* encodes a subunit of cytochrome c, which is part of the ETC.

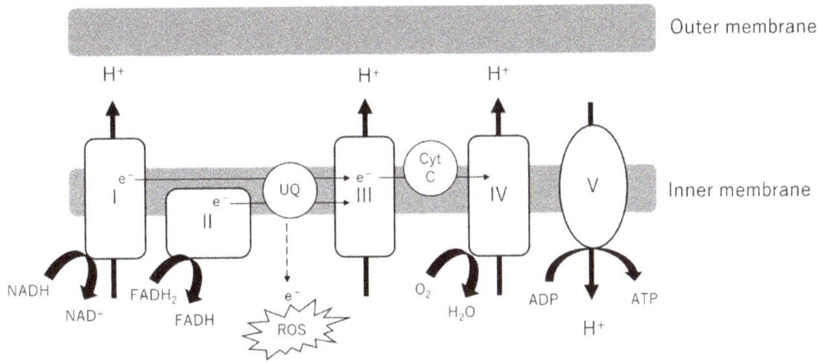

Figure 2. Electron transport chain (ETC) in mitochondrial inner membrane. NADH and $FADH_2$ from the TCA cycle donate electrons to Complexes I and II. As electrons are transported through the ETC, a proton gradient is generated, which Complex V or ATP synthase couples to ATP synthesis. Electron leakage from the ETC causes the production of ROS. ADP, adenosine diphosphate; ATP, adenosine triphosphate; Cyt C, cytochrome complex; ROS, reactive oxygen species; UQ, ubiquinone; TCA cycle, tricarboxylic acid cycle.

2.2. *Mitochondrial Reactive Oxygen Species (mtROS)*

Ever since Brownlee and colleagues proposed that hyperglycemia-induced mitochondrial reactive oxygen species (mtROS) was the unifying mechanism of diabetic microvascular complications in 2000, this paradigm has been prevalent [11,12]. Recently, the source of ROS in DKD and the pathogenic role of ROS have become controversial [13,14]. Although there may be a consensus that ROS-induced damage is increased in DKD, conflicting studies exist regarding the change in mtROS production, which can be attributed to the different methods used to detect mtROS or the varying models or timepoints of DKD. In both live and fixed db/db mouse kidneys, increased mitochondrial ROS was observed using a mitochondrial matrix-localized reduction–oxidation-sensitive green fluorescent protein probe [15]. By contrast, in streptozotocin (STZ)-injected C57BL/6J mice and Ins2-Akita mice (DBA/B6 F1 mice), decreased mitochondrial superoxide was observed upon systemic administration of dihydroethidium (DHE) both in live and fixed kidneys [16]. The latter study does not preclude ROS production in other cell compartments, including the endoplasmic reticulum (ER) or enzyme systems such as nicotinamide adenine dinucleotide phosphate oxidase (Nox). Notably, the restoration of mitochondrial biogenesis and OXPHOS activity by adenosine monophosphate-activated protein kinase (AMPK) activation increased mtROS and ameliorated the DKD phenotype, arguing against the role of mtROS in inciting DKD.

ROS does not play an exclusively detrimental role in cell biology. Mitochondrial hormesis is the concept that slightly enhanced mitochondrial superoxide at baseline can decrease susceptibility to more severe cell stress [13]. ROS also plays an essential role in certain cell signaling pathways, requiring further elucidation of the intricate characteristics of ROS.

2.3. *Biogenesis*

Cells cope with increasing energy demand by increasing mitochondrial biogenesis, in which functional mitochondria are generated by duplication of mitochondrial DNA (mtDNA) and subsequent binary fission. Peroxisome proliferator-activated receptor γ coactivator 1α (PGC1α) plays a central role in mitochondrial biogenesis [17]. PGC1α is a transcriptional regulator of mitochondrial metabolic pathways such as oxidative phosphorylation (OXPHOS), the TCA cycle and fatty acid metabolism. PGC1 peroxisome proliferator-activated receptors (PPARs) and estrogen-related receptors (ERRs) can also serve as a coactivator of PGC1α. PGC1α dimerizes with transcriptional coactivators to regulate downstream gene transcription, and these partners include cyclic AMP-responsive element-binding protein (CREB) nuclear respiratory factors 1 and 2 (NRF1 and NRF2) and activated PPARs and ERRs [4]. Nutrient-sensing pathways like the mechanistic target of rapamycin (mTOR), AMPK, sirtuin, cyclic AMP (cAMP) and cyclic guanosine monophosphate (cGMP) regulate PGC1α directly or indirectly.

In DKD, although there is some discrepancy between studies possibly due to the analysis at different disease stages, PGC1α activity is considered to be increased in the early stage of diabetes, as demonstrated in 8-week-old db/db mice, followed by a decrease in activity at later stages, as demonstrated in pretransplant patients and mice 24 weeks after diabetes induction with STZ injection [8,16,18,19]. Taurine upregulated gene 1 (*Tug1*), a long noncoding gene, was described as a regulator of PGC1α in podocytes in DKD [20]. It was demonstrated that *Tug1* binds to an element upstream of *Ppargc1a* and interacts with PGC1α binding to its own promoter, subsequently enhancing *Ppargc1a* promoter activity.

2.4. Mitochondrial Fission and Fusion

Mitochondria are dynamic organelles which undergo tightly controlled processes of fission and fusion. Mitochondrial fission is mediated by dynamin-1-like protein (DRP1) and its receptors such as fission factor 1 (FIS1), mitochondrial fission factor (MFF) and mitochondrial dynamics proteins of 49 and 51 kDa (MID49 and MID51). Mitochondrial fusion is mediated by the long isoforms of optic atrophy protein 1 (OPA1), which plays a role in inner mitochondrial membrane fusion, and the mitofusins (MFN1 and MFN2) which play a role in outer mitochondrial membrane fusion [5,6].

Although the increase in mitochondrial fission and fusion factors such as the long isoforms of OPA1, MFN1, MFN2 and MFF were observed in early DKD, mitochondria were consistently fragmented throughout early and late stage in STZ-injected rats [8]. Human kidney biopsies of patients with DKD also demonstrated fragmented mitochondria in podocytes and proximal tubular cells [21,22]. Consistent with increased fission and decreased fusion, Drp1 and FIS1 expression was increased, while MFN2 expression was shown to be decreased in tubules in the latter study.

2.5. Mitophagy

Autophagy is a pathway that degrades and recycles damaged organelles and macromolecules, and selective autophagy of mitochondria is termed as mitophagy. Mitophagy has a critical role in the maintenance of mitochondrial quality by removing damaged mitochondria. Mitophagy can be mediated by the phosphatase and tensin homolog-induced putative kinase 1 (PINK1)/parkin-mediated pathway and other outer mitochondrial membrane proteins such as BCL2/adenovirus E1B 19 kDa protein-interacting protein 3 (BNIP3) and NIP3-like protein X (NIX), or the mitophagy receptor FUN14 domain-containing protein 1 (FUNDC1).

The PINK1 and parkin-mediated pathway has been more extensively investigated than the others [23]. PINK1 has a Ser/Thr kinase domain and is found inserted into both the inner and outer mitochondrial membranes. In healthy mitochondria, PINK1 is cleaved at two points by mitochondrial proteases, leading to its dissociation from the mitochondrial membrane and degradation by the ubiquitin-proteasome system. In depolarized mitochondria, PINK1 escapes cleavage and stably resides in the outer membrane. Subsequently, PINK1 homodimerizes and autophosphorylates to recruit E3 ubiquitin ligase parkin and ubiquitin, directing the mitochondria to the mitophagy pathway. In the ubiquitin-independent pathway, outer mitochondrial membrane proteins such as BNIP3, NIX or FUNDC1 recruit microtubule-associated protein 1A/1B light chain 3 (LC3) and induce mitophagy under certain stimuli including hypoxia [24,25]. Cardiolipin, which is located in the inner mitochondrial membrane under normal conditions, is externalized by certain stimuli and detected by LC3, facilitating the engulfment of the mitochondria by autophagosomes [26]. P62 is a marker of autophagy cargo, and its accumulation can indicate stagnation in degradation via autophagic flux. In general, basal mitophagy levels of podocytes are high, which can be attributed to their terminally differentiated characteristics. In contrast, in tubular cells the mitophagy level is low at baseline but it can be induced as a consequence of stress.

Mitophagy is suppressed in DKD, which was demonstrated by low PINK1/parkin-expression levels in podocytes of STZ-induced diabetic mice and increased p62 expression levels in tubular cells of biopsy obtained from patients with DKD [27–29]. Thioredoxin-interacting protein (TXNIP) was implicated in the suppression of tubular autophagy and mitophagy induced by high glucose [27]. High glucose was also shown to inhibit the transcriptional activity of forkhead-box class O1 (FoxO1) via its phosphorylation by Akt (protein kinase B), leading to the downregulation of PINK1 [29]. The protective effect of mitoquinone on DKD was partially attributed to the restoration of PINK1 and parkin protein expression in tubular cells via NRF2 activation [30].

2.6. Urinary Mitochondrial Biomarker

It is of value to detect DKD in the early disease stage in a reliable manner, preferentially before microalbuminuria or a decrease in the estimated glomerular filtration rate (eGFR) becomes evident, to prevent further progression to end stage kidney disease. Since mitochondrial dysfunction is thought to precede overt histological changes in DKD, biomarkers of mitochondrial dysfunction have recently been vigorously investigated. Even though those studies investigated patients with established DKD, they can be useful to identify biomarkers in patients affected by DKD. A study of metabolites in urine found global suppression in mitochondrial respiration in patients with DKD compared to controls without DKD [19]. In concordance with the notion that mtDNA is released from damaged tubular cells, the same study demonstrated increased mtDNA in urine. Although urinary mtDNA had a modest but significant inverse correlation with intra-renal mtDNA and eGFR at baseline, a positive correlation with interstitial fibrosis was found. Urinary mtDNA or mtDNA of biopsy specimens did not significantly correlate with eGFR decline in the 24 months of follow up [31]. Further investigations are warranted to determine if mitochondrial biomarkers can be used clinically.

3. Glucose-Induced Mitochondrial Dysfunction in DKD

In glomerular cells, glucose is taken up by glucose transporters (GLUTs) via facilitated diffusion transport. The expression pattern of each member of the GLUTs is cell-specific. Mesangial cells express GLUT1 and 4, podocytes express GLUT1, 4 and 8 and endothelial cells express GLUT1 [32]. In contrast, tubular cells reabsorb glucose from the glomerular filtrate mainly via sodium-dependent glucose cotransporters (SGLTs). Glucose reabsorbed by tubular cells is dissipated across GLUTs in the basolateral plasma membrane and diffuses into the interstitium. Proximal tubular cells can also produce glucose via gluconeogenesis, which is increased in type 2 diabetes [33]. Hyperglycemia is the major pathogenetic factor and some intermediate metabolites of glucose metabolism have also been implicated in contributing to cell injury in DKD. In the unifying hypothesis offered by Brownlee and colleagues in 2000, it was proposed that the overproduction of mtROS due to increased influx into OXPHOS activated the nuclear DNA-repair enzyme poly(ADP-ribose) polymerase (PARP), leading to the decreased activity of glyceraldehyde 3-phosphate dehydrogenase (GAPDH) and the subsequent accumulation of toxic intermediates of glucose metabolism [11]. However, later studies have suggested that mitochondria are dysfunctional and that other mechanisms, in addition to a mere increase in substrates, contribute to increased glycolysis and the subsequent accumulation of toxic metabolites.

3.1. Warburg Effect

In general, cancer cells display enhanced glycolysis and impaired oxidative phosphorylation. While the short-time and reversible shift of this metabolic process is called the Crabtree effect, the long-term metabolic reprogramming is called the Warburg effect [34].

The Warburg effect is the term originally used to describe a shift from OXPHOS to aerobic glycolysis (in which lactate is produced as a final product from glucose) in cancer [35]. Recent studies have demonstrated that this shift also takes place in diabetic kidneys [9,19]. Mitochondria in diabetic tissues are dysfunctional, as discussed above. Transcriptomic, metabolomic and metabolite flux analysis showed increased glucose metabolism and decreased mitochondrial function in the kidney cortices of db/db mice [9]. A metabolomics analysis of urine samples demonstrated significant decreases of 13 metabolites in patients with DKD compared to healthy controls, 12 of which were associated with mitochondrial metabolism [19]. Whether mitochondrial dysfunction causes the shift to glycolysis or if increased glycolysis causes mitochondrial dysfunction remains to be established [36].

Pyruvate kinase is the enzyme which catalyzes the conversion of phosphoenolpyruvate to pyruvate, the last and irreversible step of glycolysis. Decreased pyruvate kinase M2 activity was identified as the possible mechanism of the Warburg effect in DKD by Qi and colleagues [37]. They first conducted proteomics analysis of glomeruli isolated from patients with type 1 diabetes who did not develop DKD for over 50 years (protected) and those with a histological confirmation of DKD (unprotected). Some enzymes involved in glucose metabolism and antioxidation were found to be increased in protected glomeruli, and in particular, pyruvate kinase M2 (PKM2) expression and activity were upregulated. In mechanistic studies using high-glucose-treated podocytes and STZ-injected mice, it was confirmed that PKM2 activity was decreased in the DKD mouse model, that PKM downregulation contributed to DKD exacerbation and that pharmacological activation of PKM2 reversed the elevation in toxic glucose metabolites and mitochondrial dysfunction. Similarly, decrease in pyruvate kinase activation causes the Warburg effect in cancer [38].

Other factors that are postulated to induce the Warburg effect in DKD include sphingomyelin and fumarate accumulation [39]. Matrix-assisted laser desorption/ionization mass spectrometry imaging (MALDI-MSI) revealed significant increases of ATP/AMP ratio and of a specific sphingomyelin species (SM(d18:1/16:0)) in glomeruli of DKD mice [40]. In vitro, addition of SM(d18:1/16:0) to mesangial cells activated glycolysis. Fumarate was identified as a factor that mediated Nox4-induced injury in DKD [41]. Podocyte-specific induction of Nox4 in vivo recapitulated DKD-induced glomerular injury, and metabolomic analysis demonstrated increased fumarate, which was reversed with Nox1/Nox4 inhibition. Fumarate could serve as a hypoxia-inducible factor (HIF) stabilizer, which could cause the activation of glycolysis and suppression of OXPHOS.

3.2. Toxic Metabolites of Glucose Metabolism

Four major pathways branching from glycolysis are known to produce toxic intermediate metabolites: the polyol pathway, the hexosamine pathway, the advanced glycation end-products (AGEs) pathway and the protein kinase C (PKC) pathway [12] (Figure 3). Inhibition of each one of these pathways ameliorates hyperglycemia-induced injury in preclinical models [42–45]. Notably, the activation of any of these pathways can be instantly reversed with the restoration of euglycemia.

In particular, levels of 3-deoxyglucosone (3DG) and methylglyoxal (MGO), members of reactive carbonyl species (RCS) which are produced by the degradation of glyceraldehyde in the AGE pathway, are increased with the activation of glycolysis. RCS have highly reactive carbonyl groups and can modify protein and DNA. Proteins glycated by RCS leads to formation of AGEs. Extracellular AGEs can increase the crosslinking of matrices, leading to arterial stiffening [46]. AGEs can also bind to receptors for AGE (RAGE) and transduce various signals into cells, including nuclear factor-κB (NF-κB) activation leading to ROS formation, inflammation and fibrosis [47].

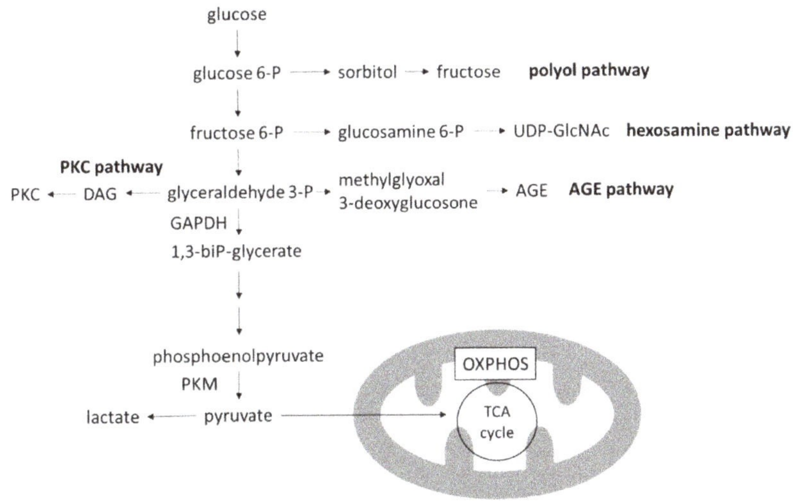

Figure 3. Glycolysis and branching pathways. In the high-glucose environment of DKD, glycolysis is increased and OXPHOS is decreased, leading to the accumulation of toxic metabolites produced in the branching pathways of glycolysis. AGE, advanced glycation end-product; DAG, diacylglycerol; GAPDH, glygeraldehyde-3-phosphate dehydrogenase; OXPHOS, oxidative phosphorylation; PKC, protein kinase C; PKM, pyruvate kinase M; TCA cycle, tricarboxylic acid cycle; UDP-GlcNAc, uridine diphosphate *N*-acetylglucosamine.

4. Non-Glucose-Induced Mitochondrial Dysfunction in DKD

Although hyperglycemia is a key factor in the development of DKD, other factors involved in DKD can also contribute to mitochondrial dysfunction. Dyslipidemia and lipid overload is a common complication in DKD, and hypoxia in the tubulointerstium occurs regardless of the cause of chronic kidney disease (CKD) [48]. Endothelin-1 (Edn1) was first identified as a downstream factor of transforming growth factor-β (TGF-β) in a model of focal segmental glomerulosclerosis (FSGS) and shown to induce albuminuria via mtROS in glomerular endothelial cells. Later, the same signaling pathway was also found to be upregulated in DKD [49,50]. In this section, we discuss the roles of these factors in the development and progression of DKD in relation to mitochondrial dysfunction.

4.1. Lipotoxicity

In kidney biopsies of patients with DKD, extensive lipid droplet accumulation was observed by electron microscopy in glomerular endothelial cells, podocytes and tubular cells compared to healthy counterparts [51,52]. Genetic analysis of these samples revealed downregulation of fatty acid oxidation (FAO)-related genes including peroxisome proliferator-activated receptor (PPAR)-α, carnitine palmitoyltransferase 1 (CPT1), acyl-CoA oxidase, and L-FABP; upregulation of cholesterol receptors including low-density lipoprotein (LDL) receptors, oxidized LDL receptors, and acetylated LDL receptors; and downregulation of cholesterol-efflux-related genes including ATP-binding cassette transporter A1 (ABCA1), ATP-binding cassette transporter G1 (ABCG1), and apolipoprotein (APOE) [51]. While downregulation of FAO-related genes suggests a decrease in mitochondrial lipid metabolism as the cause of lipid accumulation, lipid accumulation itself can induce mitochondrial dysfunction. We previously showed that human podocytes treated with the serum of patients with DKD show increased tumor necrosis factor (TNF) expression and that local rather than systemic TNF causes free cholesterol accumulation and injury via the suppression of ABCA1 in podocytes [53,54]. Notably, ABCA1 suppression induced cardiolipin accumulation and peroxidation in mitochondria, sensitizing podocytes to injury [55]. ABCA1 overexpression or inhibition of cardiolipin peroxidation by elamipretide rescued podocyte injury in experimental DKD.

Tubular cells require large amounts of ATP for solute reabsorption and depend on FAO because fatty acids yield more ATP per gram than other energy sources [56]. In the early stage of diabetes, FAO is increased in accordance with increased FA flux, and ROS production is attributed to FAO, especially to electron leakage at the electron transfer flavoprotein that shuttles electrons from acyl-CoA dehydrogenases to coenzyme Q [57]. Nevertheless, FAO is eventually decreased in established diabetes as described earlier [51,58].

Tubular cells in patients with DKD are likely exposed to fatty acid-bound albumin, since dyslipidemia and proteinuria often accompany diabetes mellitus. Whereas albumin itself can cause tubular cell damage on its reabsorption, FA-bound albumin was shown to induce more severe tubular damage [59,60]. Cytotoxicity caused by FA or glycated albumin was shown to be mediated by the uptake via the protein cluster of differentiation 36/FA translocase (CD36/FAT) in the brush border in humans, in contrast to usual albumin reabsorption via a complex of megalin, cubilin and amnionless [61]. In addition to reabsorption, synthesis of FA is also upregulated [62,63]. FAs are esterified by long-chain acyl-CoA (LC-CoA) synthetase (ACSL), which is upregulated in both mouse db/db model and human DKD kidney samples [64,65]. LC-CoA is transferred to mitochondria via CPT1 and CPT2 to produce ATP via FAO and unmetabolized LC-CoA is cleaved or stored in lipid droplets to prevent lipotoxicity. When buffering ability is saturated, LC-CoA serves as an inhibitor of the Na^+/H^+ exchanger 1 (NHE1) and phosphatidylinositol 4,5-bisphosphate [PI(4,5)P2] binding to cause apoptosis in proximal tubular cells [66].

Genetic and pharmacological enhancement of FAO could represent a new therapeutic treatment strategy for patients with DKD. Transgenic expression of PCG1α and fenofibrate treatment ameliorated tubular cell apoptosis via restoration of CPTs and/or acyl-CoA oxidases [58,67].

4.2. Hypoxia

Tubulointerstitial hypoxia is known to be a final common pathway of CKD progression [48]. Regardless of the cause of the disease, hypoxia can occur due to decreased oxygen supply via blood flow because of impaired vasodilation, loss of vasculature, fibrosis, anemia, increased oxygen demand due to increased solute reabsorption, and inefficient ATP production because of mitochondrial uncoupling, which are the shared mechanisms in CKD.

Recently, the mechanism by which acute hypoxia causes mtROS production in the ETC has been revealed [68]. Acute hypoxia prompts a conformational shift in Complex I, leading to the activation of the Na^+/Ca^{2+} exchanger in the inner mitochondrial membrane. Na^+ imported into the matrix then interacts with phospholipids to reduce membrane fluidity, resulting in the inability of free ubiquinone to move between Complex II and Complex III. Thus, Complex III produces ROS.

Hypoxia had long been deemed a direct suppressor of OXPHOS because oxygen is indispensable as a receiver of electrons in the ETC. However, to decelerate OXPHOS acutely and directly, the oxygen concentration has to be as low as 0.3% [69]. In milder and prolonged hypoxia, HIF1α acts as a mediator to suppress OXPHOS and to increase glycolysis in order to prevent ROS production. HIF1α modifies ETC complexes and promotes metabolic shifts from aerobic OXPHOS to anaerobic glycolysis [69]. Nevertheless, in chronic hypoxia, these changes limit the mitochondrial ability to produce ATP and could induce ATP deficiency, possibly leading to cytotoxicity.

4.3. Endothelin-1 (Edn1)/Edn1 Receptor Type A (Endra) Signaling

Endothelin-1 (Edn1) was first characterized as a signaling molecule released by podocytes in TGF-β-induced FSGS [49]. Surprisingly, Edn1 causes mtROS, decreases reserve respiratory capacity and mtDNA damage in endothelial cells via Edn1 receptor type A (Ednra) activation, but this is not seen in podocytes. This pathway induces glycosaminoglycan degradation in the endothelial surface layer, leading to the loss of fenestration [70]. Interestingly, EDNRA activation in endothelial cells is also required for podocyte foot process effacement and apoptosis [49].

The similar role of mtROS and EDNRA in endothelial cells and Edn1 is also described in DKD [50]. Circulating Edn1 was increased in diabetic humans and mice [50,71]. Notably, Ednra was not detectable in the glomeruli of healthy human kidneys or DKD-resistant C57BL/6J mice, but was present in those of human DKD kidneys and diabetic DBA/2J mice. Ednra could be induced in high-glucose-treated podocytes, but these podocytes did not express Edn1, implicating other cell types or stimuli in this signaling cascade [50]. Thus, Edn1/Ednra signaling has a critical role in the reciprocal crosstalk between podocytes and endothelial cells via mitochondrial dysfunction both in FSGS and DKD.

Another vasoactive pathway, the renin–angiotensin–aldosterone system (RAAS), is also known to participate in CKD progression, including in DKD. Aside from the detrimental effects of increasing systemic and intraglomerular blood pressure, angiotensin II treatment was found to exacerbate mtROS production and mitochondrial fragmentation in podocytes both in vivo and in vitro, which can be reversed by mitoquinone, a mitochondria-targeted antioxidant [72].

5. Conclusions

Mitochondrial dysfunction plays a central role in the development and progression of DKD. Thus, targeting mitochondrial dysfunction in DKD could represent a novel therapeutic strategy for patients with DKD. However, as can be learned from studies investigating the beneficial effect of systemic antioxidant administration as a treatment for DKD, it seems that intervention in mitochondrial dysfunction has to be cell type- and context-specific. Another intriguing perspective is age- and sex-related difference in mitochondrial dysfunction in DKD. Although age is an important factor that affects mitochondrial function and the development and progression of DKD, not much is known about the exact mechanisms. In terms of sex, much difference is observed between males and females in the mitochondria of certain tissues, and estrogen and possibly testosterone can mediate renal mitochondrial bioenergetics [73]. Further investigations are needed to elucidate the exact mechanisms leading to mitochondrial dysfunction in DKD, to the extent that findings can be clinically applied to patients and change their prognoses.

Author Contributions: Conceptualization, M.I., M.Z.G. and A.F.; writing—original draft preparation, M.I.; writing—review and editing, M.I., S.M. and A.F. All authors have read and agreed to the published version of the manuscript.

Funding: M.I. is supported by Manpei Suzuki Diabetes Foundation. A.F. and S.M. are supported by National Institutes of Health grants R01DK117599, R01DK104753 and R01CA227493. A.F. is also supported by U54DK083912, UM1DK100846, U01DK116101 and UL1TR000460 (Miami Clinical Translational Science Institute).

Institutional Review Board Statement: Not applicable.

Informed Consent Statement: Not applicable.

Data Availability Statement: Not applicable.

Conflicts of Interest: A.F. and S.M. are inventors of pending and issued patents (US10183038; US10052345; PCT/US2019/032215; PCT/US2019/041730; PCT/US2013/036484; US17/057247; US17/259883; Japan no. 501309/2021; Europe no. 19834217.2; China no. 201980060078.3; Canada no. 2852904; 2930119; 3012773) aimed at preventing and treating renal disease. They stand to gain royalties from the future commercialization of these patents. S.M. and A.F. hold equity interest in L&F Research and ZyVersa Therapeutics, Inc. which has licensed worldwide rights to develop and commercialize hydroxypropyl-beta-cyclodextrin from L&F Research for the treatment of kidney disease. A.F. also holds equities in Renal 3 River Corporation. A.F. and S.M. are supported by Aurinia Pharmaceuticals and Boehringer Ingelheim.

References

1. Barkoudah, E.; Skali, H.; Uno, H.; Solomon, S.D.; Pfeffer, M.A. Mortality rates in trials of subjects with type 2 diabetes. *J. Am. Heart Assoc.* **2012**, *1*, e000059. [CrossRef]
2. Pagliarini, D.J.; Calvo, S.E.; Chang, B.; Sheth, S.A.; Vafai, S.B.; Ong, S.-E.; Walford, G.A.; Sugiana, C.; Boneh, A.; Chen, W.K.; et al. A Mitochondrial Protein Compendium Elucidates Complex I Disease Biology. *Cell* **2008**, *134*, 112–123. [CrossRef]
3. O'Connor, P.M. Renal oxygen delivery: Matching delivery to metabolic demand. *Clin. Exp. Pharmacol. Physiol.* **2006**, *33*, 961–967. [CrossRef]
4. Bhargava, P.; Schnellmann, R.G. Mitochondrial energetics in the kidney. *Nat. Rev. Nephrol.* **2017**, *13*, 629–646. [CrossRef]
5. Mise, K.; Galvan, D.L.; Danesh, F.R. Shaping up Mitochondria in Diabetic Nephropathy. *Kidney360* **2020**, *1*, 982–992. [CrossRef]
6. Forbes, J.M.; Thorburn, D.R. Mitochondrial dysfunction in diabetic kidney disease. *Nat. Rev. Nephrol.* **2018**, *14*, 291–312. [CrossRef]
7. Wei, P.Z.; Szeto, C.C. Mitochondrial dysfunction in diabetic kidney disease. *Clin. Chim. Acta* **2019**, *496*, 108–116. [CrossRef]
8. Coughlan, M.T.; Nguyen, T.-V.; Penfold, S.A.; Higgins, G.C.; Thallas-Bonke, V.; Tan, S.M.; Van Bergen, N.J.; Sourris, K.C.; Harcourt, B.E.; Thorburn, D.R.; et al. Mapping time-course mitochondrial adaptations in the kidney in experimental diabetes. *Clin. Sci.* **2016**, *130*, 711–720. [CrossRef]
9. Sas, K.M.; Kayampilly, P.; Byun, J.; Nair, V.; Hinder, L.M.; Hur, J.; Zhang, H.; Lin, C.; Qi, N.R.; Michailidis, G.; et al. Tissue-specific metabolic reprogramming drives nutrient flux in diabetic complications. *JCI Insight* **2016**, *1*, e86976. [CrossRef]
10. Swan, E.J.; Salem, R.M.; Sandholm, N.; Tarnow, L.; Rossing, P.; Lajer, M.; Groop, P.H.; Maxwell, A.P.; McKnight, A.J. Genetic risk factors affecting mitochondrial function are associated with kidney disease in people with Type 1 diabetes. *Diabet. Med.* **2015**, *32*, 1104–1109. [CrossRef]
11. Nishikawa, T.; Edelstein, D.; Du, X.L.; Yamagishi, S.; Matsumura, T.; Kaneda, Y.; Yorek, M.A.; Beebe, D.; Oates, P.J.; Hammes, H.-P.; et al. Normalizing mitochondrial superoxide production blocks three pathways of hyperglycaemic damage. *Nature* **2000**, *404*, 787–790. [CrossRef]
12. Iacobini, C.; Vitale, M.; Pesce, C.; Pugliese, G.; Menini, S. Diabetic Complications and Oxidative Stress: A 20-Year Voyage Back in Time and Back to the Future. *Antioxidants* **2021**, *10*, 727. [CrossRef]
13. Coughlan, M.T.; Sharma, K. Challenging the dogma of mitochondrial reactive oxygen species overproduction in diabetic kidney disease. *Kidney Int.* **2016**, *90*, 272–279. [CrossRef] [PubMed]
14. Ram, C.; Jha, A.K.; Ghosh, A.; Gairola, S.; Syed, A.M.; Murty, U.S.; Naidu, V.G.M.; Sahu, B.D. Targeting NLRP3 inflammasome as a promising approach for treatment of diabetic nephropathy: Preclinical evidences with therapeutic approaches. *Eur. J. Pharmacol.* **2020**, *885*, 173503. [CrossRef]
15. Galvan, D.L.; Badal, S.S.; Long, J.; Chang, B.H.; Schumacker, P.T.; Overbeek, P.A.; Danesh, F.R. Real-time in vivo mitochondrial redox assessment confirms enhanced mitochondrial reactive oxygen species in diabetic nephropathy. *Kidney Int.* **2017**, *92*, 1282–1287. [CrossRef]
16. Dugan, L.L.; You, Y.-H.; Ali, S.S.; Diamond-Stanic, M.; Miyamoto, S.; DeCleves, A.-E.; Andreyev, A.; Quach, T.; Ly, S.; Shekhtman, G.; et al. AMPK dysregulation promotes diabetes-related reduction of superoxide and mitochondrial function. *J. Clin. Investig.* **2013**, *123*, 4888–4899. [CrossRef]
17. Svensson, K.; Schnyder, S.; Cardel, B.; Handschin, C. Loss of Renal Tubular PGC-1α Exacerbates Diet-Induced Renal Steatosis and Age-Related Urinary Sodium Excretion in Mice. *PLoS ONE* **2016**, *11*, e0158716. [CrossRef]
18. Wang, X.X.; Wang, D.; Luo, Y.; Myakala, K.; Dobrinskikh, E.; Rosenberg, A.Z.; Levi, J.; Kopp, J.B.; Field, A.; Hill, A.; et al. FXR/TGR5 Dual Agonist Prevents Progression of Nephropathy in Diabetes and Obesity. *J. Am. Soc. Nephrol.* **2018**, *29*, 118–137. [CrossRef]
19. Sharma, K.; Karl, B.; Mathew, A.V.; Gangoiti, J.A.; Wassel, C.L.; Saito, R.; Pu, M.; Sharma, S.; You, Y.-H.; Wang, L.; et al. Metabolomics reveals signature of mitochondrial dysfunction in diabetic kidney disease. *J. Am. Soc. Nephrol.* **2013**, *24*, 1901–1912. [CrossRef]
20. Jianyin, L.; Badal, S.S.; Zengchun, Y.; Yin, W.; Ayanga, B.A.; Galvan, D.L.; Green, N.H.; Chang, B.H.; Overbeek, P.A.; Danesh, F.R. Long noncoding RNA Tug1 regulates mitochondrial bioenergetics in diabetic nephropathy. *J. Clin. Investig.* **2016**, *126*, 4205–4218. [CrossRef]
21. Ma, Y.; Chen, Z.; Tao, Y.; Zhu, J.; Yang, H.; Liang, W.; Ding, G. Increased mitochondrial fission of glomerular podocytes in diabetic nephropathy. *Endocr. Connect.* **2019**, *8*, 1206–1212. [CrossRef] [PubMed]
22. Jiang, H.; Shao, X.; Jia, S.; Qu, L.; Weng, C.; Shen, X.; Wang, Y.; Huang, H.; Wang, C.; Wang, Y.; et al. The Mitochondria-Targeted Metabolic Tubular Injury in Diabetic Kidney Disease. *Cell Physiol. Biochem.* **2019**, *52*, 156–171. [CrossRef] [PubMed]
23. Eiyama, A.; Okamoto, K. PINK1/Parkin-mediated mitophagy in mammalian cells. *Curr. Opin. Cell Biol.* **2015**, *33*, 95–101. [CrossRef] [PubMed]
24. Ney, P.A. Mitochondrial autophagy: Origins, significance, and role of BNIP3 and NIX. *Biochim. Biophys. Acta-Mol. Cell Res.* **2015**, *1853*, 2775–2783. [CrossRef]
25. Liu, L.; Feng, D.; Chen, G.; Chen, M.; Zheng, Q.; Song, P.; Ma, Q.; Zhu, C.; Wang, R.; Qi, W.; et al. Mitochondrial outer-membrane protein FUNDC1 mediates hypoxia-induced mitophagy in mammalian cells. *Nat. Cell Biol.* **2012**, *14*, 177–185. [CrossRef]
26. Chu, C.T.; Ji, J.; Dagda, R.K.; Jiang, J.F.; Tyurina, Y.Y.; Kapralov, A.A.; Tyurin, V.A.; Yanamala, N.; Shrivastava, I.H.; Mohammadyani, D.; et al. Cardiolipin externalization to the outer mitochondrial membrane acts as an elimination signal for mitophagy in neuronal cells. *Nat. Cell Biol.* **2013**, *15*, 1197–1205. [CrossRef]

27. Huang, C.; Zhang, Y.; Kelly, D.J.; Tan, C.Y.R.; Gill, A.; Cheng, D.; Braet, F.; Park, J.-S.; Sue, C.M.; Pollock, C.A.; et al. Thioredoxin interacting protein (TXNIP) regulates tubular autophagy and mitophagy in diabetic nephropathy through the mTOR signaling pathway. *Sci. Rep.* **2016**, *6*, 29196. [CrossRef]
28. Tagawa, A.; Yasuda, M.; Kume, S.; Yamahara, K.; Nakazawa, J.; Chin-Kanasaki, M.; Araki, H.; Araki, S.; Koya, D.; Asanuma, K.; et al. Impaired Podocyte Autophagy Exacerbates Proteinuria in Diabetic Nephropathy. *Diabetes* **2016**, *65*, 755–767. [CrossRef]
29. Li, W.; Du, M.; Wang, Q.; Ma, X.; Wu, L.; Guo, F.; Ji, H.; Huang, F.; Qin, G. FoxO1 Promotes Mitophagy in the Podocytes of Diabetic Male Mice via the PINK1/Parkin Pathway. *Endocrinology* **2017**, *158*, 2155–2167. [CrossRef]
30. Xiao, L.; Xu, X.; Zhang, F.; Wang, M.; Xu, Y.; Tang, D.; Wang, J.; Qin, Y.; Liu, Y.; Tang, C.; et al. The mitochondria-targeted antioxidant MitoQ ameliorated tubular injury mediated by mitophagy in diabetic kidney disease via Nrf2/PINK1. *Redox Biol.* **2017**, *11*, 297–311. [CrossRef]
31. Wei, P.Z.; Kwan, B.C.-H.; Chow, K.M.; Cheng, P.M.-S.; Luk, C.C.-W.; Li, P.K.-T.; Szeto, C.C. Urinary mitochondrial DNA level is an indicator of intra-renal mitochondrial depletion and renal scarring in diabetic nephropathy. *Nephrol. Dial. Transplant.* **2018**, *33*, 784–788. [CrossRef] [PubMed]
32. Mather, A.; Pollock, C. Glucose handling by the kidney. *Kidney Int.* **2011**, *79*, S1–S6. [CrossRef] [PubMed]
33. Meyer, C.; Woerle, H.J.; Dostou, J.M.; Welle, S.L.; Gerich, J.E. Abnormal renal, hepatic, and muscle glucose metabolism following glucose ingestion in type 2 diabetes. *Am. J. Physiol. Endocrinol. Metab.* **2004**, *287*, E1049–E1056. [CrossRef] [PubMed]
34. Diaz-Ruiz, R.; Rigoulet, M.; Devin, A. The Warburg and Crabtree effects: On the origin of cancer cell energy metabolism and of yeast glucose repression. *Biochim. Biophys. Acta-Bioenerg.* **2011**, *1807*, 568–576. [CrossRef]
35. Warburg, O. The Metabolism of Carcinoma Cells. *J. Cancer Res.* **1925**, *9*, 148–163. [CrossRef]
36. Stanton, R.C. Role of Glucose Metabolism and Mitochondrial Function in Diabetic Kidney Disease. *Curr. Diabetes Rep.* **2021**, *21*. [CrossRef]
37. Qi, W.; Keenan, H.A.; Li, Q.; Ishikado, A.; Kannt, A.; Sadowski, T.; Yorek, M.A.; Wu, I.H.; Lockhart, S.; Coppey, L.J.; et al. Pyruvate kinase M2 activation may protect against the progression of diabetic glomerular pathology and mitochondrial dysfunction. *Nat. Med.* **2017**, *23*, 753–762. [CrossRef]
38. Christofk, H.R.; Vander Heiden, M.G.; Harris, M.H.; Ramanathan, A.; Gerszten, R.E.; Wei, R.; Fleming, M.D.; Schreiber, S.L.; Cantley, L.C. The M2 splice isoform of pyruvate kinase is important for cancer metabolism and tumour growth. *Nature* **2008**, *452*, 230–233. [CrossRef]
39. Zhang, G.; Darshi, M.; Sharma, K. The Warburg Effect in Diabetic Kidney Disease. *Semin. Nephrol.* **2018**, *38*, 111–120. [CrossRef]
40. Miyamoto, S.; Hsu, C.C.; Hamm, G.; Darshi, M.; Diamond-Stanic, M.; Declèves, A.E.; Slater, L.; Pennathur, S.; Stauber, J.; Dorrestein, P.C.; et al. Mass Spectrometry Imaging Reveals Elevated Glomerular ATP/AMP in Diabetes/obesity and Identifies Sphingomyelin as a Possible Mediator. *EBioMedicine* **2016**, *7*, 121–134. [CrossRef]
41. You, Y.H.; Quach, T.; Saito, R.; Pham, J.; Sharma, K. Metabolomics Reveals a Key Role for Fumarate in Mediating the Effects of NADPH Oxidase 4 in Diabetic Kidney Disease. *J. Am. Soc. Nephrol.* **2016**, *27*, 466–481. [CrossRef] [PubMed]
42. Lee, A.Y.W.; Chung, S.K.; Chung, S.S.M. Demonstration that polyol accumulation is responsible for diabetic cataract by the use of transgenic mice expressing the aldose reductase gene in the lens. *Proc. Natl. Acad. Sci. USA* **1995**, *92*, 2780–2784. [CrossRef] [PubMed]
43. Pang, Y.; Bounelis, P.; Chatham, J.C.; Marchase, R.B. Hexosamine pathway is responsible for inhibition by diabetes of phenylephrine-induced inotropy. *Diabetes* **2004**, *53*, 1074–1081. [CrossRef] [PubMed]
44. Sima, A.A.F.; Prashar, A.; Zhang, W.X.; Chakrabarti, S.; Greene, D.A. Preventive effect of long-term aldose reductase inhibition (ponalrestat) on nerve conduction and sural nerve structure in the spontaneously diabetic Bio-Breeding rat. *J. Clin. Investig.* **1990**, *85*, 1410–1420. [CrossRef] [PubMed]
45. Ganz, M.B.; Seftel, A. Glucose-induced changes in protein kinase C and nitric oxide are prevented by vitamin E. *Am. J. Physiol. Endocrinol. Metab.* **2000**, *278*, E146–E152. [CrossRef]
46. Kass, D.A.; Shapiro, E.P.; Kawaguchi, M.; Capriotti, A.R.; Scuteri, A.; DeGroof, R.C.; Lakatta, E.G. Improved arterial compliance by a novel advanced glycation end-product crosslink breaker. *Circulation* **2001**, *104*, 1464–1470. [CrossRef]
47. Manigrasso, M.B.; Juranek, J.; Ramasamy, R.; Schmidt, A.M. Unlocking the biology of RAGE in diabetic microvascular complications. *Trends Endocrinol. Metab.* **2014**, *25*, 15–22. [CrossRef]
48. Nangaku, M. Chronic hypoxia and tubulointerstitial injury: A final common pathway to end-stage renal failure. *J. Am. Soc. Nephrol.* **2006**, *17*, 17–25. [CrossRef]
49. Daehn, I.; Casalena, G.; Zhang, T.; Shi, S.; Fenninger, F.; Barasch, N.; Yu, L.; D'Agati, V.; Schlondorff, D.; Kriz, W.; et al. Endothelial mitochondrial oxidative stress determines podocyte depletion in segmental glomerulosclerosis. *J. Clin. Investig.* **2014**, *124*, 1608–1621. [CrossRef]
50. Qi, H.; Casalena, G.; Shi, S.; Yu, L.; Ebefors, K.; Sun, Y.; Zhang, W.; D'Agati, V.; Schlondorff, D.; Haraldsson, B.; et al. Glomerular Endothelial Mitochondrial Dysfunction Is Essential and Characteristic of Diabetic Kidney Disease Susceptibility. *Diabetes* **2017**, *66*, 763–778. [CrossRef]
51. Herman-Edelstein, M.; Scherzer, P.; Tobar, A.; Levi, M.; Gafter, U. Altered renal lipid metabolism and renal lipid accumulation in human diabetic nephropathy. *J. Lipid Res.* **2014**, *55*, 561–572. [CrossRef] [PubMed]

52. Ge, M.; Fontanesi, F.; Merscher, S.; Fornoni, A. The Vicious Cycle of Renal Lipotoxicity and Mitochondrial Dysfunction. *Front. Physiol.* **2020**, *11*, 732. [CrossRef]
53. Pedigo, C.E.; Ducasa, G.M.; Leclercq, F.; Sloan, A.; Mitrofanova, A.; Hashmi, T.; Molina-David, J.; Ge, M.; Lassenius, M.I.; Forsblom, C.; et al. Local TNF causes NFATc1-dependent cholesterol-mediated podocyte injury. *J. Clin. Investig.* **2016**, *126*, 3336–3350. [CrossRef]
54. Ducasa, G.M.; Mitrofanova, A.; Fornoni, A. Crosstalk between Lipids and Mitochondria in Diabetic Kidney Disease. *Curr. Diabetes Rep.* **2019**, *19*, 144. [CrossRef] [PubMed]
55. Ducasa, G.M.; Mitrofanova, A.; Mallela, S.K.; Liu, X.; Molina, J.; Sloan, A.; Pedigo, C.E.; Ge, M.; Santos, J.V.; Hernandez, Y.; et al. ATP-binding cassette A1 deficiency causes cardiolipin-driven mitochondrial dysfunction in podocytes. *J. Clin. Investig.* **2019**, *130*, 3387–3400. [CrossRef] [PubMed]
56. Meyer, C.; Nadkarni, V.; Stumvoll, M.; Gerich, J. Human kidney free fatty acid and glucose uptake: Evidence for a renal glucose-fatty acid cycle. *Am. J. Physiol.-Endocrinol. Metab.* **1997**, *273*, E650. [CrossRef] [PubMed]
57. Rosca, M.G.; Vazquez, E.J.; Chen, Q.; Kerner, J.; Kern, T.S.; Hoppel, C.L. Oxidation of Fatty Acids Is the Source of Increased Mitochondrial Reactive Oxygen Species Production in Kidney Cortical Tubules in Early Diabetes. *Diabetes* **2012**, *61*, 2074–2083. [CrossRef]
58. Kang, H.M.; Ahn, S.H.; Choi, P.; Ko, Y.-A.; Han, S.H.; Chinga, F.; Park, A.S.D.; Tao, J.; Sharma, K.; Pullman, J.; et al. Defective fatty acid oxidation in renal tubular epithelial cells has a key role in kidney fibrosis development. *Nat. Med.* **2015**, *21*, 37–46. [CrossRef]
59. Kamijo, A.; Kimura, K.; Sugaya, T.; Yamanouchi, M.; Hase, H.; Kaneko, T.; Hirata, Y.; Goto, A.; Fujita, T.; Omata, M. Urinary free fatty acids bound to albumin aggravate tubulointerstitial damage. *Kidney Int.* **2002**, *62*, 1628–1637. [CrossRef]
60. Ruggiero, C.; Elks, C.M.; Kruger, C.; Cleland, E.; Addison, K.; Noland, R.C.; Stadler, K. Albumin-bound fatty acids but not albumin itself alter redox balance in tubular epithelial cells and induce a peroxide-mediated redox-sensitive apoptosis. *Am. J. Physiol.-Ren. Physiol.* **2014**, *306*, 896–906. [CrossRef]
61. Susztak, K.; Ciccone, E.; McCue, P.; Sharma, K.; Böttinger, E.P. Multiple Metabolic Hits Converge on CD36 as Novel Mediator of Tubular Epithelial Apoptosis in Diabetic Nephropathy. *PLoS Med.* **2005**, *2*, e45. [CrossRef] [PubMed]
62. Sun, L.; Halaihel, N.; Zhang, W.; Rogers, H.; Levi, M. Role of sterol regulatory element-binding protein 1 in regulation of renal lipid metabolism and glomerulosclerosis in diabetes mellitus. *J. Biol. Chem.* **2002**, *277*, 18919–18927. [CrossRef] [PubMed]
63. Proctor, G.; Jiang, T.; Iwahashi, M.; Wang, Z.; Li, J.; Levi, M. Regulation of renal fatty acid and cholesterol metabolism, inflammation, and fibrosis in Akita and OVE26 mice with type 1 diabetes. *Diabetes* **2006**, *55*, 2502–2509. [CrossRef] [PubMed]
64. Mishra, R.; Emancipator, S.; Miller, C.N.; Kern, T.; Simonson, M.S. Adipose differentiation-related protein and regulators of lipid homeostasis identified by gene expression profiling in the murine db/db diabetic kidney. *Am. J. Physiol. Ren. Physiol.* **2004**, *286*, F913–F921. [CrossRef]
65. Woroniecka, K.I.; Park, A.S.D.; Mohtat, D.; Thomas, D.B.; Pullman, J.M.; Susztak, K. Transcriptome Analysis of Human Diabetic Kidney Disease. *Diabetes* **2011**, *60*, 2354–2369. [CrossRef]
66. Khan, S.; Jawdeh, B.G.A.; Goel, M.; Schilling, W.P.; Parker, M.D.; Puchowicz, M.A.; Yadav, S.P.; Harris, R.C.; El-Meanawy, A.; Hoshi, M.; et al. Lipotoxic disruption of NHE1 interaction with PI(4,5)P2 expedites proximal tubule apoptosis. *J. Clin. Investig.* **2014**, *124*, 1057–1068. [CrossRef]
67. Nishi, H.; Higashihara, T.; Inagi, R. Lipotoxicity in Kidney, Heart, and Skeletal Muscle Dysfunction. *Nutrients* **2019**, *11*, 1664. [CrossRef]
68. Hernansanz-Agustín, P.; Choya-Foces, C.; Carregal-Romero, S.; Ramos, E.; Oliva, T.; Villa-Piña, T.; Moreno, L.; Izquierdo-Álvarez, A.; Cabrera-García, J.D.; Cortés, A.; et al. Na+ controls hypoxic signalling by the mitochondrial respiratory chain. *Nature* **2020**, *586*, 287–291. [CrossRef]
69. Lee, P.; Chandel, N.S.; Simon, M.C. Cellular adaptation to hypoxia through hypoxia inducible factors and beyond. *Nat. Rev. Mol. Cell Biol.* **2020**, *21*, 268–283. [CrossRef]
70. Ebefors, K.; Wiener, R.J.; Yu, L.; Azeloglu, E.U.; Yi, Z.; Jia, F.; Zhang, W.; Baron, M.H.; He, J.C.; Haraldsson, B.; et al. Endothelin receptor-A mediates degradation of the glomerular endothelial surface layer via pathologic crosstalk between activated podocytes and glomerular endothelial cells. *Kidney Int.* **2019**, *96*, 957–970. [CrossRef]
71. Takahashi, K.; Ghatei, M.A.; Lam, H.-C.; O'Halloran, D.J.; Bloom, S.R. Elevated plasma endothelin in patients with diabetes mellitus. *Diabetologia* **1990**, *33*, 306–310. [CrossRef] [PubMed]
72. Zhu, Z.; Liang, W.; Chen, Z.; Hu, J.; Feng, J.; Cao, Y.; Ma, Y.; Ding, G. Mitoquinone Protects Podocytes from Angiotensin II-Induced Mitochondrial Dysfunction and Injury via the Keap1-Nrf2 Signaling Pathway. *Oxid. Med. Cell. Longev.* **2021**, *2021*, 1394486. [CrossRef] [PubMed]
73. Sultanova, R.F.; Schibalski, R.; Yankelevich, I.A.; Stadler, K.; Ilatovskaya, D.V. Sex differences in renal mitochondrial function: A hormone-gous opportunity for research. *Am. J. Physiol. Ren. Physiol.* **2020**, *319*, F1117–F1124. [CrossRef] [PubMed]

biomolecules

MDPI

Review

Involvement of Tricarboxylic Acid Cycle Metabolites in Kidney Diseases

Alexis Paulina Jiménez-Uribe *, Estefani Yaquelin Hernández-Cruz, Karla Jaqueline Ramírez-Magaña and José Pedraza-Chaverri

Departamento de Biología, Facultad de Química, Universidad Nacional Autónoma de México, Mexico City 04510, Mexico; estefani.hernandez@quimica.unam.mx (E.Y.H.-C.); karlaa_jaquelinee@comunidad.unam.mx (K.J.R.-M.); pedraza@unam.mx (J.P.-C.)
* Correspondence: jimenez.uribe.ap@comunidad.unam.mx

Abstract: Mitochondria are complex organelles that orchestrate several functions in the cell. The primary function recognized is energy production; however, other functions involve the communication with the rest of the cell through reactive oxygen species (ROS), calcium influx, mitochondrial DNA (mtDNA), adenosine triphosphate (ATP) levels, cytochrome c release, and also through tricarboxylic acid (TCA) metabolites. Kidney function highly depends on mitochondria; hence mitochondrial dysfunction is associated with kidney diseases. In addition to oxidative phosphorylation impairment, other mitochondrial abnormalities have been described in kidney diseases, such as induction of mitophagy, intrinsic pathway of apoptosis, and releasing molecules to communicate to the rest of the cell. The TCA cycle is a metabolic pathway whose primary function is to generate electrons to feed the electron transport system (ETS) to drives energy production. However, TCA cycle metabolites can also release from mitochondria or produced in the cytosol to exert different functions and modify cell behavior. Here we review the involvement of some of the functions of TCA metabolites in kidney diseases.

Keywords: mitochondria; TCA cycle metabolites; kidney diseases

Citation: Jiménez-Uribe, A.P.; Hernández-Cruz, E.Y.; Ramírez-Magaña, K.J.; Pedraza-Chaverri, J. Involvement of Tricarboxylic Acid Cycle Metabolites in Kidney Diseases. *Biomolecules* **2021**, *11*, 1259. https://doi.org/10.3390/biom11091259

Academic Editor: Liang-Jun Yan

Received: 31 July 2021
Accepted: 23 August 2021
Published: 24 August 2021

1. Introduction

Mitochondria are organelles that fulfill a wide variety of functions in the cell. In addition to being a bioenergetics node, they also serve as signal organelles that communicate to the rest of the cell through different mechanisms. For example, the production of reactive oxygen species (ROS) [1], calcium influx [2], adenosine triphosphate (ATP) levels that regulate adenosine monophosphate protein kinase (AMPK) activation [3], modulation of the immune response through mitochondrial DNA (mtDNA) [4], releasing of cytochrome c orchestrating apoptosis [5], and also through Krebs cycle metabolites [6]; that modulates cell adaptation to different conditions.

The tricarboxylic acid (TCA) cycle, also named the citric acid cycle and Krebs cycle (although this last name can be dissected in the three different cycles: the urea, the glyoxylate, and TCA cycles) was described by Hans Krebs and his colleagues [7]. It is known chiefly for producing electron donors, the reduced form of nicotinamide adenine dinucleotide (NADH), and the reduced form of flavin adenine dinucleotide ($FADH_2$) to feed the electrons transport system (ETS). However, their intermediates also can serve as signal molecules to drive several cell functions.

TCA cycle metabolites were discovered as signal molecules mainly in cancer cells and were defined as oncometabolites that promote tumor progression [8]. However, recent evidence suggests that they are associated with diverse pathologies, including kidney diseases.

Kidneys are highly dependent on mitochondrial function due to their energy demand, particularly by the tubular nephron section, which exerts filtration and reabsorption functions. Hence, mitochondrial alterations, such as dynamics (fusion/fission), homeostasis (biogenesis/mitophagy), and bioenergetics, impact on kidney function.

In addition, other metabolic functions of the mitochondrial, such as the TCA cycle, could also be involved in kidney diseases.

2. A Brief Overview of the TCA Cycle

TCA cycle is an amphibolic pathway; the anabolic routes involve gluconeogenesis, transamination reactions, deamination reactions, and fatty acid synthesis; whereas catabolic routes oxidized components derived from carbohydrates, proteins, and lipids to generate electron donors and guanosine triphosphate (GTP) for energy production.

The cycle requires the condensation of acetyl coenzyme A (Acetyl-CoA) with oxaloacetate (OAA) to generate citrate; this reaction is catalyzed by the citrate synthase (CS). Citrate is then isomerized to cis-aconitate and further to isocitrate; aconitase activity is necessary for these reactions.

Isocitrate is dehydrogenated and decarboxylated to alpha-ketoglutarate (AKG) by the isocitrate dehydrogenase (IDH); in this reaction CO_2 and NADH also are produced. AKG is further decarboxylated to succinyl CoA by the AKG dehydrogenase; in this reaction, as the previous, NADH and CO_2 are produced. Succinyl CoA is converted to succinate by the action of succinyl CoA synthetase (also named succinate thiokinase); in this step, a molecule of GTP is produced. Succinate is dehydrated to fumarate by the succinate dehydrogenase (SDH), in this step, $FADH_2$ also is produced. Fumarate is then hydrated to generate malate by fumarate hydratase (FH), also named fumarase. Finally, malate is dehydrogenated to generate OAA by malate dehydrogenase (MDH); in this reaction, the third molecule of NADH is also produced. The produced OAA could start the cycle again by condensing with Acetyl CoA [6,9].

The main recognized products of the TCA cycle are one molecule of GTP, three molecules of NADH, and one molecule of $FADH_2$; these last are the electron donors to feed the ETS (Figure 1). However, each of the TCA cycle intermediates can also exert different functions and may be involved in kidney diseases.

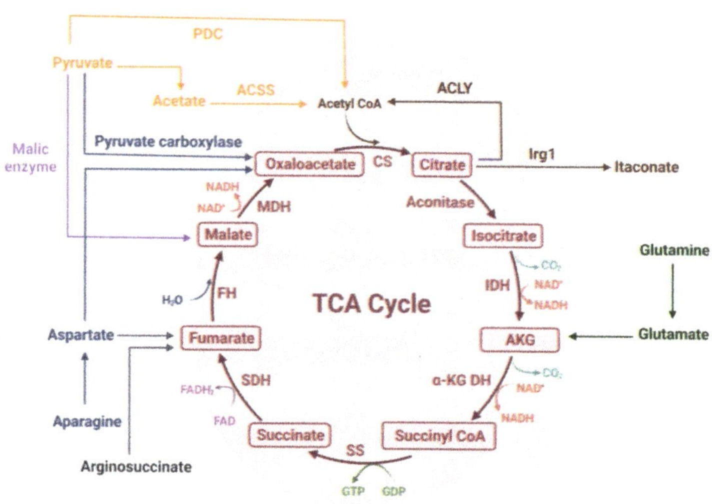

Figure 1. The tricarboxylic acid (TCA) cycle. The TCA cycle is an amphibolic route with anabolic and catabolic reactions. Anaplerotic reactions include the net input of amino acids in the cycle and the generation of oxaloacetate from pyruvate through pyruvate carboxylase. Among the main products of the TCA cycle are guanosine triphosphate (GTP), dinucleotide nicotinamide molecules (NADH),

and dinucleotide adenine flavine molecule ($FADH_2$); the latter are electron donors to feed the electron transport system (ETS). PDC: pyruvate dehydrogenase complex, ACSS: acetyl CoA synthetase, ACLY: ATP-citrate lyase, Irg1: the immune responsive gene 1 protein, CS: citrate synthase, IDH: isocitrate dehydrogenase, AKG DH: alpha-ketoglutarate dehydrogenase, SS: succinyl CoA synthetase, SDH: succinate dehydrogenase, FH: fumarate hydratase, MDH: malate dehydrogenase, NAD^+: nicotinamide adenine dinucleotide (oxidized form), FAD: flavine adenine dinucleotide (oxidized form), GDP: guanosine diphosphate. Created with Biorender.com.

3. Acetyl-CoA

Although acetyl-CoA is not inside the TCA cycle, it is a highly relevant molecule as it is required in the first step, which reacts with OAA to give rise to citrate; interestingly, citrate can be shuttled out from the mitochondrial matrix to the cytosol, where it can be re-converted to acetyl-CoA and OAA.

Acetyl-CoA is generated from different sources, such as the mentioned citrate, which is exported from mitochondria to the cytosol through the SLC25A1 transporter [10], in this compartment, by the action of the ATP-citrate lyase (ACLY) generates OAA and acetyl-CoA; from pyruvate decarboxylation by the action of pyruvate dehydrogenase complex (PDC); from acetate by the action of acetyl CoA synthetase (ACSS); and acetoacetyl-CoA by the action of thiolase. Acetyl-CoA serves as a substrate for fatty acids synthesis and protein acetylation.

Regarding acetyl-CoA involvement in lipid metabolism, fatty acid metabolism impairment has been reported in kidney diseases [11]. In diabetic nephropathy, although acetyl-CoA levels have not been directly reported, the increase in acyl-CoA levels, acetyl-CoA carboxylase, and fatty acid synthase suggest its utilization for fatty acid synthesis, whereas fatty acid oxidation is blocked [12]. The above leads to lipid accumulation and lipotoxicity; similar results were found in animal models of kidney fibrosis induced by folic acid [13], ischemia/reperfusion (I/R) [14], and unilateral ureteral obstruction (UUO) [15]. In renal cell carcinoma (RCC), fatty acid metabolism also seems to be impaired due to altered expression enzymes involved in fatty acid metabolism [16] and intracellular lipid accumulation driven by hypoxia [17]. Interestingly metabolic stress such as hypoxia and low nutrient availability has been reported to induce ACSS expression in cells from breast cancer [18]. This finding suggests that hypoxic alterations that are common in acute kidney injury (AKI) [18], chronic kidney disease (CKD) [19], and renal carcinoma [20], could also contribute to ACSS expression and activation to promote acetyl-CoA synthesis for its further utilization for de novo fatty acid synthesis in these pathologies. In summary, the evidence indicates an impairment in lipid metabolism, with increased acetyl-CoA utilization for fatty acid synthesis and decreased fatty acid oxidation in some kidney diseases.

Acetyl-CoA also serves as a substrate for protein acetylation, a post-translational modification (PTM); if acetylation occurs on lysine (K) residues of histones, this serves as an epigenetic modification for gene expression regulation. In general, histone acetylation by histone acetyltransferases (HAT) elicits genetic expression by inducing chromatin relaxation and favoring the binding of nuclear factors, whereas the absence of histone acetylation could act as a repressive mark [21].

In CKD development by high salt diet and unilateral ureteral obstruction (OUU) in rats, global acetyl-CoA levels are decreased [22,23].

In a transcriptomic analysis of kidney fibroblasts/myofibroblast derived from OUU, transforming growth factor-beta (TGF-β) signaling induces a metabolic reprogramming with reduced TCA cycle-related enzymes, including PDC [23,24], resulting in reduced acetyl-CoA synthesis and levels; moreover, the global acetylation is reduced, and as a consequence, histone acetylation also is decreased [23,24]. Interestingly, the restoration of acetyl-CoA levels partially reverses the induction of fibrotic markers alpha-smooth actin (α-SMA) and collagen [23]. Histone acetylation is a dynamic process since global acetylation of kidney fibroblasts/myofibroblast derived from OUU is decreased, impacting mainly on Histone 3 (H3) K4, K14, and K23 residues; in contrast to K18, and K27 residues that remain acetylated [24]. In the case of H3K9, acetylation is controversial; Smith et al. (2019) found a decrease in UUO rats [24], whereas Hewitson et al. (2017) found an increase in UUO [25] as occurs in diabetic nephropathy [26] in mice. These findings could be explained by the time of disease progression and the animal model. Regarding the above mention, mesangial cells in hyperglycemic conditions induce fibrotic gene expression in an ACLY-dependent manner, involving the increase of H3K9 acetylation [27]. Similarly, in kidney damage induced by obesity, ACLY, the enzyme that generates acetyl-CoA from citrate is increased, and its inhibition in vitro reduces histone acetylation and the expression of lipogenic and pro-fibrotic genes [28]. However, this effect could be in response to nutrient excess.

In AKI induced by I/R, a metabolic reprogramming also has been reported with a glycolytic shift and PDC inhibition [29], indicating a decreased activity of this enzyme complex and ergo a reduction in acetyl-CoA synthesis; moreover, in AKI induced by cisplatin, H3K27 acetylation is reduced, and the restoration of its acetylation levels decreases kidney damage [30,31]. Hence, in CKD and AKI, there appears to be a decrease in acetyl-CoA levels, which in turn impacts histone acetylation.

Contrary, in RCC, acetyl CoA synthetase 2 (ACSS2), the enzyme that produces acetyl CoA from acetate shows increased levels, and in vitro, the expression of this enzyme promotes cell migration [32,33]. In human renal cell adenocarcinoma cell lines, the expression of the snail family transcriptional repressor 1 (SNAI1) is promoted by ACSS2 [34]. SNAI1 is a transcriptional repressor involved in the progression of many types of cancer [35], including RCC [36]. Interestingly, although the global levels of H3 acetylation seems to be decreased in RCC [37], the acetylation of H3K27 seems to be necessary for SNAI1 expression, particularly under hypoglycemic conditions [34].

Based on the above data, in kidney diseases, there is a complex regulation of acetyl-CoA synthesis and its utilization as a substrate for histone acetylation, which in turn impacts on epigenetic regulation of the gene expression (Figure 2a).

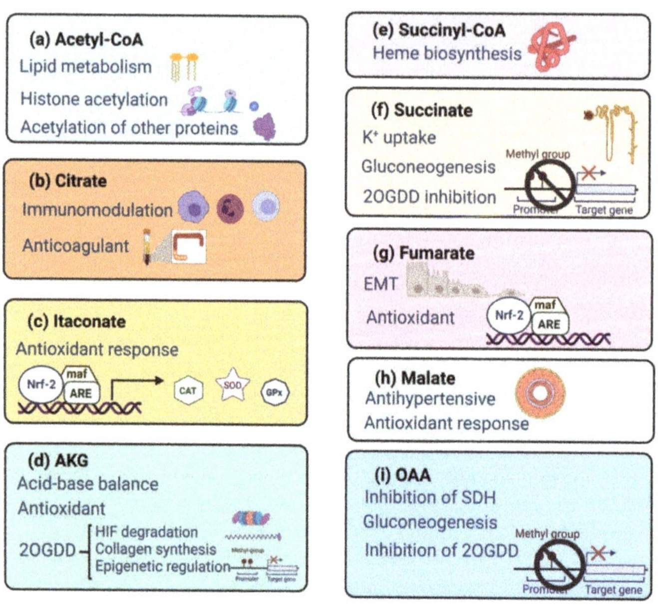

Figure 2. Involvement of TCA cycle metabolites in kidney functions. The metabolites of the TCA cycle and acetyl-CoA participate as signal molecules to promote several critical cellular functions such as epigenetic modifications, redox regulation, hypoxic response, and immunity. Nrf-2: nuclear factor erythroid 2–related factor 2, ARE: antioxidant response element, maf: musculoaponeurotic

fibrosarcoma protein, CAT: catalase, SOD: superoxide dismutase, GPx: glutathione peroxidase, 2OGDD: named 2-oxoglutarate dependent dioxygenases, HIF: hypoxia inducible factor, EMT: epithelial-mesenchymal transition, SDH: succinate dehydrogenase. Created with Biorender.com.

4. Citrate

Citrate can be obtained by uptake from dietary sources, and in mitochondria, as a result of the condensation of acetyl-CoA with OAA by the citrate CS; cytoplasmic citrate can be derived from mitochondrial export through the SLC25A1 transporter, also named citrate–malate exchanger (CIC) [10].

Physiologically, an excess of citrate also serves as a control point of glycolysis by inhibiting phosphofructokinase (PFK) [38] and PDC [39], limiting the fructose-1,6-bisphosphate and acetyl-CoA synthesis, respectively [38].

In a model of CKD by UUO or by diabetic nephropathy, urinary excretion of citrate is decreased [40,41], as occurs in patients with CKD [42–44]; in contrast, in plasma [45] and kidney tissue [46], citrate levels are increased in UUO. In addition, in cisplatin-induced CKD, CS activity is increased. In contrast, aconitase activity is reduced [47], suggesting that produced citrate is not converted to isocitrate. In non-diabetic CKD patients, the expressions of aconitase 1 and aconitase 2 are reduced; and in urine and blood, the levels of isocitrate are also decreased [42]. In addition, it is known that CS is stimulated by aldosterone [48], a hormone increased in CKD [49], suggesting that in CKD, aldosterone promotes an excess of citrate synthesis. This suggests that produced citrate (probably in excess) is not converted into isocitrate, and its retention results in the reduced urinary excretion [42], as has been demonstrated in animal models of UUO-induced CKD and I/R-induced AKI, in which kidney tissue reveals an accumulation of this metabolite [46,50].

Clinically, urinary low citrate excretion is proposed as a marker of acid retention and reduced glomerular filtration in patients with CKD [43], and plasma citrate levels correlate negatively with estimated glomerular filtration rate (eGFR) [51].

However, in diabetic nephropathy, urinary citrate excretion is controversial due to in humans being decreased [37], whereas in mice it is increased [43], although this may be the result of other metabolic disorders involved in diabetes.

Administration of citrate has been used to manage kidney diseases such as kidney stones [52], AKI, and CKD [53–55]. In kidney injury by kidney stones, citrate binds to calcium, preventing its binding to oxalate or calcium phosphate and the consequent reduction of stone formation; however, its effectiveness is still controversial [56].

Citrate administration in AKI and CKD is used as an anticoagulant during renal replacement therapy [53–55]. Moreover, in a model of AKI by I/R, citrate administration reduces plasma creatinine levels, lactate dehydrogenase activity and partially restores ATP content in tissue, reflecting improvement in kidney function [57]. Interestingly, citrate has also been associated with immunomodulatory effects. In AKI patients with continuous venovenous hemofiltration therapy, citrate administration reduces myeloperoxidase and interleukin 8 (IL-8) plasma levels [58]; in a model of CKD induced by adenine in rats, the administration of citrate reduces the production of pro-inflammatory cytokines interleukin 6 (IL-6) and interleukin 17 (IL-17), whereas it increases the anti-inflammatory cytokines interleukin 10 (IL-10) and TGF-β [59].

The immunomodulatory effects of citrate have also been reported in other cells types such as monocytes and macrophages. In these cells, ROS and pro-inflammatory cytokines were reduced in response to lipopolysaccharide (LPS) [60,61]; however, this effect could be dependent on citrate concentration [61].

In RCC, citrate levels are enriched [62], and its immunosuppressive effects could be related to the tumor progression; however, there is still no evidence of this effect. However, in RCC, citrate is re-converted to acetyl-CoA by ACLY, which in turn serves as the substrate for protein acetylation and fatty acid synthesis; as mentioned above, RCC also has elevated levels of ACLY. Interestingly, it is silencing, avoiding citrate-derived acetyl-CoA, promoting apoptosis, and reducing proliferative and migration rates in RCC cells [63].

Citrate involvement in kidney diseases includes immunomodulatory effects, regulating acetyl-CoA synthesis, and even being used in their therapeutic management (Figure 2b).

5. Isocitrate/Itaconate

Aconitase is the enzyme responsible for the conversion of citrate to cis-aconitate and later to isocitrate.

Aconitase is an iron–sulfur-containing dehydratase; its activity is sensitive to oxidation [64]. In kidney diseases, it is well-known that oxidative stress is a hallmark [65], hence suggesting that aconitase activity is reduced, as reported in CKD induced by cisplatin [47] and 5/6 nephrectomy [66]; as well in AKI induced by maleate [67] and I/R [68].

In addition, as kidney function declines in nephrectomy-induced CKD [61], aconitase activity also decreases [69].

In non-diabetic CKD, isocitrate urinary excretion, aconitase 1 (mitochondrial aconitase), and 2 (cytosolic aconitase) expression are reduced in kidney tissue [42].

At present, there is no evidence of isocitrate as a signal molecule, and its synthesis seems to be decreased in kidney diseases. On the other hand, itaconate is another intermediate of the TCA cycle derived from the decarboxylation of cis-aconitate by the immune responsive gene 1 protein (Irg1). Interestingly, itaconate seems to have immunomodulatory effects [70].

In the I/R model, Irg1 levels increase after 12 h, with a peak at 24 h; the induction of this enzyme on the different cell types depends on the stimulus. For example, renal cells respond to H_2O_2, increasing Irg1 levels, whereas macrophages respond mainly to pro-inflammatory stimulus, such as cytokines and cell lysates, and lesser extent to H_2O_2 [71]. Itaconate has protective effects since Irg1 knock-out mice exacerbate inflammatory response and reduced survival percentage induced by I/R [71]. Due to this immunomodulatory effect, this metabolite has been used for reducing damage in kidney tissue and cells.

The administration of 4-octyl itaconate (OI), a derivate of itaconate with higher fat solubility, by tail vein injection, reduces fibrotic kidney damage induced in UUO or by adenine administration in rats. This effect was partly through the reduction of the canonical signaling of TGF-β pathway and by recovering antioxidant enzyme expression in adenine-induced kidney damage or decreasing inflammatory response by reducing nuclear factor kappa B (NF-κB) activation in UUO [72]. In vitro, OI treatment also reduces fibrotic markers fibronectin, plasminogen activator inhibitor 1 (PAI-1), and α-SMA, decreases phosphorylation of p65 subunit of NF-κB; whereas stimulates antioxidant response through the increase of the nuclear factor erythroid 2-related factor (Nrf2) and reducing ROS levels in kidney epithelial cells HK-2 stimulated with TGF-β [72].

Dimethyl itaconate (DMI), another derivate of itaconate, has also been demonstrated to have a renal protective effect. The treatment of neonatal renal cells with DMI and exposure to hypoxia/reoxygenation (H/R) reduces cell death; in addition, the antioxidant response is activated due to the increase in Nrf2 nuclear translocation [71]. A similar result was demonstrated in macrophages exposed to H/R, in which DMI reduces inflammatory response by decreasing tumor necrosis factor-alpha (TNF-α) and interleukin 1-beta (IL-1β) production through reducing mitogen-activated protein kinase (MAPK) and NF-κB activation; this effect was in part due to the induction of antioxidant response mediated by Nrf2 stimulation [71].

The well-reported mechanism of action of itaconate is through its binding to Kelch-like ECH associated protein 1 (KEAP1), a negative regulator of Nrf2, the interaction of itaconate with KEAP1, elicits the dissociation of this last one from Nrf2, inducing the nuclear translocation of Nrf2 to promote antioxidant gene expression [73]. Interestingly, itaconate also inhibits SDH activity, which results in succinate accumulation and the inhibition of fumarate formation [74]; both mechanisms are demonstrated in monocytes/macrophages. Itaconate has also demonstrated antimicrobial functions, blocking the glyoxylate cycle in *Mycobacterium avium* and *Mycobacterium tuberculosis* [75] (Figure 2c).

Although itaconate immunomodulatory effects are an emerging and growing field, there is little evidence of its levels and action mechanisms on kidney diseases.

6. Alpha-Ketoglutarate

The decarboxylation of isocitrate by the IDH gives rise to 2-oxoglutarate, which is also named AKG, which can also be derived from glutaminolysis [76] and even can be produced by microorganisms [77].

At present, there is little evidence related to AKG alterations in kidney diseases, such as increased urinary excretion in diabetic nephropathy in humans [44] and mice [78]; however other report decreased urinary levels in humans and mice [42,79] and reduction of blood serum AKG concentration in diabetic nephropathy in mice [41]. On the other hand, in RCC, contrary results were found in tissue with an elevated concentration in mice [80] compared to low levels reported in human samples [81].

Even though the information related to AKG levels in kidney disease is limited and confusing, this molecule and the enzymes related to its metabolism seem to be of great relevance in different kidney conditions.

As mentioned above, IDH catalyzes the conversion of isocitrate to AKG. IDH1 is expressed in cytosol and peroxisomes, whereas IDH2 and IDH3 are expressed in mitochondria. IDH1 and IDH2 are nicotinamide adenine dinucleotide phosphate (NADP)-dependent, and each one functions as homodimers; whereas IDH3 is nicotinamide adenine dinucleotide (NAD)-dependent and is composed by three different subunits; thus resulting in the production of the reduced form of NADP (NADPH) by IDH1/2, or NADH by IDH3 in addition to the AKG synthesis [82]. It is well known that NADPH is a substrate for antioxidant defense, used for glutathione regeneration and thioredoxin activity [83–85]. In addition, the administration of AKG also has been reported to function directly as an antioxidant [86–88], as mentioned below.

In acute kidney injury by cisplatin, IDH2 levels are decreased; moreover, IDH1/2 activities are reduced, but with no IDH3 [89]; in a similar way, in I/R-induced AKI it has been reported that IDH1/2 are reduced, as well their function [83,90]; on the other hand in UUO-induced CKD, IDH2 levels also are reduced, and in diabetic nephropathy, also IDH2 activity is diminished [91]. Thus, demonstrating that IDH2 activity reduction is a common characteristic in all of these pathologies. Besides, genetic deletion of IDH2 exacerbates renal damage by increasing oxidative stress and leukocyte infiltration in I/R and cisplatin-induced AKI and UUO-induced CKD models [83,89,92], reflecting an antioxidant protective effect of this enzyme additional to its function of AKG synthesis. Additionally, an interesting finding is that in diabetic nephropathy, IDH2 deficiency also increases the expression of renin, angiotensin II type 1 receptor, angiotensinogen, and angiotensin-converting enzyme in renal tissue, as well renin and angiotensin II levels in plasma, promoting hypertension derived from oxidative stress [91].

In non-diabetic CKD patients, IDH3 expression is decreased [42], which suggests an impairment in AKG and NADH+ synthesis, resulting in low levels of electron donors for the ETS. On the other hand, in RCC, low expression of IDH1 has been associated with a poor prognosis [93]. Independent of which IDH catalyzes the synthesis of AKG, this metabolite exerts different functions in the kidney.

A new physiological function of AKG participating in the acid–base balance in the kidney has been reported. At the extracellular level, AKG can be recognized by its receptor 2-oxoglutarate receptor 1 (OXGR1), which is expressed in cells of the connecting tubule and cortical collecting tubule; once activated, this receptor acts in conjunction with pendrin, regulating the HCO_3^- excretion and NaCl reabsorption [94].

AKG also has been reported to function as a non-enzymatic antioxidant, scavenging H_2O_2 and enhancing the activity of other antioxidant molecules in liver damage induced by ethanol or acetaminophen, respectively [86,87]. Thus, in a model of hyperammonemia, liver and kidney damage were reduced by the oral administration of AKG, restoring the antioxidant status in both organs [88]. In vitro, kidney proximal tubules under hypoxic condition show mitochondrial alterations and decreased ATP levels; however, the use of AKG in combination with aspartate reduce mitochondrial structural alterations and partially restores ATP levels by replenishing TCA cycle [95]. In a similar approximation in AKI induced by I/R, the treatment with AKG plus malate did not demonstrate protective effects in the kidney, even as a deleterious effect, mean arterial blood pressure (MAP) and heart rate were decreased [96]. Although the AKG was not promising for AKI treatment, hypertension in CKD is a concomitant alteration [97], opening a new exciting research field of the effect of AKG on CKD progression.

In addition, AKG also participates in the function of a superfamily of enzymes named 2-oxoglutarate dependent dioxygenases (2OGDD). The reaction catalyzed by 2OGDD is the hydroxylation of the substrate, requiring as co-substrates O_2, Fe^{2+}, and AKG. Prolyl hydroxylases (PHD), histone demethylases (HDM), nucleic acid oxygenases, and fatty acid oxygenases are some of the 2OGDD [98]. Some structural analogs of AKG, such as pyruvate, citrate, isocitrate, succinate, fumarate, malate, OAA, R-2-hydroxyglutarate (R2HG), and L-2-hydroxyglutarate (L2HG), act as 2OGDD inhibitors [98].

PHD function hydroxylating proline residues of several proteins, such as the hypoxia-inducible factor (HIF) promoting its proteasomal degradation; and collagen, inducing its structural conformation. In I/-R-induced AKI, the pre-treatment, but no post-ischemic damage, with a PHD inhibitor (PHI), GSK1002083A, reduces the fibrotic lesions and maintains kidney function [99]. Similarly, in AKI induced by cisplatin and folic acid, the pre-treatment with other PHI, FG-4592, also decreases kidney damage, reducing the inflammatory and fibrotic responses [100,101]; however, in UUO-induced CKD, the use of PHI does not affect fibrotic or inflammatory markers [102], demonstrating that inhibition of PHD is effective only in acute damage. The action mechanism of PHI is through avoiding HIF proteasomal degradation as an acute protective response [99–101]; however, although it was not demonstrated, possible inhibition of collagen synthesis could also contribute to PHI benefits in kidney diseases. Since HIF is necessary for erythropoiesis induction, PHI roxadustat and GSK1278863 also has been proposed for anemia treatment in CKD, demonstrating the increase in erythropoietin and hematocrit levels in animal models [103,104] and even in patients [105]; hence in CKD and AKI, there is an increase in PHD activity, and their inhibition ameliorates damage associated with kidney dysfunction.

Another 2OGDD enzymes involved in kidney diseases is the ten-eleven translocation methyl-cytosine dioxygenase (TET)1-3, which catalyzes the conversion of 5-methyl cytosine (5mC) to 5-hydroxymethyl cytosine (5hmC), an epigenetic mark on DNA associated with active transcription [106]. In tissue derived from CKD patients, TET1-3 levels are increased; however, their activity is decreased [107], indicating an imbalance in 5mC/5hmC. TET2 low activity is associated with worsen acute kidney damage induced by I/R [108,109] and cisplatin [110]; whereas TET3 low activity is associated with chronic kidney damage induced by UUO [111,112]; moreover, restoration of TET function ameliorates kidney injury [107,110,112] by the hydroxymethylation of different genes such as Klotho [107] and ras protein activator like 1 (RASAL) [111] promoters, which codify for renoprotective and anti-proliferative proteins, respectively. In addition, in UUO and nephrectomy models of kidney damage, the inhibition of the histone demethylase Jumonji domain containing-3 (JMJD3), another 2OGGD, worsen fibrotic lesions through enhancing TGF-β signaling [113]. Thus, in kidney injury seems that removing DNA and histone methylation by 2OGDD demethylases has protective effects.

It is crucial to notice that the 2OGDDs function depends on their expression, the availability of co-substrates AKG, O_2, and Fe^{2+}, and the presence of structural inhibitors. For example, in RCC, despite the increase of the 2OGDD histone and DNA demethylases, there is a reduction in 5hmC levels. This finding is probably secondary to the low availability of AKG and the presence of its structural analog L2HG [114].

In summary, the AKG, the enzymes involved in its synthesis, and the enzymes that require it for their function seem to be highly relevant in kidney diseases development.

Here we mention only a few examples of 2OGDD; however, these enzymes are involved in a great variety of biological functions (Figure 2d).

7. Succinyl-CoA

Succinyl-CoA is derived from AKG by the action of the AKG dehydrogenase, also named 2-oxoglutarate dehydrogenase (2OGDH); succinyl-CoA can also be derived from succinate by the action of the succinyl-CoA synthetase.

In addition to its role in the TCA cycle, succinyl-CoA and glycine participate in the biosynthetic heme pathway, necessary for different kidney hemoproteins synthesis.

2OGDH is decreased in CKD induced by high-salt diet and aristocholic acid administration [22,115], as has been reported in non-diabetic CKD patients [42]. On the other hand, in RCC 2OGDH lower expression was associated with poor outcome [116]. Together, this indicates a decreased synthesis of succinyl-CoA in some progressive kidney diseases.

P450 cytochromes (CPY) are hemoproteins that have been associated with kidney dysfunction. In end-stage renal failure, a decrease in CYP1A2, CYP2C9, CYP2C19, and CYP3A4 expression in blood samples has been reported [117]; these CPYs are associated with drug excretion. On the other hand, CPY24, an enzyme that regulates vitamin D levels, seems to be increased in adenine-induced CKD [118]. However, there is no evidence that succinyl-CoA levels in kidney diseases drive altered expression or synthesis of CPY.

Other hemoproteins include catalase, nitric oxide synthase, and prostaglandin synthase; however, in kidney diseases, these enzymes are evaluated by their function rather than their synthesis derived by the succinyl-dependent biosynthetic heme pathway (Figure 2e).

Currently, there is little evidence related to the direct role of succinyl-CoA in kidney pathologies.

8. Succinate

Succinate is derived from succinyl-CoA by the reaction of succinyl CoA synthetase. In high-diet-salt-induced CKD, succinate levels decrease in tissue [22], as has been reported in non-diabetic patients with CKD with decreased levels in kidney biopsies and low urinary excretion [42]. In diabetic nephropathy in rodents, succinate levels are increased in urine [41,78,119], whereas in kidney tissue are reported both, decreased [119] and increase levels [120]. In contrast, in UUO, succinate levels increase in plasma and kidney tissue [45,46], as happens in I/R-induced AKI [50,121] and polycystic kidney disease [122]. In RCC also increased succinate levels has been found [62].

Hence, it seems that there is a dynamic regulation of succinate levels and its excretion depending on kidney damage injury.

An exciting finding from almost two decades ago was the discovering of succinate receptor 1 (SUCNR1) in the kidney, which is found mainly in the proximal tubules [123].

It is known that in the proximal tubule, succinate stimulates gluconeogenesis [124] and induces membrane hyperpolarization by increasing K^+ uptake [125], although it is uncertain if these functions depend on SUCNR1. However, a well-documented function of succinate/SUCNR1 signaling is the stimulation of arachidonic acid, prostaglandin E2, and prostaglandin I2 release, which in turn stimulates renin release [120,123,126,127]. Moreover, in diabetic nephropathy in mice, hyperglycemia induces renin release through SUCNR1 [120]. In CKD, there are alterations in the renin/angiotensin/aldosterone axis; however, currently is unknown if succinate and its receptor are participating.

Another reported function of succinate is its inhibitory effect on 2OGDD mentioned above, particularly inhibiting PHD and indirectly stabilizing HIF [128]. Regarding above mentioned, succinate has also been described as a pro-inflammatory signal promoting the expression IL-1β via HIF activation in macrophages [129], opening a new panorama of the participation of this metabolite during the inflammatory response in kidney diseases (Figure 2f).

9. Fumarate

SDH catalyzes the reaction that transforms succinate to fumarate; this reaction also can be in the direction of fumarate to succinate. Interestingly, SDH also is named complex II and is part of the ETS.

This metabolite also can be derived from arginosuccinate as a part of the urea cycle. Fumarate increased levels in the plasma of diabetic and non-diabetic CKD in mice [41,42], whereas increased urinary excretion and levels in the renal cortex in diabetic mice [78] are reported. In contrast, in kidney damage induced by cisplatin, decreased urinary excretion has been found [130], similarly in injury induced by I/R and adenine, decreased tissue fumarate levels [50,131] has been reported. In RCC, fumarate also appears to be reduced in tissue [62]. As happens with other TCA cycle metabolites, its regulation seems to be dynamic.

Besides, SDH activity is reduced in CKD models induced by potassium dichromate [132], sulfasalazine [133], cisplatin [134], and UUO [135]. In addition, the exposure of proximal tubular epithelial cells to uremic toxins decreases SDH activity [136], suggesting that fumarate is not synthesized and the ETS is not working fully. In comparison, during acute injury by I/R, SDH blockade with malonate has protective effects in the kidney [137]; however, also reduction in its activity has been reported in this model [138]; hence, deeper studies are necessary to understand the molecular mechanism of the SDH under the specific condition of renal damage.

In addition, fumarate, as succinate, is a 2-OGDD inhibitor and has exceptional attention in a subtype of RCC (FH-deficient RCC), in which an FH mutation avoids the conversion of fumarate into malate, leading to an excessive fumarate accumulation [139]. Fumarate accumulation has been demonstrated to induce epithelial-mesenchymal transition (EMT) through epigenetic regulation inhibiting TET demethylase. The above, finally provides phenotypic mesenchymal characteristics and migratory capacities to the cells, thus is highly relevant in the progression of RCC [140]. In other kidney disorders, such as CKD-induced fibrosis, EMT is a phenomenon also observed [141,142] in which the fumarate role has not been elucidated.

In addition, fumarate seems to have protective effects in the kidney, as demonstrated in kidney damage induced by ciclosporin, cisplatin, folic acid, and I/R [143–145], in which dimethyl fumarate administration reduce kidney damage by enhancing the antioxidant response driven by Nrf2 (Figure 2g). Moreover, dimethyl fumarate is already approved by the food and drug administration (FDA) as an immunomodulatory drug for the therapeutic management of multiple sclerosis [146].

10. Malate

Malate is raised from fumarate by FH action and from pyruvate by the action of the malic enzyme.

There is little evidence of malate alterations in kidney diseases, such as increased levels in serum and urine in diabetic nephropathy in mice [41,78] and reduced levels in kidney tissue from RCC and I/R injury [50,62]. In fact, the reduction of FH activity has been proposed as a biomarker of acute kidney injury [147]. The silencing of FH in HK-2 renal epithelial cells increases fumarate levels, whereas it decreases malate levels as expected; interestingly, it also reduces nitric oxide levels and the activity of nitric oxide synthase (NOS) [148], which is known to induce vascular relaxation. In a model of hypertension in rats, malate administration increased NOS levels and activity, and alleviated hypertension, reducing the MAP [148]. Similar results were obtained in a model of I/R in which malate administration plus AKG causes hypotension reducing the MAP [96].

In addition, malate synthesis by the malic enzyme is highly relevant due to the formation of NADPH for glutathione and thioredoxin antioxidant activities [149]. Related to the above, in kidney damage the cisplatin malic enzyme increases its activity [145], probably as a reparative mechanism (Figure 2h).

11. Oxaloacetate

OAA can be synthesized from malate by the MDH. It can also be derived from pyruvate catalyzed by the pyruvate decarboxylase, or aspartate by the glutamic oxaloacetate transaminase (GOT). OAAs can be condensed with acetyl-CoA to start the cycle again and also can be used for gluconeogenesis.

Currently, there is no information related to OAA levels in kidney diseases, probably by the difficulties in its measurement [150]. However, in kidney injury induced by toxic compounds potassium dichromate [151], gentamicin [152], melamine/cyanuric acid, and in diabetic nephropathy [153], MDH activity reduction has been reported; suggesting a decrease in OAA synthesis. Moreover, GOT serum levels in CKD patients are reduced and correlated with advanced stages of the disease [154].

Contrary, in RCC, MDH and GOT expression are increased [155], suggesting an increase in OAA synthesis. Furthermore, OAA inhibits SDH [156], thus impacting ETS activity and promoting succinate accumulation, which can inhibit 2OGDD as mentioned above (Figure 2i).

12. Clinical Significance of TCA Metabolites

Clinically, kidney function is evaluated by indirect measurement of glomerular filtration by serum creatinine levels, albuminuria, proteinuria, and eGFR. Recently, the use of mass spectrometry (MS) as a tool with proteomics [157], peptidomics [158], and metabolomics [159] approaches to the discovery of new biomarkers in urine and serum, has increased, showing a large number of molecules with potential use in the clinic. Some examples of molecules identified by mass spectrometry currently useful as biomarkers in clinics include cystatin C [160,161], neutrophil gelatinase-associated lipocalin (NGAL), and kidney injury molecule 1 (KIM1) [162–165]. Hence, the use of new biomarkers in conjunction with the classical method of kidney function evaluation could be helpful in a more accurate diagnosis or prognosis of different kidney diseases.

Due to the involvement of TCA cycle metabolites in kidney physiology and pathophysiology, identifying these in biofluids, such as serum and urine by metabolomics, could give insights into their use as potential biomarkers in different kidney diseases.

Acetyl-CoA. Currently, acetyl-CoA has not been identified as a biomarker in kidney diseases, probably by its multiple sources and its implication in diverse biochemical pathways. However, as mentioned above, one of its functions is in the fatty acid metabolism, which seems to be impaired in kidney diseases [11]. Carnitine can react with acetyl-CoA to form acetyl-carnitine during fatty acid metabolism by the carnitine acetyltransferase (CAT).

In CKD, serum levels of acetyl-carnitine increase along with disease progression, whereas in urine are decreased; even more, serum acetyl-carnitine shows a negative correlation with eGRF [166,167]. In AKI patients, serum levels of acetyl-carnitine levels also are increased [168]. In biopsies of renal cell carcinoma, acetyl-carnitine is increased; moreover, there are differences between clear cell, papillary, and chromophobe subtypes, with a more noticeable increase in clear cell RCC subtype [169].

Currently, acetyl-carnitine has been proposed as a biomarker for hepatocellular carcinoma, in which it is increased [170,171]; and in major depressive disorder, in which levels are decreased in serum. In kidney diseases, the use of this metabolite and the eGFR could help evaluate kidney function. However, more in-depth studies are necessary to determine its utility in discriminating against different kidney diseases.

Citrate. As mentioned above, urinary excretion of citrate is decreased in patients with CKD [42–44]. Clinically, urinary low citrate excretion is proposed as a marker of acid retention and reduced glomerular filtration in patients with CKD [43]. The meaning of plasma citrate is not clear enough since both negative and positive correlations with estimated glomerular filtration rate (eGFR) have been proposed [51,172]; in addition, the ratio of myo-inositol:citrate in urine seems to predict active renal vasculitis [173]. Similarly, in AKI pediatric patients, urinary citrate levels were found to be reduced [174]. In RCC, citrate levels decreased in urine [175], but these are enriched in tissue [62]. Hence, reduced citrate levels in urine seem to be a promising biomarker of altered kidney function.

Isocitrate. There is scarce information related to isocitrate alteration in serum or urine levels in kidney diseases. This metabolite and its derivate cis-aconitate are decreased in the urine of CKD patients [42]; additionally, plasma isocitrate correlates negatively with eGFR [51], being a possible predictor of disease progression. On the other hand, in RCC, low expression of IDH1, the enzyme responsible for isocitrate conversion to AKG, has been associated with a poor prognosis [93]. More in-depth studies are necessary to understand the clinical significance of this metabolite.

AKG. In CKD patients, decreased urinary levels of AKG have been reported [42], whereas there are no differences in AKI patients [176]. In RCC, AKG urinary excretion is increased [177], whereas, in biopsies, reduced levels have been reported; even more, tissue levels of this metabolite could be helpful in the prognosis of this neoplasia [81].

Succinate. In CKD patients, decreased urinary levels of succinate have been reported [42], and plasma succinate correlates negatively with eGFR [51]. In RCC, succinate levels are increased in tissue [62] and decreased in urine [175]. However, there is no information related to alterations of this metabolite in urine or serum from AKI patients. Currently, increased urinary levels of succinate and AKG have been proposed as biomarkers of major depressive disorder [178], opening a new panorama for the use of these metabolites in the clinic.

Fumarate. Increased urinary levels of fumarate in CKD patients have been described [44]; also, plasma fumarate correlates positively with eGFR [51] and has been associated with mortality [179]. In RCC, tissue fumarate levels are decreased [62], and no altered levels are reported in AKI patients. However, in an animal model, FH activity in urine and plasma has been proposed as a biomarker of AKI [147].

Malate. As fumarate, increased malate levels in the urine of CKD patients have been reported [44], and plasma malate also correlates negatively with eGFR [51]. In RCC, tissue malate levels are decreased [62]. In AKI patients, there are no reported altered levels of this metabolite.

OAA. As mentioned above, OAA is difficult to detect by MS; hence, its potential as biomarker is limited. In CKD, GOT serum levels are reduced and correlated with advanced stages of the disease [154]. In RCC tissue, GOT expression is increased. GOT is currently used for clinical evaluation of liver function; its use in conjunction with other parameters in the diagnosis and prognosis of kidney diseases could be helpful in clinics.

A summary of alterations of some TCA cycle metabolites in kidney diseases in humans and their potential as biomarkers is showed in Table 1.

Table 1. TCA cycle metabolites alterations in kidney diseases with potential use as biomarkers.

Metabolite	Kidney Disease			References
	CKD	AKI	RCC	
Acetyl-carnitine	Δ serum ∇ urine	Δ serum	Δ tissue	[166–169]
Citrate	∇ urine	∇ urine	∇ urine	[43,62,174,175]
Isocitrate	∇ urine	-	-	[42]
AKG	∇ urine	-	∇ urine Δ tissue	[42,81,177]
Succinate	∇ urine	-	∇ urine Δ tissue	[42,62,175]
Fumarate	Δ urine	-	∇ tissue	[44,62]
Malate	Δ urine	-	∇ tissue	[44,62]

Summary of tricarboxylic citric acid (TCA) cycle alterations in different kidney diseases in humans. Δ = increased levels, ∇ = decreased levels; CKD, chronic kidney disease; AKI, acute kidney injury; RCC, renal cell carcinoma; AKG, alpha-ketoglutarate.

13. Concluding Remarks and Future Directions

Mitochondria perform several functions, including metabolic pathways and communicating to the rest of the cell to drive its behavior. The TCA cycle occurs in the mitochondrial matrix, and the metabolites that compose it are dependent on each other; hence the excess or lack of the TCA cycle metabolites are regulated by their release from mitochondria or can be replenished from cytosolic precursors.

As we review, TCA cycle metabolites are involved in several kidney functions in health and disease. Moreover, in kidney diseases, there are alterations in the levels of TCA cycle metabolites and in the enzymes involved in their synthesis, which drive cell fate impacting kidney function.

Knowing the role of these metabolites in kidney diseases is of great relevance to understanding the pathophysiology and for their possible application in future therapeutic options and for clinical use as prognosis/diagnosis biomarkers.

Author Contributions: Conceptualization, A.P.J.-U.; writing—original draft preparation, A.P.J.-U.; writing—review and editing, E.Y.H.-C., K.J.R.-M. and J.P.-C.; figures preparation, E.Y.H.-C. and A.P.J.-U.; funding acquisition, J.P.-C. All authors have read and agreed to the published version of the manuscript.

Funding: This research was funded by the Consejo Nacional de Ciencia y Tecnología (CONACYT), grant number A1-S-7495 and by the Dirección General de Asuntos del Personal Académico (DGAPA), grant numbers IN202219 and IN200922.

Institutional Review Board Statement: Not applicable.

Informed Consent Statement: Not applicable.

Acknowledgments: A.P.J.-U. is a Ph.D. student from Posgrado en Ciencias Bioquímicas at the Universidad Nacional Autónoma de México.

Conflicts of Interest: The authors declare no conflict of interest.

References

1. Hamanaka, R.B.; Chandel, N.S. Mitochondrial reactive oxygen species regulate cellular signaling and dictate biological outcomes. *Trends Biochem. Sci.* **2010**, *35*, 505–513. [CrossRef]
2. Bononi, A.; Missiroli, S.; Poletti, F.; Suski, J.M.; Agnoletto, C.; Bonora, M.; De Marchi, E.; Giorgi, C.; Marchi, S.; Patergnani, S.; et al. Mitochondria-associated membranes (MAMs) as hotspot Ca(2+) signaling units. *Adv. Exp. Med. Biol.* **2012**, *740*, 411–437. [CrossRef] [PubMed]
3. Herzig, S.; Shaw, R.J. AMPK: Guardian of metabolism and mitochondrial homeostasis. *Nat. Rev. Mol. Cell Biol.* **2018**, *19*, 121–135. [CrossRef] [PubMed]
4. Riley, J.S.; Tait, S.W. Mitochondrial DNA in inflammation and immunity. *EMBO Rep.* **2020**, *21*, e49799. [CrossRef] [PubMed]
5. Garrido, C.; Galluzzi, L.; Brunet, M.; Puig, P.E.; Didelot, C.; Kroemer, G. Mechanisms of cytochrome c release from mitochondria. *Cell Death Differ.* **2006**, *13*, 1423–1433. [CrossRef]
6. Martinez-Reyes, I.; Chandel, N.S. Mitochondrial TCA cycle metabolites control physiology and disease. *Nat. Commun.* **2020**, *11*, 102. [CrossRef] [PubMed]
7. Kornberg, H. Krebs and his trinity of cycles. *Nat. Rev. Mol. Cell Biol.* **2000**, *1*, 225–228. [CrossRef]
8. Sciacovelli, M.; Frezza, C. Oncometabolites: Unconventional triggers of oncogenic signalling cascades. *Free Radic. Biol. Med.* **2016**, *100*, 175–181. [CrossRef]
9. Akram, M. Citric acid cycle and role of its intermediates in metabolism. *Cell Biochem. Biophys.* **2014**, *68*, 475–478. [CrossRef]
10. Catalina-Rodriguez, O.; Kolukula, V.K.; Tomita, Y.; Preet, A.; Palmieri, F.; Wellstein, A.; Byers, S.; Giaccia, A.J.; Glasgow, E.; Albanese, C.; et al. The mitochondrial citrate transporter, CIC, is essential for mitochondrial homeostasis. *Oncotarget* **2012**, *3*, 1220–1235. [CrossRef]
11. Jang, H.S.; Noh, M.R.; Kim, J.; Padanilam, B.J. Defective Mitochondrial Fatty Acid Oxidation and Lipotoxicity in Kidney Diseases. *Front. Med.* **2020**, *7*, 65. [CrossRef]
12. Afshinnia, F.; Nair, V.; Lin, J.; Rajendiran, T.M.; Soni, T.; Byun, J.; Sharma, K.; Fort, P.E.; Gardner, T.W.; Looker, H.C.; et al. Increased lipogenesis and impaired beta-oxidation predict type 2 diabetic kidney disease progression in American Indians. *JCI Insight* **2019**, *4*. [CrossRef] [PubMed]
13. Kang, H.M.; Ahn, S.H.; Choi, P.; Ko, Y.A.; Han, S.H.; Chinga, F.; Park, A.S.; Tao, J.; Sharma, K.; Pullman, J.; et al. Defective fatty acid oxidation in renal tubular epithelial cells has a key role in kidney fibrosis development. *Nat. Med.* **2015**, *21*, 37–46. [CrossRef] [PubMed]

14. Ke, Q.; Yuan, Q.; Qin, N.; Shi, C.; Luo, J.; Fang, Y.; Xu, L.; Sun, Q.; Zen, K.; Jiang, L.; et al. UCP2-induced hypoxia promotes lipid accumulation and tubulointerstitial fibrosis during ischemic kidney injury. *Cell Death Dis.* **2020**, *11*, 26. [CrossRef]
15. Yan, Q.; Song, Y.; Zhang, L.; Chen, Z.; Yang, C.; Liu, S.; Yuan, X.; Gao, H.; Ding, G.; Wang, H. Autophagy activation contributes to lipid accumulation in tubular epithelial cells during kidney fibrosis. *Cell Death Discov.* **2018**, *4*, 2. [CrossRef]
16. Zhao, Z.; Liu, Y.; Liu, Q.; Wu, F.; Liu, X.; Qu, H.; Yuan, Y.; Ge, J.; Xu, Y.; Wang, H. The mRNA Expression Signature and Prognostic Analysis of Multiple Fatty Acid Metabolic Enzymes in Clear Cell Renal Cell Carcinoma. *J. Cancer* **2019**, *10*, 6599–6607. [CrossRef]
17. Du, W.; Zhang, L.; Brett-Morris, A.; Aguila, B.; Kerner, J.; Hoppel, C.L.; Puchowicz, M.; Serra, D.; Herrero, L.; Rini, B.I.; et al. HIF drives lipid deposition and cancer in ccRCC via repression of fatty acid metabolism. *Nat. Commun.* **2017**, *8*, 1769. [CrossRef]
18. Schug, Z.T.; Peck, B.; Jones, D.T.; Zhang, Q.; Grosskurth, S.; Alam, I.S.; Goodwin, L.M.; Smethurst, E.; Mason, S.; Blyth, K.; et al. Acetyl-CoA synthetase 2 promotes acetate utilization and maintains cancer cell growth under metabolic stress. *Cancer Cell* **2015**, *27*, 57–71. [CrossRef]
19. Fu, Q.; Colgan, S.P.; Shelley, C.S. Hypoxia: The Force that Drives Chronic Kidney Disease. *Clin. Med. Res.* **2016**, *14*, 15–39. [CrossRef]
20. Choueiri, T.K.; Bauer, T.M.; Papadopoulos, K.P.; Plimack, E.R.; Merchan, J.R.; McDermott, D.F.; Michaelson, M.D.; Appleman, L.J.; Thamake, S.; Perini, R.F.; et al. Inhibition of hypoxia-inducible factor-2alpha in renal cell carcinoma with belzutifan: A phase 1 trial and biomarker analysis. *Nat. Med.* **2021**, *27*, 802–805. [CrossRef]
21. Bannister, A.J.; Kouzarides, T. Regulation of chromatin by histone modifications. *Cell Res.* **2011**, *21*, 381–395. [CrossRef]
22. Tanada, Y.; Okuda, J.; Kato, T.; Minamino-Muta, E.; Murata, I.; Soga, T.; Shioi, T.; Kimura, T. The metabolic profile of a rat model of chronic kidney disease. *PeerJ* **2017**, *5*, e3352. [CrossRef]
23. Smith, E.R.; Hewitson, T.D. TGF-beta1 is a regulator of the pyruvate dehydrogenase complex in fibroblasts. *Sci. Rep.* **2020**, *10*, 17914. [CrossRef] [PubMed]
24. Smith, E.R.; Wigg, B.; Holt, S.; Hewitson, T.D. TGF-beta1 modifies histone acetylation and acetyl-coenzyme A metabolism in renal myofibroblasts. *Am. J. Physiol. Renal Physiol.* **2019**, *316*, F517–F529. [CrossRef] [PubMed]
25. Hewitson, T.D.; Holt, S.G.; Tan, S.J.; Wigg, B.; Samuel, C.S.; Smith, E.R. Epigenetic Modifications to H3K9 in Renal Tubulointerstitial Cells after Unilateral Ureteric Obstruction and TGF-beta1 Stimulation. *Front. Pharmacol.* **2017**, *8*, 307. [CrossRef] [PubMed]
26. Sayyed, S.G.; Gaikwad, A.B.; Lichtnekert, J.; Kulkarni, O.; Eulberg, D.; Klussmann, S.; Tikoo, K.; Anders, H.J. Progressive glomerulosclerosis in type 2 diabetes is associated with renal histone H3K9 and H3K23 acetylation, H3K4 dimethylation and phosphorylation at serine 10. *Nephrol. Dial. Transplant.* **2010**, *25*, 1811–1817. [CrossRef] [PubMed]
27. Deb, D.K.; Chen, Y.; Sun, J.; Wang, Y.; Li, Y.C. ATP-citrate lyase is essential for high glucose-induced histone hyperacetylation and fibrogenic gene upregulation in mesangial cells. *Am. J. Physiol. Renal Physiol.* **2017**, *313*, F423–F429. [CrossRef]
28. Chen, Y.; Deb, D.K.; Fu, X.; Yi, B.; Liang, Y.; Du, J.; He, L.; Li, Y.C. ATP-citrate lyase is an epigenetic regulator to promote obesity-related kidney injury. *FASEB J.* **2019**, *33*, 9602–9615. [CrossRef]
29. Lan, R.; Geng, H.; Singha, P.K.; Saikumar, P.; Bottinger, E.P.; Weinberg, J.M.; Venkatachalam, M.A. Mitochondrial Pathology and Glycolytic Shift during Proximal Tubule Atrophy after Ischemic AKI. *J. Am. Soc. Nephrol.* **2016**, *27*, 3356–3367. [CrossRef]
30. Jiang, W.; Yuan, X.; Zhu, H.; He, C.; Ge, C.; Tang, Q.; Xu, C.; Hu, B.; Huang, C.; Ma, T. Inhibition of Histone H3K27 Acetylation Orchestrates Interleukin-9-Mediated and Plays an Anti-Inflammatory Role in Cisplatin-Induced Acute Kidney Injury. *Front. Immunol.* **2020**, *11*, 231. [CrossRef]
31. Zhu, H.; Jiang, W.; Zhao, H.; He, C.; Tang, X.; Xu, S.; Xu, C.; Feng, R.; Li, J.; Ma, T.; et al. PSTPIP2 inhibits cisplatin-induced acute kidney injury by suppressing apoptosis of renal tubular epithelial cells. *Cell Death Dis.* **2020**, *11*, 1057. [CrossRef] [PubMed]
32. Zhang, S.; He, J.; Jia, Z.; Yan, Z.; Yang, J. Acetyl-CoA synthetase 2 enhances tumorigenesis and is indicative of a poor prognosis for patients with renal cell carcinoma. *Urol. Oncol.* **2018**, *36*, 243.e9–243.e20. [CrossRef] [PubMed]
33. Yao, L.; Guo, X.; Gui, Y. Acetyl-CoA Synthetase 2 Promotes Cell Migration and Invasion of Renal Cell Carcinoma by Upregulating Lysosomal-Associated Membrane Protein 1 Expression. *Cell Physiol. Biochem.* **2018**, *45*, 984–992. [CrossRef] [PubMed]
34. Yao, L.; Jiang, L.; Zhang, F.; Li, M.; Yang, B.; Zhang, F.; Guo, X. Acetate promotes SNAI1 expression by ACSS2-mediated histone acetylation under glucose limitation in renal cell carcinoma cell. *Biosci. Rep.* **2020**, *40*. [CrossRef]
35. Wang, Y.; Shi, J.; Chai, K.; Ying, X.; Zhou, B.P. The Role of Snail in EMT and Tumorigenesis. *Curr. Cancer Drug Targets* **2013**, *13*, 963–972. [CrossRef]
36. Cai, J. Roles of transcriptional factor Snail and adhesion factor E-cadherin in clear cell renal cell carcinoma. *Exp. Ther. Med.* **2013**, *6*, 1489–1493. [CrossRef] [PubMed]
37. Kanao, K.; Mikami, S.; Mizuno, R.; Shinojima, T.; Murai, M.; Oya, M. Decreased acetylation of histone H3 in renal cell carcinoma: A potential target of histone deacetylase inhibitors. *J. Urol.* **2008**, *180*, 1131–1136. [CrossRef]
38. Usenik, A.; Legisa, M. Evolution of allosteric citrate binding sites on 6-phosphofructo-1-kinase. *PLoS ONE* **2010**, *5*, e15447. [CrossRef]
39. Taylor, W.M.; Halperin, M.L. Regulation of pyruvate dehydrogenase in muscle: Inhibition by citrate. *J. Biol. Chem.* **1973**, *248*, 6080–6083. [CrossRef]
40. MacLellan, D.L.; Mataija, D.; Doucette, A.; Huang, W.; Langlois, C.; Trottier, G.; Burton, I.W.; Walter, J.A.; Karakach, T.K. Alterations in urinary metabolites due to unilateral ureteral obstruction in a rodent model. *Mol. Biosyst.* **2011**, *7*, 2181–2188. [CrossRef] [PubMed]

41. Li, M.; Wang, X.; Aa, J.; Qin, W.; Zha, W.; Ge, Y.; Liu, L.; Zheng, T.; Cao, B.; Shi, J.; et al. GC/TOFMS analysis of metabolites in serum and urine reveals metabolic perturbation of TCA cycle in db/db mice involved in diabetic nephropathy. *Am. J. Physiol. Renal Physiol.* **2013**, *304*, F1317–F1324. [CrossRef]
42. Hallan, S.; Afkarian, M.; Zelnick, L.R.; Kestenbaum, B.; Sharma, S.; Saito, R.; Darshi, M.; Barding, G.; Raftery, D.; Ju, W.; et al. Metabolomics and Gene Expression Analysis Reveal Down-regulation of the Citric Acid (TCA) Cycle in Non-diabetic CKD Patients. *EBioMedicine* **2017**, *26*, 68–77. [CrossRef]
43. Goraya, N.; Simoni, J.; Sager, L.N.; Madias, N.E.; Wesson, D.E. Urine citrate excretion as a marker of acid retention in patients with chronic kidney disease without overt metabolic acidosis. *Kidney Int.* **2019**, *95*, 1190–1196. [CrossRef] [PubMed]
44. Liu, J.J.; Liu, S.; Gurung, R.L.; Ching, J.; Kovalik, J.P.; Tan, T.Y.; Lim, S.C. Urine Tricarboxylic Acid Cycle Metabolites Predict Progressive Chronic Kidney Disease in Type 2 Diabetes. *J. Clin. Endocrinol. Metab.* **2018**, *103*, 4357–4364. [CrossRef]
45. Chen, L.; Chen, D.Q.; Liu, J.R.; Zhang, J.; Vaziri, N.D.; Zhuang, S.; Chen, H.; Feng, Y.L.; Guo, Y.; Zhao, Y.Y. Unilateral ureteral obstruction causes gut microbial dysbiosis and metabolome disorders contributing to tubulointerstitial fibrosis. *Exp. Mol. Med.* **2019**, *51*, 1–18. [CrossRef] [PubMed]
46. Liu, H.; Li, W.; He, Q.; Xue, J.; Wang, J.; Xiong, C.; Pu, X.; Nie, Z. Mass Spectrometry Imaging of Kidney Tissue Sections of Rat Subjected to Unilateral Ureteral Obstruction. *Sci. Rep.* **2017**, *7*, 41954. [CrossRef] [PubMed]
47. Mapuskar, K.A.; Wen, H.; Holanda, D.G.; Rastogi, P.; Steinbach, E.; Han, R.; Coleman, M.C.; Attanasio, M.; Riley, D.P.; Spitz, D.R.; et al. Persistent increase in mitochondrial superoxide mediates cisplatin-induced chronic kidney disease. *Redox Biol.* **2019**, *20*, 98–106. [CrossRef] [PubMed]
48. Ullian, M.E.; Robinson, C.J.; Evans, C.T.; Melnick, J.Z.; Fitzgibbon, W.R. Role of citrate synthase in aldosterone-mediated sodium reabsorption. *Hypertension* **2000**, *35*, 875–879. [CrossRef]
49. Minakuchi, H.; Wakino, S.; Urai, H.; Kurokochi, A.; Hasegawa, K.; Kanda, T.; Tokuyama, H.; Itoh, H. The effect of aldosterone and aldosterone blockade on the progression of chronic kidney disease: A randomized placebo-controlled clinical trial. *Sci. Rep.* **2020**, *10*, 16626. [CrossRef]
50. Wei, Q.; Xiao, X.; Fogle, P.; Dong, Z. Changes in metabolic profiles during acute kidney injury and recovery following ischemia/reperfusion. *PLoS ONE* **2014**, *9*, e106647. [CrossRef]
51. Toyohara, T.; Akiyama, Y.; Suzuki, T.; Takeuchi, Y.; Mishima, E.; Tanemoto, M.; Momose, A.; Toki, N.; Sato, H.; Nakayama, M.; et al. Metabolomic profiling of uremic solutes in CKD patients. *Hypertens Res.* **2010**, *33*, 944–952. [CrossRef]
52. Phillips, R.; Hanchanale, V.S.; Myatt, A.; Somani, B.; Nabi, G.; Biyani, C.S. Citrate salts for preventing and treating calcium containing kidney stones in adults. *Cochrane Database Syst. Rev.* **2015**, CD010057. [CrossRef]
53. Fiaccadori, E.; Regolisti, G.; Cademartiri, C.; Cabassi, A.; Picetti, E.; Barbagallo, M.; Gherli, T.; Castellano, G.; Morabito, S.; Maggiore, U. Efficacy and safety of a citrate-based protocol for sustained low-efficiency dialysis in AKI using standard dialysis equipment. *Clin. J. Am. Soc. Nephrol.* **2013**, *8*, 1670–1678. [CrossRef]
54. Hanevold, C.; Lu, S.; Yonekawa, K. Utility of citrate dialysate in management of acute kidney injury in children. *Hemodial. Int.* **2010**, *14* (Suppl. S1), S2–S6. [CrossRef] [PubMed]
55. Mariano, F.; Bergamo, D.; Gangemi, E.N.; Hollo, Z.; Stella, M.; Triolo, G. Citrate anticoagulation for continuous renal replacement therapy in critically ill patients: Success and limits. *Int. J. Nephrol.* **2011**, *2011*, 748320. [CrossRef] [PubMed]
56. Krieger, N.S.; Asplin, J.R.; Frick, K.K.; Granja, I.; Culbertson, C.D.; Ng, A.; Grynpas, M.D.; Bushinsky, D.A. Effect of Potassium Citrate on Calcium Phosphate Stones in a Model of Hypercalciuria. *J. Am. Soc. Nephrol.* **2015**, *26*, 3001–3008. [CrossRef] [PubMed]
57. Bienholz, A.; Reis, J.; Sanli, P.; de Groot, H.; Petrat, F.; Guberina, H.; Wilde, B.; Witzke, O.; Saner, F.H.; Kribben, A.; et al. Citrate shows protective effects on cardiovascular and renal function in ischemia-induced acute kidney injury. *BMC Nephrol.* **2017**, *18*, 130. [CrossRef]
58. Tiranathanagul, K.; Jearnsujitwimol, O.; Susantitaphong, P.; Kijkriengkraikul, N.; Leelahavanichkul, A.; Srisawat, N.; Praditpornsilpa, K.; Eiam-Ong, S. Regional citrate anticoagulation reduces polymorphonuclear cell degranulation in critically ill patients treated with continuous venovenous hemofiltration. *Ther. Apher. Dial.* **2011**, *15*, 556–564. [CrossRef]
59. Ou, Y.; Li, S.; Zhu, X.; Gui, B.; Yao, G.; Ma, L.; Zhu, D.; Fu, R.; Ge, H.; Wang, L.; et al. Citrate Attenuates Adenine-Induced Chronic Renal Failure in Rats by Modulating the Th17/Treg Cell Balance. *Inflammation* **2016**, *39*, 79–86. [CrossRef]
60. Choi, E.Y.; Kim, H.J.; Han, J.S. Anti-inflammatory effects of calcium citrate in RAW 264.7cells via suppression of NF-kappaB activation. *Environ. Toxicol. Pharmacol.* **2015**, *39*, 27–34. [CrossRef]
61. Ashbrook, M.J.; McDonough, K.L.; Pituch, J.J.; Christopherson, P.L.; Cornell, T.T.; Selewski, D.T.; Shanley, T.P.; Blatt, N.B. Citrate modulates lipopolysaccharide-induced monocyte inflammatory responses. *Clin. Exp. Immunol.* **2015**, *180*, 520–530. [CrossRef]
62. Hakimi, A.A.; Reznik, E.; Lee, C.H.; Creighton, C.J.; Brannon, A.R.; Luna, A.; Aksoy, B.A.; Liu, E.M.; Shen, R.; Lee, W.; et al. An Integrated Metabolic Atlas of Clear Cell Renal Cell Carcinoma. *Cancer Cell* **2016**, *29*, 104–116. [CrossRef] [PubMed]
63. Teng, L.; Chen, Y.; Cao, Y.; Wang, W.; Xu, Y.; Wang, Y.; Lv, J.; Li, C.; Su, Y. Overexpression of ATP citrate lyase in renal cell carcinoma tissues and its effect on the human renal carcinoma cells in vitro. *Oncol. Lett.* **2018**, *15*, 6967–6974. [CrossRef] [PubMed]
64. Lushchak, O.V.; Piroddi, M.; Galli, F.; Lushchak, V.I. Aconitase post-translational modification as a key in linkage between Krebs cycle, iron homeostasis, redox signaling, and metabolism of reactive oxygen species. *Redox Rep.* **2014**, *19*, 8–15. [CrossRef] [PubMed]
65. Gyuraszova, M.; Gurecka, R.; Babickova, J.; Tothova, L. Oxidative Stress in the Pathophysiology of Kidney Disease: Implications for Noninvasive Monitoring and Identification of Biomarkers. *Oxid. Med. Cell Longev.* **2020**, *2020*, 5478708. [CrossRef]

66. Correa, F.; Buelna-Chontal, M.; Hernandez-Resendiz, S.; Garcia-Nino, W.R.; Roldan, F.J.; Soto, V.; Silva-Palacios, A.; Amador, A.; Pedraza-Chaverri, J.; Tapia, E.; et al. Curcumin maintains cardiac and mitochondrial function in chronic kidney disease. *Free Radic. Biol. Med.* **2013**, *61*, 119–129. [CrossRef]
67. Tapia, E.; Sanchez-Lozada, L.G.; Garcia-Nino, W.R.; Garcia, E.; Cerecedo, A.; Garcia-Arroyo, F.E.; Osorio, H.; Arellano, A.; Cristobal-Garcia, M.; Loredo, M.L.; et al. Curcumin prevents maleate-induced nephrotoxicity: Relation to hemodynamic alterations, oxidative stress, mitochondrial oxygen consumption and activity of respiratory complex I. *Free Radic. Res.* **2014**, *48*, 1342–1354. [CrossRef] [PubMed]
68. Nilakantan, V.; Liang, H.L.; Rajesh, S.; Mortensen, J.; Chandran, K. Time-dependant protective effects of mangenese(III) tetrakis (1-methyl-4-pyridyl) porphyrin on mitochondrial function following renal ischemia-reperfusion injury. *Free Radic. Res.* **2010**, *44*, 773–782. [CrossRef] [PubMed]
69. Yarian, C.S.; Toroser, D.; Sohal, R.S. Aconitase is the main functional target of aging in the citric acid cycle of kidney mitochondria from mice. *Mech. Ageing Dev.* **2006**, *127*, 79–84. [CrossRef]
70. Hooftman, A.; O'Neill, L.A.J. The Immunomodulatory Potential of the Metabolite Itaconate. *Trends Immunol.* **2019**, *40*, 687–698. [CrossRef] [PubMed]
71. Zhu, D.; Zhao, Y.; Luo, Y.; Qian, X.; Zhang, Z.; Jiang, G.; Guo, F. Irg1-itaconate axis protects against acute kidney injury via activation of Nrf2. *Am. J. Transl. Res.* **2021**, *13*, 1155–1169. [PubMed]
72. Tian, F.; Wang, Z.; He, J.; Zhang, Z.; Tan, N. 4-Octyl itaconate protects against renal fibrosis via inhibiting TGF-beta/Smad pathway, autophagy and reducing generation of reactive oxygen species. *Eur J. Pharmacol.* **2020**, *873*, 172989. [CrossRef] [PubMed]
73. Mills, E.L.; Ryan, D.G.; Prag, H.A.; Dikovskaya, D.; Menon, D.; Zaslona, Z.; Jedrychowski, M.P.; Costa, A.S.H.; Higgins, M.; Hams, E.; et al. Itaconate is an anti-inflammatory metabolite that activates Nrf2 via alkylation of KEAP1. *Nature* **2018**, *556*, 113–117. [CrossRef] [PubMed]
74. Cordes, T.; Wallace, M.; Michelucci, A.; Divakaruni, A.S.; Sapcariu, S.C.; Sousa, C.; Koseki, H.; Cabrales, P.; Murphy, A.N.; Hiller, K.; et al. Immunoresponsive Gene 1 and Itaconate Inhibit Succinate Dehydrogenase to Modulate Intracellular Succinate Levels. *J. Biol. Chem.* **2016**, *291*, 14274–14284. [CrossRef]
75. Honer Zu Bentrup, K.; Miczak, A.; Swenson, D.L.; Russell, D.G. Characterization of activity and expression of isocitrate lyase in Mycobacterium avium and Mycobacterium tuberculosis. *J. Bacteriol* **1999**, *181*, 7161–7167. [CrossRef]
76. Zdzisinska, B.; Zurek, A.; Kandefer-Szerszen, M. Alpha-Ketoglutarate as a Molecule with Pleiotropic Activity: Well-Known and Novel Possibilities of Therapeutic Use. *Arch. Immunol. Ther. Exp.* **2017**, *65*, 21–36. [CrossRef]
77. Otto, C.; Yovkova, V.; Barth, G. Overproduction and secretion of alpha-ketoglutaric acid by microorganisms. *Appl. Microbiol. Biotechnol.* **2011**, *92*, 689–695. [CrossRef]
78. You, Y.H.; Quach, T.; Saito, R.; Pham, J.; Sharma, K. Metabolomics Reveals a Key Role for Fumarate in Mediating the Effects of NADPH Oxidase 4 in Diabetic Kidney Disease. *J. Am. Soc. Nephrol.* **2016**, *27*, 466–481. [CrossRef]
79. Salek, R.M.; Maguire, M.L.; Bentley, E.; Rubtsov, D.V.; Hough, T.; Cheeseman, M.; Nunez, D.; Sweatman, B.C.; Haselden, J.N.; Cox, R.D.; et al. A metabolomic comparison of urinary changes in type 2 diabetes in mouse, rat, and human. *Physiol. Genom.* **2007**, *29*, 99–108. [CrossRef]
80. Shroff, E.H.; Eberlin, L.S.; Dang, V.M.; Gouw, A.M.; Gabay, M.; Adam, S.J.; Bellovin, D.I.; Tran, P.T.; Philbrick, W.M.; Garcia-Ocana, A.; et al. MYC oncogene overexpression drives renal cell carcinoma in a mouse model through glutamine metabolism. *Proc. Natl. Acad. Sci. USA* **2015**, *112*, 6539–6544. [CrossRef]
81. Chen, S.; Wang, Y.; Xiong, Y.; Peng, T.; Lu, M.; Zhang, L.; Guo, Z. Wild-type IDH1 inhibits the tumor growth through degrading HIF-alpha in renal cell carcinoma. *Int. J. Biol. Sci.* **2021**, *17*, 1250–1262. [CrossRef] [PubMed]
82. Al-Khallaf, H. Isocitrate dehydrogenases in physiology and cancer: Biochemical and molecular insight. *Cell Biosci.* **2017**, *7*, 37. [CrossRef] [PubMed]
83. Han, S.J.; Jang, H.S.; Noh, M.R.; Kim, J.; Kong, M.J.; Kim, J.I.; Park, J.W.; Park, K.M. Mitochondrial NADP(+)-Dependent Isocitrate Dehydrogenase Deficiency Exacerbates Mitochondrial and Cell Damage after Kidney Ischemia-Reperfusion Injury. *J. Am. Soc. Nephrol.* **2017**, *28*, 1200–1215. [CrossRef] [PubMed]
84. Jo, S.H.; Son, M.K.; Koh, H.J.; Lee, S.M.; Song, I.H.; Kim, Y.O.; Lee, Y.S.; Jeong, K.S.; Kim, W.B.; Park, J.W.; et al. Control of mitochondrial redox balance and cellular defense against oxidative damage by mitochondrial NADP+-dependent isocitrate dehydrogenase. *J. Biol. Chem.* **2001**, *276*, 16168–16176. [CrossRef]
85. Hanschmann, E.M.; Godoy, J.R.; Berndt, C.; Hudemann, C.; Lillig, C.H. Thioredoxins, glutaredoxins, and peroxiredoxins–molecular mechanisms and health significance: From cofactors to antioxidants to redox signaling. *Antioxid. Redox Signal.* **2013**, *19*, 1539–1605. [CrossRef]
86. Velvizhi, S.; Nagalashmi, T.; Essa, M.M.; Dakshayani, K.B.; Subramanian, P. Effects of alpha-ketoglutarate on lipid peroxidation and antioxidant status during chronic ethanol administration in Wistar rats. *Pol. J. Pharmacol.* **2002**, *54*, 231–236.
87. Mehra, L.; Hasija, Y.; Mittal, G. Therapeutic potential of alpha-ketoglutarate against acetaminophen-induced hepatotoxicity in rats. *J. Pharm. Bioallied. Sci.* **2016**, *8*, 296–299. [CrossRef] [PubMed]
88. Velvizhi, S.; Dakshayani, K.B.; Subramanian, P. Effects of alpha-ketoglutarate on antioxidants and lipid peroxidation products in rats treated with ammonium acetate. *Nutrition* **2002**, *18*, 747–750. [CrossRef]
89. Kong, M.J.; Han, S.J.; Kim, J.I.; Park, J.W.; Park, K.M. Mitochondrial NADP(+)-dependent isocitrate dehydrogenase deficiency increases cisplatin-induced oxidative damage in the kidney tubule cells. *Cell Death Dis.* **2018**, *9*, 488. [CrossRef]

90. Kim, J.; Kim, K.Y.; Jang, H.S.; Yoshida, T.; Tsuchiya, K.; Nitta, K.; Park, J.W.; Bonventre, J.V.; Park, K.M. Role of cytosolic NADP+-dependent isocitrate dehydrogenase in ischemia-reperfusion injury in mouse kidney. *Am. J. Physiol. Renal Physiol.* **2009**, *296*, F622–F633. [CrossRef]
91. Noh, M.R.; Kong, M.J.; Han, S.J.; Kim, J.I.; Park, K.M. Isocitrate dehydrogenase 2 deficiency aggravates prolonged high-fat diet intake-induced hypertension. *Redox Biol.* **2020**, *34*, 101548. [CrossRef] [PubMed]
92. Kim, J.I.; Noh, M.R.; Yoon, G.E.; Jang, H.S.; Kong, M.J.; Park, K.M. IDH2 gene deficiency accelerates unilateral ureteral obstruction-induced kidney inflammation through oxidative stress and activation of macrophages. *Korean J. Physiol. Pharmacol.* **2021**, *25*, 139–146. [CrossRef] [PubMed]
93. Laba, P.; Wang, J.; Zhang, J. Low level of isocitrate dehydrogenase 1 predicts unfavorable postoperative outcomes in patients with clear cell renal cell carcinoma. *BMC Cancer* **2018**, *18*, 852. [CrossRef]
94. Tokonami, N.; Morla, L.; Centeno, G.; Mordasini, D.; Ramakrishnan, S.K.; Nikolaeva, S.; Wagner, C.A.; Bonny, O.; Houillier, P.; Doucet, A.; et al. alpha-Ketoglutarate regulates acid-base balance through an intrarenal paracrine mechanism. *J. Clin. Investig.* **2013**, *123*, 3166–3171. [CrossRef]
95. Weinberg, J.M.; Venkatachalam, M.A.; Roeser, N.F.; Nissim, I. Mitochondrial dysfunction during hypoxia/reoxygenation and its correction by anaerobic metabolism of citric acid cycle intermediates. *Proc. Natl. Acad. Sci. USA* **2000**, *97*, 2826–2831. [CrossRef]
96. Bienholz, A.; Petrat, F.; Wenzel, P.; Ickerott, P.; Weinberg, J.M.; Witzke, O.; Kribben, A.; de Groot, H.; Feldkamp, T. Adverse effects of alpha-ketoglutarate/malate in a rat model of acute kidney injury. *Am. J. Physiol. Renal Physiol.* **2012**, *303*, F56–F63. [CrossRef]
97. Peralta, C.A.; Hicks, L.S.; Chertow, G.M.; Ayanian, J.Z.; Vittinghoff, E.; Lin, F.; Shlipak, M.G. Control of hypertension in adults with chronic kidney disease in the United States. *Hypertension* **2005**, *45*, 1119–1124. [CrossRef] [PubMed]
98. Losman, J.A.; Koivunen, P.; Kaelin, W.G., Jr. 2-Oxoglutarate-dependent dioxygenases in cancer. *Nat. Rev. Cancer* **2020**, *20*, 710–726. [CrossRef]
99. Kapitsinou, P.P.; Jaffe, J.; Michael, M.; Swan, C.E.; Duffy, K.J.; Erickson-Miller, C.L.; Haase, V.H. Preischemic targeting of HIF prolyl hydroxylation inhibits fibrosis associated with acute kidney injury. *Am. J. Physiol. Renal Physiol.* **2012**, *302*, F1172–F1179. [CrossRef] [PubMed]
100. Yang, Y.; Yu, X.; Zhang, Y.; Ding, G.; Zhu, C.; Huang, S.; Jia, Z.; Zhang, A. Hypoxia-inducible factor prolyl hydroxylase inhibitor roxadustat (FG-4592) protects against cisplatin-induced acute kidney injury. *Clin. Sci.* **2018**, *132*, 825–838. [CrossRef]
101. Li, X.; Zou, Y.; Xing, J.; Fu, Y.Y.; Wang, K.Y.; Wan, P.Z.; Zhai, X.Y. Pretreatment with Roxadustat (FG-4592) Attenuates Folic Acid-Induced Kidney Injury through Antiferroptosis via Akt/GSK-3beta/Nrf2 Pathway. *Oxid. Med. Cell Longev.* **2020**, *2020*, 6286984. [CrossRef]
102. Kabei, K.; Tateishi, Y.; Nozaki, M.; Tanaka, M.; Shiota, M.; Osada-Oka, M.; Nishide, S.; Uchida, J.; Nakatani, T.; Tomita, S.; et al. Role of hypoxia-inducible factor-1 in the development of renal fibrosis in mouse obstructed kidney: Special references to HIF-1 dependent gene expression of profibrogenic molecules. *J. Pharmacol. Sci.* **2018**, *136*, 31–38. [CrossRef]
103. Del Balzo, U.; Signore, P.E.; Walkinshaw, G.; Seeley, T.W.; Brenner, M.C.; Wang, Q.; Guo, G.; Arend, M.P.; Flippin, L.A.; Chow, F.A.; et al. Nonclinical Characterization of the Hypoxia-Inducible Factor Prolyl Hydroxylase Inhibitor Roxadustat, a Novel Treatment of Anemia of Chronic Kidney Disease. *J. Pharmacol. Exp. Ther.* **2020**, *374*, 342–353. [CrossRef] [PubMed]
104. Kabei, K.; Tateishi, Y.; Shiota, M.; Osada-Oka, M.; Nishide, S.; Uchida, J.; Nakatani, T.; Matsunaga, S.; Yamaguchi, T.; Tomita, S.; et al. Effects of orally active hypoxia inducible factor alpha prolyl hydroxylase inhibitor, FG4592 on renal fibrogenic potential in mouse unilateral ureteral obstruction model. *J. Pharmacol. Sci.* **2020**, *142*, 93–100. [CrossRef] [PubMed]
105. Brigandi, R.A.; Johnson, B.; Oei, C.; Westerman, M.; Olbina, G.; de Zoysa, J.; Roger, S.D.; Sahay, M.; Cross, N.; McMahon, L.; et al. A Novel Hypoxia-Inducible Factor-Prolyl Hydroxylase Inhibitor (GSK1278863) for Anemia in CKD: A 28-Day, Phase 2A Randomized Trial. *Am. J. Kidney Dis.* **2016**, *67*, 861–871. [CrossRef] [PubMed]
106. Seethy, A.; Pethusamy, K.; Chattopadhyay, I.; Sah, R.; Chopra, A.; Dhar, R.; Karmakar, S. TETology: Epigenetic Mastermind in Action. *Appl. Biochem. Biotechnol.* **2021**, *193*, 1701–1726. [CrossRef]
107. Gu, Y.; Chen, J.; Zhang, H.; Shen, Z.; Liu, H.; Lv, S.; Yu, X.; Zhang, D.; Ding, X.; Zhang, X. Hydrogen sulfide attenuates renal fibrosis by inducing TET-dependent DNA demethylation on Klotho promoter. *FASEB J.* **2020**, *34*, 11474–11487. [CrossRef]
108. Huang, N.; Tan, L.; Xue, Z.; Cang, J.; Wang, H. Reduction of DNA hydroxymethylation in the mouse kidney insulted by ischemia reperfusion. *Biochem. Biophys. Res. Commun.* **2012**, *422*, 697–702. [CrossRef]
109. Yan, H.; Tan, L.; Liu, Y.; Huang, N.; Cang, J.; Wang, H. Ten-eleven translocation methyl-cytosine dioxygenase 2 deficiency exacerbates renal ischemia-reperfusion injury. *Clin. Epigenetics* **2020**, *12*, 98. [CrossRef]
110. Bao, Y.; Bai, M.; Zhu, H.; Yuan, Y.; Wang, Y.; Zhang, Y.; Wang, J.; Xie, X.; Yao, X.; Mao, J.; et al. DNA demethylase Tet2 suppresses cisplatin-induced acute kidney injury. *Cell Death Discov.* **2021**, *7*, 167. [CrossRef]
111. Tampe, B.; Tampe, D.; Zeisberg, E.M.; Muller, G.A.; Bechtel-Walz, W.; Koziolek, M.; Kalluri, R.; Zeisberg, M. Induction of Tet3-dependent Epigenetic Remodeling by Low-dose Hydralazine Attenuates Progression of Chronic Kidney Disease. *EBioMedicine* **2015**, *2*, 19–36. [CrossRef]
112. Tampe, B.; Tampe, D.; Muller, C.A.; Sugimoto, H.; LeBleu, V.; Xu, X.; Muller, G.A.; Zeisberg, E.M.; Kalluri, R.; Zeisberg, M. Tet3-mediated hydroxymethylation of epigenetically silenced genes contributes to bone morphogenic protein 7-induced reversal of kidney fibrosis. *J. Am. Soc. Nephrol.* **2014**, *25*, 905–912. [CrossRef]

113. Yu, C.; Xiong, C.; Tang, J.; Hou, X.; Liu, N.; Bayliss, G.; Zhuang, S. Histone demethylase JMJD3 protects against renal fibrosis by suppressing TGFbeta and Notch signaling and preserving PTEN expression. *Theranostics* **2021**, *11*, 2706–2721. [CrossRef] [PubMed]
114. Shenoy, N.; Bhagat, T.D.; Cheville, J.; Lohse, C.; Bhattacharyya, S.; Tischer, A.; Machha, V.; Gordon-Mitchell, S.; Choudhary, G.; Wong, L.F.; et al. Ascorbic acid-induced TET activation mitigates adverse hydroxymethylcytosine loss in renal cell carcinoma. *J. Clin. Investig.* **2019**, *129*, 1612–1625. [CrossRef] [PubMed]
115. Harzandi, A.; Lee, S.; Bidkhori, G.; Saha, S.; Hendry, B.M.; Mardinoglu, A.; Shoaie, S.; Sharpe, C.C. Acute kidney injury leading to CKD is associated with a persistence of metabolic dysfunction and hypertriglyceridemia. *iScience* **2021**, *24*, 102046. [CrossRef] [PubMed]
116. Zhang, Y.; Chen, M.; Liu, M.; Xu, Y.; Wu, G. Glycolysis-Related Genes Serve as Potential Prognostic Biomarkers in Clear Cell Renal Cell Carcinoma. *Oxid. Med. Cell Longev.* **2021**, *2021*, 6699808. [CrossRef] [PubMed]
117. Deri, M.T.; Kiss, A.F.; Toth, K.; Paulik, J.; Sarvary, E.; Kobori, L.; Monostory, K. End-stage renal disease reduces the expression of drug-metabolizing cytochrome P450s. *Pharmacol. Rep.* **2020**, *72*, 1695–1705. [CrossRef]
118. Helvig, C.F.; Cuerrier, D.; Hosfield, C.M.; Ireland, B.; Kharebov, A.Z.; Kim, J.W.; Ramjit, N.J.; Ryder, K.; Tabash, S.P.; Herzenberg, A.M.; et al. Dysregulation of renal vitamin D metabolism in the uremic rat. *Kidney Int.* **2010**, *78*, 463–472. [CrossRef]
119. Zhao, L.; Gao, H.; Lian, F.; Liu, X.; Zhao, Y.; Lin, D. (1)H-NMR-based metabonomic analysis of metabolic profiling in diabetic nephropathy rats induced by streptozotocin. *Am. J. Physiol. Renal Physiol.* **2011**, *300*, F947–F956. [CrossRef]
120. Toma, I.; Kang, J.J.; Sipos, A.; Vargas, S.; Bansal, E.; Hanner, F.; Meer, E.; Peti-Peterdi, J. Succinate receptor GPR91 provides a direct link between high glucose levels and renin release in murine and rabbit kidney. *J. Clin. Investig.* **2008**, *118*, 2526–2534. [CrossRef]
121. Kamarauskaite, J.; Baniene, R.; Trumbeckas, D.; Strazdauskas, A.; Trumbeckaite, S. Increased Succinate Accumulation Induces ROS Generation in In Vivo Ischemia/Reperfusion-Affected Rat Kidney Mitochondria. *Biomed. Res. Int.* **2020**, *2020*, 8855585. [CrossRef] [PubMed]
122. Kocyigit, I.; Taheri, S.; Eroglu, E.; Sener, E.F.; Zararsiz, G.; Uzun, I.; Tufan, E.; Mehmetbeyoglu, E.; Korkmaz Bayramov, K.; Sipahioglu, M.H.; et al. Systemic Succinate, Hypoxia-Inducible Factor-1 Alpha, and IL-1beta Gene Expression in Autosomal Dominant Polycystic Kidney Disease with and without Hypertension. *Cardiorenal. Med.* **2019**, *9*, 370–381. [CrossRef] [PubMed]
123. He, W.; Miao, F.J.; Lin, D.C.; Schwandner, R.T.; Wang, Z.; Gao, J.; Chen, J.L.; Tian, H.; Ling, L. Citric acid cycle intermediates as ligands for orphan G-protein-coupled receptors. *Nature* **2004**, *429*, 188–193. [CrossRef]
124. Gullans, S.R.; Brazy, P.C.; Dennis, V.W.; Mandel, L.J. Interactions between gluconeogenesis and sodium transport in rabbit proximal tubule. *Am. J. Physiol.* **1984**, *246*, F859–F869. [CrossRef] [PubMed]
125. Gullans, S.R.; Kone, B.C.; Avison, M.J.; Giebisch, G. Succinate alters respiration, membrane potential, and intracellular K+ in proximal tubule. *Am. J. Physiol.* **1988**, *255*, F1170–F1177. [CrossRef] [PubMed]
126. Robben, J.H.; Fenton, R.A.; Vargas, S.L.; Schweer, H.; Peti-Peterdi, J.; Deen, P.M.; Milligan, G. Localization of the succinate receptor in the distal nephron and its signaling in polarized MDCK cells. *Kidney Int.* **2009**, *76*, 1258–1267. [CrossRef] [PubMed]
127. Vargas, S.L.; Toma, I.; Kang, J.J.; Meer, E.J.; Peti-Peterdi, J. Activation of the succinate receptor GPR91 in macula densa cells causes renin release. *J. Am. Soc. Nephrol.* **2009**, *20*, 1002–1011. [CrossRef]
128. Koivunen, P.; Hirsila, M.; Remes, A.M.; Hassinen, I.E.; Kivirikko, K.I.; Myllyharju, J. Inhibition of hypoxia-inducible factor (HIF) hydroxylases by citric acid cycle intermediates: Possible links between cell metabolism and stabilization of HIF. *J. Biol. Chem.* **2007**, *282*, 4524–4532. [CrossRef] [PubMed]
129. Tannahill, G.M.; Curtis, A.M.; Adamik, J.; Palsson-McDermott, E.M.; McGettrick, A.F.; Goel, G.; Frezza, C.; Bernard, N.J.; Kelly, B.; Foley, N.H.; et al. Succinate is an inflammatory signal that induces IL-1beta through HIF-1alpha. *Nature* **2013**, *496*, 238–242. [CrossRef]
130. Won, A.J.; Kim, S.; Kim, Y.G.; Kim, K.B.; Choi, W.S.; Kacew, S.; Kim, K.S.; Jung, J.H.; Lee, B.M.; Kim, S.; et al. Discovery of urinary metabolomic biomarkers for early detection of acute kidney injury. *Mol. Biosyst.* **2016**, *12*, 133–144. [CrossRef]
131. Khattri, R.B.; Thome, T.; Ryan, T.E. Tissue-Specific (1)H-NMR Metabolomic Profiling in Mice with Adenine-Induced Chronic Kidney Disease. *Metabolites* **2021**, *11*, 45. [CrossRef]
132. Molina-Jijon, E.; Tapia, E.; Zazueta, C.; El Hafidi, M.; Zatarain-Barron, Z.L.; Hernandez-Pando, R.; Medina-Campos, O.N.; Zarco-Marquez, G.; Torres, I.; Pedraza-Chaverri, J. Curcumin prevents Cr(VI)-induced renal oxidant damage by a mitochondrial pathway. *Free Radic. Biol. Med.* **2011**, *51*, 1543–1557. [CrossRef] [PubMed]
133. Niknahad, H.; Heidari, R.; Mohammadzadeh, R.; Ommati, M.M.; Khodaei, F.; Azarpira, N.; Abdoli, N.; Zarei, M.; Asadi, B.; Rasti, M.; et al. Sulfasalazine induces mitochondrial dysfunction and renal injury. *Ren. Fail.* **2017**, *39*, 745–753. [CrossRef] [PubMed]
134. Tanabe, K.; Tamura, Y.; Lanaspa, M.A.; Miyazaki, M.; Suzuki, N.; Sato, W.; Maeshima, Y.; Schreiner, G.F.; Villarreal, F.J.; Johnson, R.J.; et al. Epicatechin limits renal injury by mitochondrial protection in cisplatin nephropathy. *Am. J. Physiol. Renal Physiol.* **2012**, *303*, F1264–F1274. [CrossRef] [PubMed]
135. Jimenez-Uribe, A.P.; Bellido, B.; Aparicio-Trejo, O.E.; Tapia, E.; Sanchez-Lozada, L.G.; Hernandez-Santos, J.A.; Fernandez-Valverde, F.; Hernandez-Cruz, E.Y.; Orozco-Ibarra, M.; Pedraza-Chaverri, J. Temporal characterization of mitochondrial impairment in the unilateral ureteral obstruction model in rats. *Free Radic. Biol. Med.* **2021**, *172*, 358–371. [CrossRef]
136. Mutsaers, H.A.; Wilmer, M.J.; Reijnders, D.; Jansen, J.; van den Broek, P.H.; Forkink, M.; Schepers, E.; Glorieux, G.; Vanholder, R.; van den Heuvel, L.P.; et al. Uremic toxins inhibit renal metabolic capacity through interference with glucuronidation and mitochondrial respiration. *Biochim. Biophys. Acta* **2013**, *1832*, 142–150. [CrossRef]

137. Beach, T.E.; Prag, H.A.; Pala, L.; Logan, A.; Huang, M.M.; Gruszczyk, A.V.; Martin, J.L.; Mahbubani, K.; Hamed, M.O.; Hosgood, S.A.; et al. Targeting succinate dehydrogenase with malonate ester prodrugs decreases renal ischemia reperfusion injury. *Redox Biol.* **2020**, *36*, 101640. [CrossRef]
138. Mohan, D.; Balasubramanian, E.D.; Ravindran, S.; Kurian, G.A. Renal mitochondria can withstand hypoxic/ischemic injury secondary to renal failure in uremic rats pretreated with sodium thiosulfate. *Indian J. Pharmacol.* **2017**, *49*, 317–321. [CrossRef]
139. Sun, G.; Zhang, X.; Liang, J.; Pan, X.; Zhu, S.; Liu, Z.; Armstrong, C.M.; Chen, J.; Lin, W.; Liao, B.; et al. Integrated Molecular Characterization of Fumarate Hydratase-deficient Renal Cell Carcinoma. *Clin. Cancer Res.* **2021**, *27*, 1734–1743. [CrossRef]
140. Sciacovelli, M.; Goncalves, E.; Johnson, T.I.; Zecchini, V.R.; da Costa, A.S.; Gaude, E.; Drubbel, A.V.; Theobald, S.J.; Abbo, S.R.; Tran, M.G.; et al. Fumarate is an epigenetic modifier that elicits epithelial-to-mesenchymal transition. *Nature* **2016**, *537*, 544–547. [CrossRef]
141. Lovisa, S.; LeBleu, V.S.; Tampe, B.; Sugimoto, H.; Vadnagara, K.; Carstens, J.L.; Wu, C.C.; Hagos, Y.; Burckhardt, B.C.; Pentcheva-Hoang, T.; et al. Epithelial-to-mesenchymal transition induces cell cycle arrest and parenchymal damage in renal fibrosis. *Nat. Med.* **2015**, *21*, 998–1009. [CrossRef] [PubMed]
142. Grande, M.T.; Sanchez-Laorden, B.; Lopez-Blau, C.; De Frutos, C.A.; Boutet, A.; Arevalo, M.; Rowe, R.G.; Weiss, S.J.; Lopez-Novoa, J.M.; Nieto, M.A. Snail1-induced partial epithelial-to-mesenchymal transition drives renal fibrosis in mice and can be targeted to reverse established disease. *Nat. Med.* **2015**, *21*, 989–997. [CrossRef] [PubMed]
143. Takasu, C.; Vaziri, N.D.; Li, S.; Robles, L.; Vo, K.; Takasu, M.; Pham, C.; Liu, S.; Farzaneh, S.H.; Foster, C.E., 3rd; et al. Treatment With Dimethyl Fumarate Attenuates Calcineurin Inhibitor-induced Nephrotoxicity. *Transplantation* **2015**, *99*, 1144–1150. [CrossRef]
144. Yang, Y.; Cai, F.; Zhou, N.; Liu, S.; Wang, P.; Zhang, S.; Zhang, Y.; Zhang, A.; Jia, Z.; Huang, S. Dimethyl fumarate prevents ferroptosis to attenuate acute kidney injury by acting on NRF2. *Clin. Transl. Med.* **2021**, *11*, e382. [CrossRef]
145. Sasaki, A.; Koike, N.; Murakami, T.; Suzuki, K. Dimethyl fumarate ameliorates cisplatin-induced renal tubulointerstitial lesions. *J. Toxicol. Pathol.* **2019**, *32*, 79–89. [CrossRef] [PubMed]
146. Valencia-Sanchez, C.; Carter, J.L. An evaluation of dimethyl fumarate for the treatment of relapsing remitting multiple sclerosis. *Expert. Opin. Pharmacother.* **2020**, *21*, 1399–1405. [CrossRef]
147. Nielsen, P.M.; Eldirdiri, A.; Bertelsen, L.B.; Jorgensen, H.S.; Ardenkjaer-Larsen, J.H.; Laustsen, C. Fumarase activity: An in vivo and in vitro biomarker for acute kidney injury. *Sci. Rep.* **2017**, *7*, 40812. [CrossRef]
148. Hou, E.; Sun, N.; Zhang, F.; Zhao, C.; Usa, K.; Liang, M.; Tian, Z. Malate and Aspartate Increase L-Arginine and Nitric Oxide and Attenuate Hypertension. *Cell Rep.* **2017**, *19*, 1631–1639. [CrossRef]
149. Bradshaw, P.C. Cytoplasmic and Mitochondrial NADPH-Coupled Redox Systems in the Regulation of Aging. *Nutrients* **2019**, *11*, 504. [CrossRef]
150. Al Kadhi, O.; Melchini, A.; Mithen, R.; Saha, S. Development of a LC-MS/MS Method for the Simultaneous Detection of Tricarboxylic Acid Cycle Intermediates in a Range of Biological Matrices. *J. Anal. Methods Chem.* **2017**, *2017*, 5391832. [CrossRef]
151. Shil, K.; Pal, S. Metabolic adaptability in hexavalent chromium-treated renal tissue: An in vivo study. *Clin. Kidney J.* **2018**, *11*, 222–229. [CrossRef]
152. Khan, S.A.; Priyamvada, S.; Farooq, N.; Khan, S.; Khan, M.W.; Yusufi, A.N. Protective effect of green tea extract on gentamicin-induced nephrotoxicity and oxidative damage in rat kidney. *Pharmacol. Res.* **2009**, *59*, 254–262. [CrossRef]
153. Rony, K.A.; Ajith, T.A.; Kuttikadan, T.A.; Blaze, R.; Janardhanan, K.K. Phellinus rimosus improves mitochondrial energy status and attenuates nephrotoxicity in diabetic rats. *J. Basic Clin. Physiol. Pharmacol.* **2017**, *28*, 455–461. [CrossRef] [PubMed]
154. Sette, L.H.; Lopes, E.P. The reduction of serum aminotransferase levels is proportional to the decline of the glomerular filtration rate in patients with chronic kidney disease. *Clinics* **2015**, *70*, 346–349. [CrossRef]
155. Pandey, N.; Lanke, V.; Vinod, P.K. Network-based metabolic characterization of renal cell carcinoma. *Sci. Rep.* **2020**, *10*, 5955. [CrossRef] [PubMed]
156. Fink, B.D.; Bai, F.; Yu, L.; Sheldon, R.D.; Sharma, A.; Taylor, E.B.; Sivitz, W.I. Oxaloacetic acid mediates ADP-dependent inhibition of mitochondrial complex II-driven respiration. *J. Biol. Chem.* **2018**, *293*, 19932–19941. [CrossRef]
157. Mischak, H.; Delles, C.; Vlahou, A.; Vanholder, R. Proteomic biomarkers in kidney disease: Issues in development and implementation. *Nat. Rev. Nephrol.* **2015**, *11*, 221–232. [CrossRef] [PubMed]
158. Sirolli, V.; Pieroni, L.; Di Liberato, L.; Urbani, A.; Bonomini, M. Urinary Peptidomic Biomarkers in Kidney Diseases. *Int. J. Mol. Sci.* **2019**, *21*, 96. [CrossRef] [PubMed]
159. Dubin, R.F.; Rhee, E.P. Proteomics and Metabolomics in Kidney Disease, including Insights into Etiology, Treatment, and Prevention. *Clin. J. Am. Soc. Nephrol.* **2020**, *15*, 404–411. [CrossRef]
160. Inker, L.A.; Okparavero, A. Cystatin C as a marker of glomerular filtration rate: Prospects and limitations. *Curr. Opin Nephrol. Hypertens* **2011**, *20*, 631–639. [CrossRef]
161. Peralta, C.A.; Katz, R.; Sarnak, M.J.; Ix, J.; Fried, L.F.; De Boer, I.; Palmas, W.; Siscovick, D.; Levey, A.S.; Shlipak, M.G. Cystatin C identifies chronic kidney disease patients at higher risk for complications. *J. Am. Soc. Nephrol.* **2011**, *22*, 147–155. [CrossRef]
162. Nielsen, S.E.; Schjoedt, K.J.; Astrup, A.S.; Tarnow, L.; Lajer, M.; Hansen, P.R.; Parving, H.H.; Rossing, P. Neutrophil Gelatinase-Associated Lipocalin (NGAL) and Kidney Injury Molecule 1 (KIM1) in patients with diabetic nephropathy: A cross-sectional study and the effects of lisinopril. *Diabet. Med.* **2010**, *27*, 1144–1150. [CrossRef]
163. Devarajan, P. Proteomics for biomarker discovery in acute kidney injury. *Semin. Nephrol.* **2007**, *27*, 637–651. [CrossRef] [PubMed]

164. Malyszko, J.; Bachorzewska-Gajewska, H.; Sitniewska, E.; Malyszko, J.S.; Poniatowski, B.; Dobrzycki, S. Serum neutrophil gelatinase-associated lipocalin as a marker of renal function in non-diabetic patients with stage 2-4 chronic kidney disease. *Ren. Fail.* **2008**, *30*, 625–628. [CrossRef] [PubMed]
165. Han, W.K.; Bailly, V.; Abichandani, R.; Thadhani, R.; Bonventre, J.V. Kidney Injury Molecule-1 (KIM-1): A novel biomarker for human renal proximal tubule injury. *Kidney Int.* **2002**, *62*, 237–244. [CrossRef] [PubMed]
166. Miyamoto, Y.; Miyazaki, T.; Honda, A.; Shimohata, H.; Hirayama, K.; Kobayashi, M. Retention of acetylcarnitine in chronic kidney disease causes insulin resistance in skeletal muscle. *J. Clin. Biochem. Nutr.* **2016**, *59*, 199–206. [CrossRef] [PubMed]
167. Chen, D.Q.; Cao, G.; Chen, H.; Argyopoulos, C.P.; Yu, H.; Su, W.; Chen, L.; Samuels, D.C.; Zhuang, S.; Bayliss, G.P.; et al. Identification of serum metabolites associating with chronic kidney disease progression and anti-fibrotic effect of 5-methoxytryptophan. *Nat. Commun.* **2019**, *10*, 1476. [CrossRef]
168. Sun, J.; Shannon, M.; Ando, Y.; Schnackenberg, L.K.; Khan, N.A.; Portilla, D.; Beger, R.D. Serum metabolomic profiles from patients with acute kidney injury: A pilot study. *J. Chromatogr. B Analyt. Technol. Biomed. Life Sci.* **2012**, *893–894*, 107–113. [CrossRef]
169. Jing, L.; Guigonis, J.M.; Borchiellini, D.; Durand, M.; Pourcher, T.; Ambrosetti, D. LC-MS based metabolomic profiling for renal cell carcinoma histologic subtypes. *Sci. Rep.* **2019**, *9*, 15635. [CrossRef]
170. Lu, Y.; Li, N.; Gao, L.; Xu, Y.J.; Huang, C.; Yu, K.; Ling, Q.; Cheng, Q.; Chen, S.; Zhu, M.; et al. Acetylcarnitine Is a Candidate Diagnostic and Prognostic Biomarker of Hepatocellular Carcinoma. *Cancer Res.* **2016**, *76*, 2912–2920. [CrossRef]
171. Takaya, H.; Namisaki, T.; Kitade, M.; Shimozato, N.; Kaji, K.; Tsuji, Y.; Nakanishi, K.; Noguchi, R.; Fujinaga, Y.; Sawada, Y.; et al. Acylcarnitine: Useful biomarker for early diagnosis of hepatocellular carcinoma in non-steatohepatitis patients. *World J. Gastrointest. Oncol.* **2019**, *11*, 887–897. [CrossRef] [PubMed]
172. Lee, J.; Choi, J.Y.; Kwon, Y.K.; Lee, D.; Jung, H.Y.; Ryu, H.M.; Cho, J.H.; Ryu, D.H.; Kim, Y.L.; Hwang, G.S. Changes in serum metabolites with the stage of chronic kidney disease: Comparison of diabetes and non-diabetes. *Clin. Chim. Acta* **2016**, *459*, 123–131. [CrossRef] [PubMed]
173. Al-Ani, B.; Fitzpatrick, M.; Al-Nuaimi, H.; Coughlan, A.M.; Hickey, F.B.; Pusey, C.D.; Savage, C.; Benton, C.M.; O'Brien, E.C.; O'Toole, D.; et al. Changes in urinary metabolomic profile during relapsing renal vasculitis. *Sci. Rep.* **2016**, *6*, 38074. [CrossRef] [PubMed]
174. Muhle-Goll, C.; Eisenmann, P.; Luy, B.; Kolker, S.; Tonshoff, B.; Fichtner, A.; Westhoff, J.H. Urinary NMR Profiling in Pediatric Acute Kidney Injury-A Pilot Study. *Int. J. Mol. Sci.* **2020**, *21*, 1187. [CrossRef]
175. Falegan, O.S.; Arnold Egloff, S.A.; Zijlstra, A.; Hyndman, M.E.; Vogel, H.J. Urinary Metabolomics Validates Metabolic Differentiation Between Renal Cell Carcinoma Stages and Reveals a Unique Metabolic Profile for Oncocytomas. *Metabolites* **2019**, *9*, 155. [CrossRef] [PubMed]
176. Gisewhite, S.; Stewart, I.J.; Beilman, G.; Lusczek, E. Urinary metabolites predict mortality or need for renal replacement therapy after combat injury. *Crit. Care* **2021**, *25*, 119. [CrossRef] [PubMed]
177. Falegan, O.S.; Ball, M.W.; Shaykhutdinov, R.A.; Pieroraio, P.M.; Farshidfar, F.; Vogel, H.J.; Allaf, M.E.; Hyndman, M.E. Urine and Serum Metabolomics Analyses May Distinguish between Stages of Renal Cell Carcinoma. *Metabolites* **2017**, *7*, 6. [CrossRef]
178. Zheng, P.; Wang, Y.; Chen, L.; Yang, D.; Meng, H.; Zhou, D.; Zhong, J.; Lei, Y.; Melgiri, N.D.; Xie, P. Identification and validation of urinary metabolite biomarkers for major depressive disorder. *Mol. Cell Proteom.* **2013**, *12*, 207–214. [CrossRef] [PubMed]
179. Hu, J.R.; Coresh, J.; Inker, L.A.; Levey, A.S.; Zheng, Z.; Rebholz, C.M.; Tin, A.; Appel, L.J.; Chen, J.; Sarnak, M.J.; et al. Serum metabolites are associated with all-cause mortality in chronic kidney disease. *Kidney Int.* **2018**, *94*, 381–389. [CrossRef] [PubMed]

biomolecules

MDPI

Review

Antioxidant Roles of SGLT2 Inhibitors in the Kidney

Carmen Llorens-Cebrià [1,†], Mireia Molina-Van den Bosch [1,†], Ander Vergara [1,2], Conxita Jacobs-Cachá [1,2,*] and Maria José Soler [1,2,*]

1 Nephrology and Transplantation Research Group, Vall d'Hebron Institut de Recerca (VHIR), Vall d'Hebron Barcelona Hospital Campus, Vall d'Hebron Hospital Universitari, 08035 Barcelona, Spain; carmen.llorens@vhir.org (C.L.-C.); mireia.molina@vhir.org (M.M.-V.d.B.); vergara.ander@gmail.com (A.V.)
2 Redes de Investigación Cooperativa Orientadas a Resultados en Salud (RICORS), RD21/0005/0016, Instituto de Salud Carlos III, 28029 Madrid, Spain
* Correspondence: conxita.jacobs@vhir.org (C.J.-C.); m.soler@vhebron.net (M.J.S.)
† These authors contributed equally to this work.

Abstract: The reduction-oxidation (redox) system consists of the coupling and coordination of various electron gradients that are generated thanks to serial reduction-oxidation enzymatic reactions. These reactions happen in every cell and produce radical oxidants that can be mainly classified into reactive oxygen species (ROS) and reactive nitrogen species (RNS). ROS and RNS modulate cell-signaling pathways and cellular processes fundamental to normal cell function. However, overproduction of oxidative species can lead to oxidative stress (OS) that is pathological. Oxidative stress is a main contributor to diabetic kidney disease (DKD) onset. In the kidney, the proximal tubular cells require a high energy supply to reabsorb proteins, metabolites, ions, and water. In a diabetic milieu, glucose-induced toxicity promotes oxidative stress and mitochondrial dysfunction, impairing tubular function. Increased glucose level in urine and ROS enhance the activity of sodium/glucose co-transporter type 2 (SGLT2), which in turn exacerbates OS. SGLT2 inhibitors have demonstrated clear cardiovascular benefits in DKD which may be in part ascribed to the generation of a beneficial equilibrium between oxidant and antioxidant mechanisms.

Keywords: redox; diabetic kidney disease; oxidative stress; mitochondrial dysfunction; SGLT2

Citation: Llorens-Cebrià, C.; Molina-Van den Bosch, M.; Vergara, A.; Jacobs-Cachá, C.; Soler, M.J. Antioxidant Roles of SGLT2 Inhibitors in the Kidney. *Biomolecules* **2022**, *12*, 143. https://doi.org/10.3390/biom12010143

Academic Editor: Liang-Jun Yan

Received: 16 December 2021
Accepted: 13 January 2022
Published: 16 January 2022

1. Introduction: The Redox System

In general terms, the reduction-oxidation (redox) system consists of the coupling and coordination of various electron gradients that are generated by serial reduction-oxidation enzymatic reactions. These reduction-oxidation reactions happen in fundamental biological processes in every cell [1] and lead to the production of radical oxidants or free radicals [2]. Free radicals are in fact oxygen or nitrogen metabolites that contain an unpaired electron and are thus partially reduced [3]. These free radicals have strong oxidizing capacity and can be classified into reactive oxygen species (ROS) and reactive nitrogen species (RNS) [4] (Table 1).

Table 1. Main known reactive oxygen species (ROS) and reactive nitrogen species (RNS).

Reactive Oxygen Species	Superoxide	O_2^-
	Hydrogen peroxide	H_2O_2
	Hydroxyl radical	OH
Reactive nitrogen species	Nitric oxide	NO
	Peroxynitrite	$ONOO^-$
	Dinitrogen trioxide	N_2O_3
	Dinitrogen tetraoxide	N_2O_4
	Nitrogen dioxide	NO_2
	S-nitrosothiols	RSNO

The redox system is especially important during energy metabolism to generate adenosine triphosphate (ATP). The most important source of energy is glucose which is transformed into pyruvate via glycolysis. Pyruvate enters the tricarboxylic acid or Krebs cycle and reduces nicotinamide adenine dinucleotide (NAD) to NADH. Further, fatty acids (FA) can undergo β-oxidation to generate acetyl-CoA which then enters the Krebs cycle [5,6]. The NADH produced during the Krebs cycle is oxidized again in the mitochondrial respiratory chain [7]. The complexes that constitute the mitochondrial respiratory chain are NADH dehydrogenase (complex I), succinate dehydrogenase (complex II), cytochrome c reductase (complex III), cytochrome c oxidase (complex IV), and ATP synthase (complex V). These enzymes are able to oxidize NADH, starting the electron transport across the mitochondrial membrane and generating an electrochemical potential that allows complex V to generate ATP. Additionally, the electrons transported from complex I, through II and III, to complex IV are used to reduce oxygen to water [8,9]. ROS are generated as a byproduct of electron transfer. Therefore, mitochondria are considered the main source of ROS [10]. Approximately, 1–2% of the O_2 consumed in the mitochondria is incompletely metabolized [11]. Here, the predominant origin of ROS is the respiratory chain (mainly complexes I and III), necessary for ATP synthesis. However, other mitochondrial enzymes have been reported to be sources of ROS, such as pyruvate dehydrogenase, α-ketoglutarate dehydrogenase, or succinate dehydrogenase, which are a part of the Krebs cycle [12]. Additionally, p66shc, an adaptor protein with proapoptotic activity, is involved in the production of ROS [13]. In vitro and in vivo experiments have shown that this protein can generate ROS by oxidizing cytochrome C [14,15]. The NADPH oxidase (NOX) family of enzymes are another source of ROS. NOX are multiunit enzymes that use NADPH as an electron donor to reduce oxygen, leading to the generation of superoxide. There are seven members of the NOX family, NOX1–5 and DUOX1 and 2. NOX enzymes are expressed in different tissues and differ in the domains that constitute them, for example, DUOX1 and 2 have an additional peroxidase domain at their N-terminal end [16,17] (Figure 1).

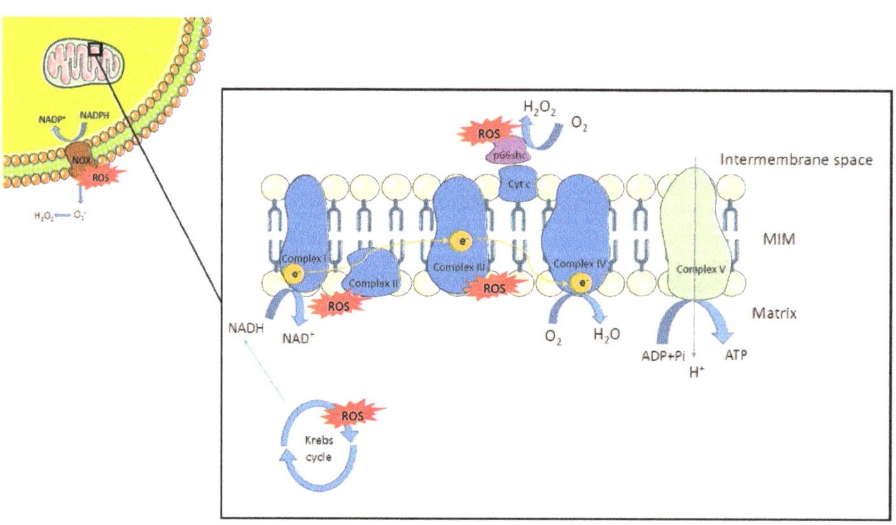

Figure 1. Sources of reactive oxygen species. Reactive oxygen species (ROS) are generated at different sites within the cell. Respiratory chain complexes I and III (in blue) produce ROS as a byproduct of electron transfer in mitochondria. Several enzymes involved in the Krebs cycle are also reported as sources of ROS. p66shc (in purple), by oxidizing cytochrome c (Cyt C), is able to generate ROS within the mitochondria intermembrane space. The NADPH oxidase (NOX) family of enzymes (in brown) are another important source of ROS by reducing oxygen to generate superoxide. MIM: mitochondria inner membrane.

Regarding RNS, these comprise nitric oxide (NO) and its derivatives, such as peroxynitrite ($ONOO^-$), dinitrogen trioxide (N_2O_3), dinitrogen tetraoxide (N_2O_4), nitrogen dioxide (NO_2), or S-nitrosothiols (RSNO) [18,19] (Table 1). NO is produced from the metabolism of L-arginine [20]. The enzyme responsible for its synthesis is nitric oxide synthase (NOS) that converts L-arginine into L-citrulline, forming NO during the reaction. There are three forms of NOS: endothelial NOS (eNOS), neuronal NOS (nNOS) and inducible NOS (iNOS). eNOS and nNOS are constitutively expressed while iNOS is only expressed under specific stimuli like infection or trauma. The different NOS isoforms are expressed in different types of cells in which NO is involved in signaling and regulation of several cellular functions [21]. In mitochondria, NO reacts with complex III of the respiratory chain, inhibiting the electron transfer and enhancing the production of O_2^-. O_2^- is highly reactive and may combine with NO, leading to the formation of peroxynitrite ($ONOO^-$). $ONOO^-$ is a powerful RNS that can irreversibly inhibit electron transport, which is pathological [18,22–24].

Both ROS and RNS are necessary for normal cell function as they modulate cell-signaling pathways and cellular processes. However, overproduction of any of them is pathological since it can damage macromolecules like proteins, lipids, or DNA [25]. DNA damage can lead to defective complexes I and III, which may result in a shutdown of mitochondrial energy production [26] and further increase ROS levels [27,28]. This eventually ends up causing mitochondrial damage and apoptosis [29,30]. Therefore, to regulate the generated ROS, there is a network of antioxidant systems to alleviate this stress. Superoxide dismutase (SOD) is an enzyme that contributes to the elimination of some oxidants. According to its subcellular location, we can differentiate between SOD1 (in the cytosol and mitochondrial intermembrane space), SOD2 (in the mitochondrial matrix), and SOD3 (in the extracellular matrix) [31]. SOD enzymes convert the superoxide radical into hydrogen peroxide [32,33] that is finally detoxified by catalase or glutathione peroxidase [34,35]. Catalases decompose H_2O_2 into water and oxygen, and glutathione peroxidases reduce H_2O_2 to water [36]. Glutathione has also a role in the antioxidant system, as it can either react directly with ROS and RNS or act as a cofactor for various enzymes helping the cell to maintain its redox status [37]. Further, exogenous and synthetic antioxidants can also be effective to prevent oxidative stress [38,39].

In this review, we will focus on oxidative stress-mediated mechanisms of kidney tubular dysfunction. Further, the contribution of the sodium/glucose co-transporter type 2 (SGLT2) in diabetic kidney disease (DKD) onset and the antioxidant potential of the SGLT2 inhibitors will be discussed.

2. Oxidative Stress-Induced Tubular Impairment

The tubular system has a very precise and highly regulated transport capacity between the tubule lumen and the bloodstream. A broad range of tubular transporters contribute to ion, amino acid, glucose, and other solute reabsorption. Therefore, tubular cells require a sufficient energy supply to sustain the high rate of exchange within the two compartments. Tubular cells are the most energy-demanding cells of the body after cardiac cells [40]. Their metabolism is highly dependent on oxygen consumption to produce ATP by oxidative phosphorylation. The oxidation of a glucose molecule provides a high energy yield (30–32 ATP) and is used as a substrate in every cell. However, in the kidney, each section of the tubule has its own preferences in terms of energy substrates, including amino acids, ketone bodies, or FA. Strikingly, FAs are an important energy fuel in tubular cells, providing 106–129 ATP through β-oxidation within the mitochondria [41]. Proximal tubular cells (PTC), located at the first section of the tubular system (segments S1, S2, and S3), are particularly energy-consuming, as 80% of the biomolecules filtered by the glomerulus are reabsorbed by these cells [42,43]. In PTC, ATP generation relies mostly on oxidative phosphorylation of FA rather than of glucose. However, under pathological conditions, such as acute kidney injury (AKI), which is associated with hypoxia [44], PTCs undergo a metabolic shift towards glycolysis [45].

As mentioned above, mitochondrial metabolism is a source of ROS that directly modulates cell-signaling pathways, protein function through post-translational modifications, and influences cell survival, proliferation, and apoptosis [46]. However, increased ROS production results in oxidative stress that causes damage to lipids, proteins, and nucleic acids, eventually disrupting cellular homeostasis. Overall, elevated ROS levels can lead to mitochondrial dysfunction, bioenergetics defects, altered gene expression, and cell death. PTCs have a high mitochondrial density that makes them susceptible to ROS-induced cell damage [42]. In vitro experiments performed using the HK-2 cell line (human PTCs) demonstrated an increase of oxidative stress, mitochondrial destabilization, DNA damage, apoptosis, and cell senescence when treated with hydrogen peroxide (H_2O_2) [47]. Aragno et al. showed that rats subjected to unilateral ischemia/reperfusion (I/R) injury exhibit high hydrogen peroxide levels in the cytosol obtained from kidney homogenate and high levels of nitrite/nitrate in serum compared to control and sham-operated animals. Consequently, the proximal tubules showed dilation and cell debris and cast formation in the lumen. These effects were attenuated in rats supplemented with dehydroepiandrosterone (DHEA), a steroid with antioxidant properties [48]. Beyond the respiratory chain, there are other sources of oxidants in the kidney. NOX enzymes have been recognized as a primary source of ROS in the kidney, especially NOX4. NOX4-derived H_2O_2 mediates several cell functions, but excessive levels can induce inflammation, apoptosis, fibrosis, and cell damage. NOX4 has constitutive activity, hence, the amount of H_2O_2 depends on the expression level. As a result, it can be damaging or beneficial depending on its abundance at the given time [49]. Normally, the levels of ROS increase in some AKI and chronic kidney disease (CKD) models, secondary to the overexpression of NOX4 [50]. Upregulation of renal NOX4 plays an important role in several pathologies, like DKD, hypertensive nephropathy, and polycystic kidney disease by increasing ROS levels and mitochondrial damage [51–55]. It has been described that mitochondrial ROS and NOX-produced ROS can enhance damage and depolarization of the mitochondrial membrane potential [46,56,57]. The total absence of NOX4 has also been reported to be pathological since it results in fibrosis and oxidative stress [58]. This could be in part related to the fact that H_2O_2 generated by NOX4 enhances nuclear factor erythroid 2-related factor 2 (Nrf2) stability, a regulator of antioxidant activity [59].

Excessive ROS production can damage mitochondria [60]. Mitochondrial damage has been recognized as a main contributor to tubular necrosis and apoptosis. Tubular cell necrosis involves disruption of respiration complexes, loss of mitochondrial membrane potential, and mitochondrial membrane transition, while apoptosis is caused by mitochondrial outer membrane permeabilization and release of apoptogenic factors [61]. Mitochondrial damage secondary to necrosis or apoptosis is especially problematic for the tubular cells as it may compromise its normal function and ultimately lead to renal dysfunction. Despite the effects of oxidative stress in kidney impairment having been studied extensively [42,46], its complexity and interconnection with a broad range of metabolic and cellular pathways make it puzzling to decipher its full implications in the onset, progression, and development of renal pathology.

3. Oxidative Stress in Diabetic Kidney Disease: Contribution of Sodium Glucose Co-Transporter Type 2 (SGLT2)

Albeit glucose consumption is low in the proximal tubule (except for segment S3 [62]), PTCs are primarily responsible for its reabsorption via the sodium/glucose co-transporter type 2 (SGLT2) [44]. SGLT2 is located at the brush border of PTCs in the S1 and S2 segments of the proximal tubule, where 90% of the total filtered glucose is reabsorbed [63]. Its transport is coupled with sodium (Na^+/Glucose, 1:1) and it accounts for 5.7% of total renal Na^+ reabsorption in healthy individuals [64]. Although SGLT2 activity has a role in maintaining the intrarenal osmolarity, the main sodium exchanger at the tubular level is the Na^+/H^+ Exchanger-3 (NHE3). Interestingly, NHE3 and SGLT2 have been shown to be interdependent [65]. Therefore, SGLT2 activity has a huge impact on both glucose and sodium balance. The Na^+ gradient created by the sodium-potassium pump (Na^+/K^+-ATPase), located in the basolateral membrane of PTC, is harnessed by SGLT2 to transport glucose and sodium across the plasma membrane down-gradient [65]. At the same time, the basolateral glucose transporter, GLUT2, maintains the glucose gradient by passively easing its backflow into the bloodstream [66].

SGLT2 plays a crucial role in glucose transport and intrarenal osmolarity. For that reason, the dysfunction of this co-transporter, which happens in diabetes, has a direct effect on intrarenal glucose metabolism and also on the redox environment. In diabetic patients, SGLT2 is overactivated, leading to increased glucose and sodium reabsorption [67]. The deleterious effects in kidney function and fluid homeostasis include the alteration of tubular-glomerular feedback (a self-regulated system of the kidney) and the increase of the glomerular filtration rate and the tubular reabsorption capacity. Systematically, the plasma volume and blood pressure increase, leading to hypertension, a common feature of these patients [68]. At the molecular level, this has drastic consequences for the activity of other transporters located in the tubule, such as NHE3, which is sensitive to small changes in sodium concentration. Packer et al. reviewed that Na^+/H^+ Exchanger-1 (NHE1) in the heart and vasculature and NHE3—the kidney isoform—are upregulated in heart failure and type 2 diabetes mellitus (T2DM) [69]. Additionally, when sodium levels drop in the luminal filtrate, Na^+/K^+-ATPase activity increases to restore the sodium gradient. As this transporter is ATP-dependent, its overactivation implies an increase in energy demand and oxygen consumption. It is worth mentioning that under normal conditions, 60% of kidney ATP consumption is intended for sodium uptake which is mostly attributed to basal Na^+/K^+-ATPase activity [70]. Therefore, the energy consumption of this transporter cannot be underestimated, and its impact on mitochondrial function should be further explored. The enhanced mitochondrial phosphorylation produces substantial accumulation of ROS products. In addition, it has been described that interleukin 6 (IL-6) activates SGLT2 through ROS in cultured tubular cells [71], which suggests that an inflammatory context would further promote ROS generation and SGLT2 overactivation. Thus, SGLT2 activity, oxidative stress, and inflammation actively contribute to the onset of DKD. By definition, DKD is a microvascular complication of diabetes mellitus that affects the kidney. These patients are diagnosed by the presence of persistent albuminuria, starting at first stage by hyperfiltration and ending with a progressive decrease in GFR at later stages. The Renal Pathology Society divide DKD pathology into glomerular, tubulointerstitial, and vascular lesions [72]. Focusing on the tubule, tubular membrane thickening, interstitial fibrosis, and tubular atrophy are usually observed. Coughlan et al. mapped the changes in mitochondrial adaptation in experimental rat's kidney mitochondria at 4, 8, 16, and 32 weeks after induction of diabetes by streptozotocin intravenous injection [73]. The authors demonstrated that DKD begins with decreased ATP production in the renal cortex, mitochondrial fragmentation, which is accompanied by increased ROS, and early renal hyperfiltration. Their results suggest that mitochondrial dynamics and bioenergetic function worsen over time, preceding the development of albuminuria and histological lesions. An in vitro experiment with HK-2 cells found out that under high glucose conditions, HK-2 cells were susceptible to mitochondrial fragmentation. These effects were ameliorated with a SGLT2 inhibitor (empagliflozin), by regulating proteins of mitochondrial fission and fusion [74]. Mitochondrial fusion and fission are essential for mitochondrial maintenance but also for mitochondrial DNA (mtDNA) integrity, which can participate in the regulation of cell survival, metabolic processes, and redox-sensitive signals [75]. In addition, high glucose conditions increase superoxide formation in mitochondria, which combined with NO (released by endothelial cells under stress) generates peroxynitrite [76]. Peroxynitrite is a cytotoxic oxidant that induces eNOS uncoupling, restricting vascular relaxation, and promoting diabetic vascular complications. As mentioned before, NOX enzymes have been shown to mediate ROS production. NOX-derived ROS regulate many physiological mechanisms of the kidney, including tubular-glomerular feedback, glucose metabolism and transport, kidney hemodynamics, and electrolyte transport [77]. NOX4 is the major source of ROS production in podocytes, and it is upregulated under high-glucose conditions [78]. For this reason, NOX4 has been largely linked to DKD [53,79]. Induction of NOX4 in the Akita model of DKD results in the emergence of the typical structural changes seen in the diabetic kidney: glomerular hyperfiltration, mesangial matrix accumulation, glomerular membrane thickening, albuminuria, and podocyte loss [53]. In addition, metabolomic analysis performed on these mice revealed alterations in tubular energy metabolism with the accumulation Krebs cycle-related urinary metabolites [53]. The Krebs cycle is the central axis of energy production in mitochondria, thus, disproportion of the cycle substrates leads to mitochondrial dysfunction in advanced DKD [80]. Accordingly, a cross-sectional study in patients with and without DKD found that injured tubular cells showed several signs of mitochondrial damage, including (1) changes in mitochondrial dynamics, (2) decreased mtDNA copy number, and (3) increased release of mtDNA into the extracellular space [81]. Paradoxically, Cao et al. reported that plasma mtDNA content was decreased in patients with T2DM with clinically significant proteinuria (24 h urinary protein level >0.5 g) but not mild proteinuria (24 h urinary protein level $\leq$0.5 g), however, mtDNA abundance was increased in the urine of the former [82]. They propose that hyperglycemia reduces intracellular mtDNA and facilitates its extracellular release, so that circulating mtDNA may be filtered by the kidneys and end up in the urine [82]. In concordance, Kafaji et al. reported that a decreased mtDNA copy number in blood was associated with the severity and the presence of DKD [83]. Their group found out that mtDNA was lower in diabetic patients with macroalbuminuria than in patients with microalbuminuria or normalbuminuria. Therefore, altered mtDNA content can predict the occurrence of DKD and the oxidative stress environment [84]. Interestingly, release of mtDNA under cellular stress has been previously considered as a damage-associated molecular pattern (DAMP) [85]. mtDNA can be recognized by Toll-like receptor 9 (TLR9) and cystosolic Cgas-stimulator of interferon genes (STING) and activate the inflammasome, leading to kidney inflammation and fibrosis [86]. In this line, recent evidence showed an increase in macrophage infiltration in a mouse model of type 2 diabetes [87] and in human progressive DKD [88], which correlates with the disease state and renal injury. Other than the direct effect of mtDNA as DAMP, oxidative stress can induce inflammation in the kidney by stimulating cytokine production [89]. ROS derivates, which participate in cellular signaling, can activate the transcription factors nuclear factor kappa B (NFκB) and activator protein-1 (AP-1), promoting the transcription of cytokines, growth factors, and extracellular membrane proteins [89] (Figure 2).

Adding the immune response to the theme above reaffirms that mitochondrial dysfunction and redox imbalance lead to the emergence of the main precipitating factors: altered metabolism, oxidative stress, and inflammation. These three elements combine and reciprocally feed back into each other. As in the chicken-and-egg dilemma, determining which of these events triggers kidney disease in the first place entails a great difficulty. However, it is clear that pharmacological approaches focused on blocking/stimulating the involved pathways mentioned in this section and targeting the SGLT2 transporter are of particular interest.

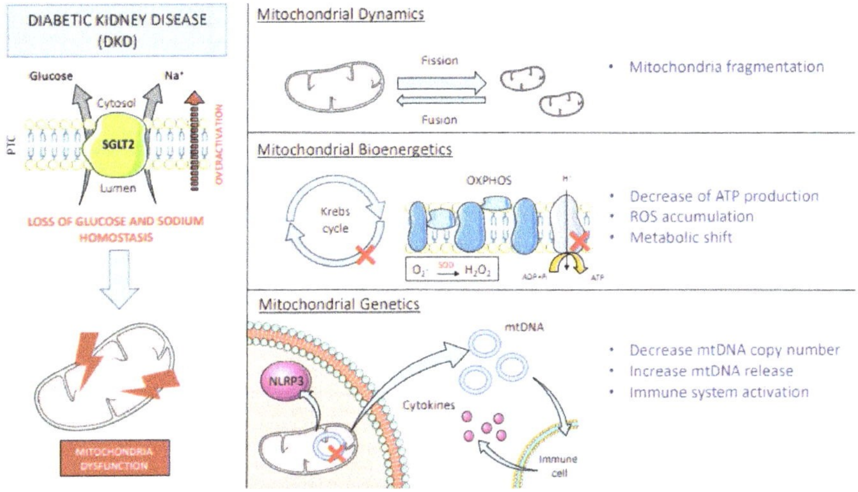

Figure 2. Mitochondrial dysfunction in DKD. The sodium/glucose co-transporter type 2 (SGLT2) is located in the apical membrane of the proximal tubular cells (PTC) of the kidney and is responsible for glucose and sodium uptake. Diabetic patients have overactivation of SGLT2 that alters glucose and sodium homeostasis, which directly affects mitochondrial function at different levels; imbalance of fission and fusion together with mitochondrial fragmentation (mitochondrial dynamics). Alteration and accumulation of Krebs cycle substrates, uncoupling of the oxidative phosphorylation chain (OXPHOS) with generation of ROS (e.g., formation of H_2O_2 by the enzyme superoxide dismutase (SOD)), leading to ATP depletion, oxidative stress, and a metabolic shift to oxygen-independent energy sources (mitochondrial bioenergetics). Mitochondrial DNA (mtDNA) is also damaged, leading to a reduction in mtDNA copy number and increased release of mtDNA into the cytosol, which activates the NLR family pyrin domain-containing 3 (NLRP3) inflammasome, and into the extracellular space, triggering the recruitment of immune cells and the onset of the inflammatory response (mitochondrial genetics). DKD: diabetic kidney disease; ROS: reactive oxygen species; ADP: adenosine diphosphate; ATP: adenosine triphosphate; NLR: NOD-like receptor.

4. Antioxidant Properties of SGLT2 Inhibitors

SGLT2 inhibitors (SGLT2is) are hypoglycemic drugs that target SGLT2 producing glycosuria. Consequently, SGLT2is decrease blood glucose in an insulin-independent manner, improving insulin resistance in diabetes [90,91]. To date, different SGLT2is, such as canagliflozin, dapagliflozin, ertugluflozin, and empagliflozin, have been approved to treat T2DM and DKD. Large clinical trials have demonstrated that these drugs (mainly canagliflozin, dapagliflozin, and empagliflozin) delay the progression of DKD on top of the standard of care with renin-angiotensin system (RAS) blockers [92–95]. The protective effect of the SGLT2is has been widely attributed to normalization of glycemia, natriuresis, body weight reduction, and decrease of intraglomerular pressure [63,96,97]. However, the use of SGLT2is has been associated with reduction of oxidative stress and inflammation as well (Table 2). In type 2 diabetic patients, twelve-week treatment with dapagliflozin significantly reduced the urinary excretion of 8-oxo-7,8-dihydro-2′-deoxyguanosine (8-oxo-dG), a marker of DNA oxidation [98]. In addition, empagliflozin treatment increased 2,2¢-azino-bis-(3-ethylbenzthiazoline-6-sulphonic acid) (ABTS) radical scavenging capacity (a measure of antioxidant capacity) in diabetic patients, suggesting a beneficial equilibrium between oxidant and antioxidant mechanisms despite empagliflozin also increasing serum levels of the thiobarbituric acid reactive substances (TBARS) and malondialdehyde (MDA). The authors suggest that the increased levels of TBARS and MDA (both indicators of lipid peroxidation) can be explained by the ketogenesis induced by SGLT2is that may contribute to lipid peroxidation [99]. Contrarily, Nabrdalik-Leśniak et al. have described that therapy of at least one month with either canagliflozin or empagliflozin decreased total antioxidant capacity (estimated by ABTS urine levels) in diabetic patients as compared to that of diabetic patients not treated with SGLT2is and healthy controls. However, the urinary activity of the antioxidant enzymes SOD and MnSOD was increased in diabetic patients treated with SGLT2is, suggesting that these drugs can also promote the activation of antioxidant mechanisms [100]. In this line, Iannantuoni et al. have described that 24-week long treatment with empagliflozin reduces superoxide production, increases glutathione content (an antioxidant), and enhances the expression of glutathione s-reductase and catalase in leukocytes of diabetic patients. In this study, the authors found that empagliflozin increased the serum levels of the anti-inflammatory interleukin 10 (IL-10) and decreased C reactive protein and myeloperoxidase serum levels suggesting, an antioxidant and anti-inflammatory role of SGLT2is [101]. Similarly, CD34+ endothelial progenitors (markers of endothelial function) isolated from peripheral blood of diabetic patients treated with canagliflozin showed increased SOD2, catalase, and glutathione peroxidase gene expression, which suggests that SGLT2is induce a beneficial antioxidant profile on these cells [102].

Table 2. Summary of the antioxidant effects of SGLT2is observed in clinical and preclinical studies.

Publication	Study Design	SGLT2i Tested	Relevant Antioxidant Effects of the SGLT2i
van Bommel et al., 2020 [98]	**Clinical Trial:** Randomized, double-blind RED trial with 44 T2DM patients on metformin monotherapy (hemoglobin A1c 7.4%, mGFR 113 mL/min) treated for 12 weeks.	Dapagliflozin	Dapagliflozin reduced the urinary excretion of 8-oxo-7,8-dihydro-2′-deoxyguanosine (8-oxo-dG), a DNA oxidation marker.
Lambadiari et al., 2021 [99]	**Clinical Trial:** 160 T2DM patients with high cardiovascular risk (SCORE $\geq$ 10%). Follow-up duration was 12 months.	Empagliflozin	Empagliflozin treatment increased 2,2¢-azino-bis-(3-ethylbenzthiazoline-6-sulphonic acid) (ABTS) radical scavenging capacity, a measure of antioxidant capacity, and serum levels of the thiobarbituric acid reactive substances (TBARS) and malondialdehyde (MDA), indicators of lipid peroxidation.
Nabrdalik-Leśniak et al., 2021 [100]	**Clinical Trial:** Observational study of a total of 101 subjects, 33 healthy and 68 with T2DM, treated (37) or not (31) with an SGLT2 inhibitor for at least one month.	SGLT2 inhibitors (not specified)	SGLT2 inhibitors improve the superoxide dismutase (SOD) antioxidant defense.

Table 2. *Cont.*

Publication	Study Design	SGLT2i Tested	Relevant Antioxidant Effects of the SGLT2i
Iannantuoni et al., 2019 [101]	**Clinical Trial:** Prospective follow-up study of 15 T2DM patients who received treatment with empagliflozin for 24 weeks.	Empagliflozin	Empagliflozin reduced superoxide production in leukocytes of diabetic patients and increased glutathione content, prominently after 24 weeks of empagliflozin treatment. Leukocyte expression of glutathione s-reductase and catalase, and serum levels of IL-10 were enhanced at 24 weeks of empagliflozin treatment.
Nandula et al., 2021 [102]	**Clinical Trial:** Double-blind, randomized trial of a total of 29 T2DM subjects (HbA1c of 7–10%) taking metformin and/or insulin were treated with canagliflozin or placebo for 16 weeks.	Canagliflozin	Canagliflozin increase the levels of the antioxidants superoxide dismutase 2 (SOD2), catalase and glutathione peroxidase in CD34+ vs cells of diabetic patients.
Kamezaki et al., 2018 [103]	**Preclinical study:** 8-weeks-old male type 2 diabetes (db/db) mice (n = 5), type 1 diabetes (BALB/c mice injected with STZ) mice (n = 5) and nondiabetic heterozygote (db/m) mice (n = 5) were administrated either ipragliflozin or placebo for 8 weeks.	Ipragliflozin	Ipragliflozin reduced cortical hypoxia and NADPH oxidase 4 expression, and subsequent oxidative stress, in early diabetic nephropathy.
Sun et al., 2020 [104]	**Preclinical study:** 6-week-old C57BL/6J mice and Sestrin 2 knockout mice were fed with normal chow diet or high-fat diet (HFD) for 12 weeks and then treated with or without empagliflozin for 8 weeks.	Empagliflozin	Empagliflozin reduced mitochondrial injury, and increased Sestrin 2 levels and AMPK and endothelial nitric oxide synthase phosphorylation, but inhibited Akt and mTOR phosphorylation. Additionally, empagliflozin enhanced the Nrf2/HO-1–mediated oxidative stress response.
Rahadian et al., 2020 [105]	**Preclinical study:** Male, 8-week-old ApoE−/− mice injected with STZ were treated with canagliflozin or placebo for 12 or 8 weeks.	Canagliflozin	Canagliflozin reduced the expressions of NADPH oxidase subunits (NOX2 and p22phox) and reduced urinary excretion of 8-oxo-dG.

In diabetic experimental models, similar results have been found (Table 2). The administration of SGLT2is in rat and mice models decreases glycemia [106], and in experimental models of diabetic nephropathy, these drugs have demonstrated cardiorenal protection [103,107–110]. In a high-fat diet (HFD)-induced obesity mouse model, 8 weeks of empagliflozin administration improved cardiac dysfunction via Sestrin2-mediated AMPK-mammalian target of rapamycin (mTOR) pathways by maintaining the redox stability [104]. Similarly, canagliflozin administration for 12 weeks improved atherosclerosis lesions and endothelial dysfunction in streptozotocin-induced diabetic apolipoprotein E-deficient (ApoE−/−) mice. These beneficial effects could be in part ascribed to the reduction of oxidative stress, as canagliflozin decreased the expression of NOX2 and p22phox (both NADPH oxidase subunits) and the urinary excretion of 8-OHdG [105]. Regarding the kidney, Kamezaki et al. have demonstrated that ipragliflozin reduced renal NOX4 expression and oxidative stress in both the tubular epithelium and glomerular podocytes using a model of early diabetic nephropathy (db/db mice) [103]. In obese prediabetic rats, dapagliflozin suppressed renal gluconeogenesis and oxidative stress [111]. Thus, in a diabetic milieu, SGLT2is seem to promote antioxidant effects that could participate in the cardiorenal protective mechanisms of these drugs. Interestingly, blockade of SGLT2 (with SGLT2is or by silencing with siRNAs) ameliorates oxidative stress in endothelial [112,113], tubular [114–116], and mesangial cells [117] in culture. This suggests that these drugs have direct antioxidant effects on kidney cells and possibly on other organs (even off-target [113]).

The beneficial antioxidant effects of SGLT2is observed in DKD can in part be attributed to improvement of glycemia and blood pressure control, as other hypoglycemic drugs also reduce oxidative stress [98]. However, other protective mechanisms cannot be ruled out. Recent data obtained from EMPEROR-Reduced and DAPA-CKD studies have confirmed the efficacy of SGLT2is in patients with heart failure or CKD without a history of diabetes [118]. In the EMPEROR-Reduced double-blind trial, patients receiving empagliflozin had a lower risk of cardiovascular death or hospitalization and lesser GFR decline over time, regardless of the presence or absence of diabetes [119]. In addition, among CKD patients participating in the DAPA-CKD clinical trial, those treated with dapagliflozin had better stabilization of GFR and a lower risk of progression to end-stage renal disease or death, again, independent of the presence of diabetes [120]. Although the renal antioxidant effect of the SGLT2is in nondiabetic human CKD needs to be assessed in future research, studies in nondiabetic experimental models of kidney damage have already shown that these drugs protect the kidney by decreasing oxidative stress [121,122]. Furthermore, SGLT2is also have antioxidant properties in organs other than the kidney. Olgar et al. demonstrated that the increase of SGLT2 activity in cardiomyocytes of old Wistar male rats (24-month age) triggers ROS accumulation, alters mitochondrial dynamics, and promotes the loss of mitochondrial membrane potential. Interestingly, these cardiac events were mitigated by dapagliflozin treatment [123].

It has also been demonstrated that targeting SGLT2 with canagliflozin reduces cancer cell proliferation by inhibiting mitochondrial complex I-mediated respiration [124]. In the same line, dapagliflozin and canagliflozin produced cell growth arrest in breast cancer cells both in vitro and in vivo via the AMPK-mTOR pathway. The activation of the AMPK-mTOR pathway basically inhibits energy-consuming processes such as the mitochondrial phosphorylation and increases catabolic processes to restore cellular energy homeostasis [125]. Thus, as SGLT2is have antioxidant effects in nondiabetic CKD and other pathologies not related to the kidney, it is possible that the antioxidant properties of SGLT2is observed in DKD patients happen due to a combination of several factors: (1) glycemia and blood pressure control, (2) reduction of glucose concentration in the target cells, and (3) activation of pathways that improve glucose utilization. Further research is needed to ascertain the exact mechanism by which SGLT2is decrease ROS, but it is a fact that these drugs promote the restoration of the oxidant/antioxidant balance in the diabetic kidney as well as in other scenarios.

5. Conclusions

In conclusion, oxidative stress has an important role in DKD onset. High levels of glucose together with increased ROS production overactivate the SGLT2 transporter in tubular cells, which, in turn, exacerbates oxidative stress. The use of SGLT2is has demonstrated clear cardiovascular benefits which may be in part ascribed by a beneficial balance between oxidant and antioxidant pathways.

Author Contributions: Writing—original draft preparation, C.L.-C., M.M.-V.d.B. and A.V.; writing—review and editing, C.J.-C. and M.J.S.; figures, C.L.-C. and M.M.-V.d.B. All authors have read and agreed to the published version of the manuscript.

Funding: The authors are current recipients of grants from FONDO DE INVESTIGACIÓN SANITARIA-FEDER, ISCIII (PI17/00257 and RICORS RD21/0005/0016), and Fundació la Marató de TV3 (421/C/2020, 759/U/2020 and 215/C/2021).

Conflicts of Interest: A.V. reports personal fees from MUNDIPHARMA, and nonfinancial support from MUNDIPHARMA, SANOFI, and NOVONORDISK outside this. C.J.C. declares travel support and a research grant from TRAVERE THERAPEUTICS outside this work. M.J.S. reports grants from BOEHRINGER, personal fees from NOVONORDISK, JANSSEN, BOHERINGER ASTRAZENECA, FRESENIUS, MUNDIPHARMA, PFIZER, ICU, GE Healthcare BAYER, TRAVERE THERAPEUTICS, and VIFOR, and nonfinancial support from ELI LILLY and ESTEVE, outside this work.

References

1. Sies, H. Oxidative stress: A concept in redox biology and medicine. *Redox Biol.* **2015**, *4*, 180–183. [CrossRef]
2. Davies, M.J. The oxidative environment and protein damage. *Biochim. Biophys. Acta -Proteins Proteomics* **2005**, *1703*, 93–109. [CrossRef]
3. Lander, H.M. An essential role for free radicals and derived species in signal transduction. *FASEB J.* **1997**, *11*, 118–124. [CrossRef]
4. Mittal, M.; Siddiqui, M.R.; Tran, K.; Reddy, S.P.; Malik, A.B. Reactive oxygen species in inflammation and tissue injury. *Antioxid. Redox Signal.* **2014**, *20*, 1126–1167. [CrossRef] [PubMed]
5. Houten, S.M.; Wanders, R.J.A. A general introduction to the biochemistry of mitochondrial fatty acid β-oxidation. *J. Inherit. Metab. Dis.* **2010**, *33*, 469–477. [CrossRef] [PubMed]
6. Stahl, A. A current review of fatty acid transport proteins (SLC27). *Pflug. Arch. Eur. J. Physiol.* **2004**, *447*, 722–727. [CrossRef] [PubMed]
7. Romano, A.H.; Conway, T. Evolution of carbohydrate metabolic pathways. *Res. Microbiol.* **1996**, *147*, 448–455. [CrossRef]
8. Napolitano, G.; Fasciolo, G.; Venditti, P. Mitochondrial Management of Reactive Oxygen Species. *Antioxidants* **2021**, *10*, 1824. [CrossRef] [PubMed]
9. Schägger, H.; De Coo, R.; Bauer, M.F.; Hofmann, S.; Godino, C.; Brandt, U. Significance of respirasomes for the assembly/stability of human respiratory chain complex I. *J. Biol. Chem.* **2004**, *279*, 36349–36353. [CrossRef]
10. Peoples, J.N.; Saraf, A.; Ghazal, N.; Pham, T.T.; Kwong, J.Q. Mitochondrial dysfunction and oxidative stress in heart disease. *Exp. Mol. Med.* **2019**, *51*, 1–13. [CrossRef]
11. Boveris, A.; Chance, B. The mitochondrial generation of hydrogen peroxide. General properties and effect of hyperbaric oxygen. *Biochem. J.* **1973**, *134*, 707–716. [CrossRef] [PubMed]
12. Starkov, A.A.; Fiskum, G.; Chinopoulos, C.; Lorenzo, B.J.; Browne, S.E.; Patel, M.S.; Beal, M.F. Mitochondrial α-ketoglutarate dehydrogenase complex generates reactive oxygen species. *J. Neurosci.* **2004**, *24*, 7779–7788. [CrossRef]
13. Pellegrini, M.; Pacini, S.; Baldari, C.T. p66SHC: The apoptotic side of Sch proteins. *Apoptosis* **2005**, *10*, 13–18. [CrossRef] [PubMed]
14. Galimov, E.R. The Role of p66shc in Oxidative Stress and Apoptosis. *Acta Nat.* **2010**, *2*, 44–51. [CrossRef]
15. Giorgio, M.; Migliaccio, E.; Orsini, F.; Paolucci, D.; Moroni, M.; Contursi, C.; Pelliccia, G.; Luzi, L.; Minucci, S.; Marcaccio, M.; et al. Electron transfer between cytochrome c and p66Shc generates reactive oxygen species that trigger mitochondrial apoptosis. *Cell* **2005**, *122*, 221–233. [CrossRef] [PubMed]
16. Lambeth, J.D. NOX enzymes and the biology of reactive oxygen. *Nat. Rev. Immunol.* **2004**, *4*, 181–189. [CrossRef] [PubMed]
17. Taylor, J.P.; Tse, H.M. The role of NADPH oxidases in infectious and inflammatory diseases. *Redox Biol.* **2020**, *48*, 102159. [CrossRef]
18. Borutaite, V.; Budriunaite, A.; Brown, G.C. Reversal of nitric oxide-, peroxynitrite-and S-nitrosothiol-induced inhibition of mitochondrial respiration or complex I activity by light and thiols. *Biochim. Biophys. Acta (BBA)-Bioenerg.* **2000**, *1459*, 405–412. [CrossRef]
19. Patel, R.P.; McAndrew, J.; Sellak, H.; White, C.R.; Jo, H.; Freeman, B.A.; Darley-Usmar, V.M. Biological aspects of reactive nitrogen species. *Biochim. Biophys. Acta -Bioenerg.* **1999**, *1411*, 385–400. [CrossRef]
20. Wu, G.; Morris, S.M. Arginine metabolism: Nitric oxide and beyond. *Biochem. J.* **1998**, *336*, 1–17. [CrossRef]
21. Paolocci, N.; Ekelund, U.E.G.; Isoda, T.; Ozaki, M.; Vandegaer, K.; Georgakopoulos, D.; Harrison, R.W.; Kass, D.A.; Hare, J.M. cGMP-independent inotropic effects of nitric oxide and peroxynitrite donors: Potential role for nitrosylation. *Am. J. Physiol. -Heart Circ. Physiol.* **2000**, *279*, H1982–H1988. [CrossRef]
22. Di Meo, S.; Reed, T.T.; Venditti, P.; Victor, V.M. Role of ROS and RNS Sources in Physiological and Pathological Conditions. *Oxid. Med. Cell. Longev.* **2016**, *2016*, 1245049. [CrossRef]

23. Poderoso, J.J.; Peralta, J.G.; Lisdero, C.L.; Carreras, M.C.; Radisic, M.; Schöpfer, F.; Scho¨pfer, S.; Cadenas, E.; Boveris, A.; Schö, F.-C. Nitric oxide regulates oxygen uptake and hydrogen peroxide release by the isolated beating rat heart. *Am. J. Physiol. -Cell Physiol.* **1998**, *274*, C112–C119. [CrossRef]
24. Simon, D.I.; Mullinst, M.E.; Jiat, L.; Gaston, B.; Singeli, D.J.; Stamlerti, J.S.; Reilly, T. Polynitrosylated proteins: Characterization, bioactivity, and functional consequences. *Proc. Natl. Acad. Sci. USA* **1996**, *93*, 4736–4741. [CrossRef]
25. Van Houten, B.; Woshner, V.; Santos, J.H. Role of mitochondrial DNA in toxic responses to oxidative stress. *DNA Repair* **2006**, *5*, 145–152. [CrossRef]
26. Verkaart, S.; Koopman, W.J.H.; van Emst-de Vries, S.E.; Nijtmans, L.G.J.; van den Heuvel, L.W.P.J.; Smeitink, J.A.M.; Willems, P.H.G.M. Superoxide production is inversely related to complex I activity in inherited complex I deficiency. *Biochim. Biophys. Acta -Mol. Basis Dis.* **2007**, *1772*, 373–381. [CrossRef] [PubMed]
27. Castro, M.d.R.; Suarez, E.; Kraiselburd, E.; Isidro, A.; Paz, J.; Ferder, L.; Ayala-Torres, S. Aging increases mitochondrial DNA damage and oxidative stress in liver of rhesus monkeys. *Exp. Gerontol.* **2012**, *47*, 29–37. [CrossRef]
28. Voets, A.M.; Huigsloot, M.; Lindsey, P.J.; Leenders, A.M.; Koopman, W.J.H.; Willems, P.H.G.M.; Rodenburg, R.J.; Smeitink, J.A.M.; Smeets, H.J.M. Transcriptional changes in OXPHOS complex I deficiency are related to anti-oxidant pathways and could explain the disturbed calcium homeostasis. *Biochim. Biophys. Acta -Mol. Basis Dis.* **2012**, *1822*, 1161–1168. [CrossRef] [PubMed]
29. Sarsour, E.H.; Kumar, M.G.; Chaudhuri, L.; Kalen, A.L.; Goswami, P.C. Redox Control of the Cell Cycle in Health and Disease. *Antioxid. Redox Signal.* **2009**, *11*, 2985–3011. [CrossRef] [PubMed]
30. Sauer, H.; Wartenberg, M.; Hescheler, J. Reactive Oxygen Species as Intracellular Messengers During Cell Growth and Differentiation. *Cell. Physiol. Biochem.* **2001**, *11*, 173–186. [CrossRef] [PubMed]
31. Fukai, T.; Ushio-Fukai, M. Superoxide dismutases: Role in redox signaling, vascular function, and diseases. *Antioxid. Redox Signal.* **2011**, *15*, 1583–1606. [CrossRef] [PubMed]
32. Kawamata, H.; Manfredi, G. Import, maturation, and function of SOD1 and its copper chaperone CCS in the mitochondrial intermembrane space. *Antioxid. Redox Signal.* **2010**, *13*, 1375–1384. [CrossRef]
33. Suzuki, Y.; Ali, M.; Fischer, M.; Riemer, J. Human copper chaperone for superoxide dismutase 1 mediates its own oxidation-dependent import into mitochondria. *Nat. Commun.* **2013**, *4*, 2430. [CrossRef]
34. Gaetani, G.F.; Galiano, S.; Canepa, L.; Ferraris, A.M.; Kirkman, H.N. Catalase and Glutathione Peroxidase Are Equally Active in Detoxification of Hydrogen Peroxide in Human Erythrocytes. *Blood* **1989**, *73*, 334–339. [CrossRef]
35. Radi, R.; Turrens, J.F.; Chang, L.Y.; Bush, K.M.; Crapo, J.D.; Freeman, B.A. Detection of catalase in rat heart mitochondria. *J. Biol. Chem.* **1991**, *266*, 22028–22034. [CrossRef]
36. Irshad, M.; Chaudhuri, P.S. Oxidant-antioxidant system: Role and significance in human body. *Indian J. Exp. Biol.* **2002**, *40*, 1233–1239.
37. Lushchak, V.I. Glutathione Homeostasis and Functions: Potential Targets for Medical Interventions. *J. Amino Acids* **2012**, *2012*, 736837. [CrossRef]
38. Forman, H.J.; Davies, K.J.A.; Ursini, F. How Do Nutritional Antioxidants Really Work: Nucleophilic Tone and Para-Hormesis Versus Free Radical Scavenging in vivo Henry. *Free Radic. Biol. Med.* **2014**, *66*, 24–35. [CrossRef] [PubMed]
39. Mathers, J.; Fraser, J.A.; McMahon, M.; Saunders, R.D.C.; Hayes, J.D.; McLellan, L.I. Antioxidant and cytoprotective responses to redox stress. *Biochem. Soc. Symp.* **2004**, *71*, 157–176. [CrossRef]
40. Simon, N.; Hertig, A. Alteration of fatty acid oxidation in tubular epithelial cells: From acute kidney injury to renal fibrogenesis. *Front. Med.* **2015**, *2*, 52. [CrossRef] [PubMed]
41. Schaub, J.A.; Venkatachalam, M.A.; Weinberg, J.M. Proximal Tubular Oxidative Metabolism in Acute Kidney Injury and the Transition to CKD. *Kidney360* **2021**, *2*, 355–364. [CrossRef]
42. Khan, M.A.; Wang, X.; Giuliani, K.T.K.; Nag, P.; Grivei, A.; Ungerer, J.; Hoy, W.; Healy, H.; Gobe, G.; Kassianos, A.J. Underlying histopathology determines response to oxidative stress in cultured human primary proximal tubular epithelial cells. *Int. J. Mol. Sci.* **2020**, *21*, 560. [CrossRef]
43. Ratliff, B.B.; Abdulmahdi, W.; Pawar, R.; Wolin, M.S. Oxidant Mechanisms in Renal Injury and Disease. *Antioxid. Redox Signal.* **2016**, *25*, 119. [CrossRef] [PubMed]
44. Faivre, A.; Verissimo, T.; Auwerx, H.; Legouis, D.; de Seigneux, S. Tubular Cell Glucose Metabolism Shift During Acute and Chronic Injuries. *Front. Med.* **2021**, *8*, 742072. [CrossRef] [PubMed]
45. Zager, R.A.; Johnson, A.C.M.; Becker, K. Renal cortical pyruvate depletion during AKI. *J. Am. Soc. Nephrol.* **2014**, *25*, 998–1012. [CrossRef] [PubMed]
46. Aranda-Rivera, A.K.; Cruz-Gregorio, A.; Aparicio-Trejo, O.E.; Pedraza-Chaverri, J. Mitochondrial redox signaling and oxidative stress in kidney diseases. *Biomolecules* **2021**, *11*, 1144. [CrossRef]
47. Small, D.M.; Bennett, N.C.; Roy, S.; Gabrielli, B.G.; Johnson, D.W.; Gobe, G.C. Oxidative stress and cell senescence combine to cause maximal renal tubular epithelial cell dysfunction and loss in an in vitro model of kidney disease. *Nephron. Exp. Nephrol.* **2012**, *122*, 123–130. [CrossRef]
48. Aragno, M.; Cutrin, J.C.; Mastrocola, R.; Perrelli, M.G.; Restivo, F.; Poli, G.; Danni, O.; Boccuzzi, G. Oxidative stress and kidney dysfunction due to ischemia/reperfusion in rat: Attenuation by dehydroepiandrosterone. *Kidney Int.* **2003**, *64*, 836–843. [CrossRef]
49. Irazabal, M.V.; Torres, V.E. ROS and Redox signaling in Chronic kidney disease. *Cells* **2020**, *9*, 1342. [CrossRef]

50. Molina-Jijón, E.; Aparicio-Trejo, O.E.; Rodríguez-Muñoz, R.; León-Contreras, J.C.; del Carmen Cárdenas-Aguayo, M.; Medina-Campos, O.N.; Tapia, E.; Sánchez-Lozada, L.G.; Hernández-Pando, R.; Reyes, J.L.; et al. The nephroprotection exerted by curcumin in maleate-induced renal damage is associated with decreased mitochondrial fission and autophagy. *BioFactors* **2016**, *42*, 686–702. [CrossRef]
51. Sedeek, M.; Nasrallah, R.; Touyz, R.M.; Hébert, R.L. NADPH oxidases, reactive oxygen species, and the kidney: Friend and foe. *J. Am. Soc. Nephrol.* **2013**, *24*, 1512–1518. [CrossRef] [PubMed]
52. Thallas-Bonke, V.; Jandeleit-Dahm, K.A.M.; Cooper, M.E. Nox-4 and progressive kidney disease. *Curr. Opin. Nephrol. Hypertens.* **2015**, *24*, 74–80. [CrossRef] [PubMed]
53. You, Y.H.; Quach, T.; Saito, R.; Pham, J.; Sharma, K. Metabolomics reveals a key role for fumarate in mediating the effects of NADPH oxidase 4 in diabetic kidney disease. *J. Am. Soc. Nephrol.* **2016**, *27*, 466–481. [CrossRef] [PubMed]
54. Cowley, A.W., Jr.; Yang, C.; Zheleznova, N.N.; Staruschenko, A.; Kurth, T.; Rein, L.; Kumar, V.; Sadovnikov, K.; Dayton, A.; Hoffman, M.; et al. Evidence of the importance of nox4 in production of hypertension in dahl salt-sensitive rats. *Hypertension* **2016**, *67*, 440–450. [CrossRef]
55. Kahveci, A.S.; Barnatan, T.T.; Kahveci, A.; Adrian, A.E.; Arroyo, J.; Eirin, A.; Harris, P.C.; Lerman, A.; Lerman, L.O.; Torres, V.E.; et al. Oxidative stress and mitochondrial abnormalities contribute to decreased endothelial nitric oxide synthase expression and renal disease progression in early experimental polycystic kidney disease. *Int. J. Mol. Sci.* **2020**, *21*, 1994. [CrossRef] [PubMed]
56. Dikalov, S. Crosstalk between mitochondria and NADPH oxidases Sergey. *Free Radic. Biol. Med.* **2011**, *51*, 1289–1301. [CrossRef] [PubMed]
57. Wenzel, P.; Mollnau, H.; Oelze, M.; Schulz, E.; Wickramanayake, J.M.D.; Müller, J.; Schuhmacher, S.; Hortmann, M.; Baldus, S.; Gori, T.; et al. First evidence for a crosstalk between mitochondrial and NADPH oxidase-derived reactive oxygen species in nitroglycerin-triggered vascular dysfunction. *Antioxid. Redox Signal.* **2008**, *10*, 1435–1447. [CrossRef]
58. Khodo, S.N.; Dizin, E.; Sossauer, G.; Szanto, I.; Martin, P.Y.; Feraille, E.; Krause, K.H.; De Seigneux, S. NADPH-oxidase 4 protects against kidney fibrosis during chronic renal injury. *J. Am. Soc. Nephrol.* **2012**, *23*, 1967–1976. [CrossRef]
59. Brewer, A.C.; Murray, T.V.A.; Arno, M.; Zhang, M.; Anilkumar, N.P.; Mann, G.E.; Shah, A.M. Nox4 regulates Nrf2 and glutathione redox in cardiomyocytes in vivo. *Free Radic. Biol. Med.* **2011**, *51*, 205–215. [CrossRef]
60. Yang, Y.; Liu, H.; Liu, F.; Dong, Z. Mitochondrial Dysregulation and Protection in Cisplatin Nephrotoxicity. *Arch Toxicol.* **2014**, *88*, 1249–1256. [CrossRef]
61. Brooks, C.; Wei, Q.; Cho, S.G.; Dong, Z. Regulation of mitochondrial dynamics in acute kidney injury in cell culture and rodent models. *J. Clin. Investig.* **2009**, *119*, 1275–1285. [CrossRef]
62. Uchida, S.; Endou, H. Substrate specificity to maintain cellular ATP along the mouse nephron. *Am. J. Physiol.* **1988**, *255*, F977–F983. [CrossRef]
63. García-Carro, C.; Vergara, A.; Agraz, I.; Jacobs-Cachá, C.; Espinel, E.; Seron, D.; Soler, M.J. The New Era for Reno-Cardiovascular Treatment in Type 2 Diabetes. *J. Clin. Med.* **2019**, *8*, 864. [CrossRef] [PubMed]
64. Brady, J.A.; Hallow, K.M. Model-Based Evaluation of Proximal Sodium Reabsorption Through SGLT2 in Health and Diabetes and the Effect of Inhibition With Canagliflozin. *J. Clin. Pharmacol.* **2018**, *58*, 377–385. [CrossRef]
65. Onishi, A.; Fu, Y.; Patel, R.; Darshi, M.; Crespo-Masip, M.; Huang, W.; Song, P.; Freeman, B.; Kim, Y.C.; Soleimani, M.; et al. A role for tubular Na+/H+ exchanger NHE3 in the natriuretic effect of the SGLT2 inhibitor empagliflozin. *Am. J. Physiol. -Ren. Physiol.* **2020**, *319*, F712–F728. [CrossRef] [PubMed]
66. Ghezzi, C.; Loo, D.D.F.; Wright, E.M. Physiology of renal glucose handling via SGLT1, SGLT2 and GLUT2. *Diabetologia* **2018**, *61*, 2087–2097. [CrossRef]
67. Gnudi, L.; Coward, R.J.M.; Long, D.A. Diabetic Nephropathy: Perspective on Novel Molecular Mechanisms. *Trends Endocrinol. Metab.* **2016**, *27*, 820–830. [CrossRef]
68. Lastra, G.; Syed, S.; Kurukulasuriya, L.R.; Manrique, C.; Sowers, J.R. Type 2 diabetes mellitus and hypertension: An update. *Endocrinol. Metab. Clin. N. Am.* **2014**, *43*, 103–122. [CrossRef]
69. Packer, M. Activation and Inhibition of Sodium-Hydrogen Exchanger Is a Mechanism That Links the Pathophysiology and Treatment of Diabetes Mellitus With That of Heart Failure. *Circulation* **2017**, *136*, 1548–1559. [CrossRef] [PubMed]
70. Kogot-Levin, A.; Hinden, L.; Riahi, Y.; Israeli, T.; Tirosh, B.; Cerasi, E.; Mizrachi, E.B.; Tam, J.; Mosenzon, O.; Leibowitz, G. Proximal Tubule mTORC1 Is a Central Player in the Pathophysiology of Diabetic Nephropathy and Its Correction by SGLT2 Inhibitors. *Cell Rep.* **2020**, *32*, 107954. [CrossRef] [PubMed]
71. Lee, Y.J.; Heo, J.S.; Suh, H.N.; Lee, M.Y.; Han, H.J. Interleukin-6 stimulates α-MG uptake in renal proximal tubule cells: Involvement of STAT3, PI3K/Akt, MAPKs, and NF-κB. *Am. J. Physiol. -Ren. Physiol.* **2007**, *293*, 1036–1046. [CrossRef] [PubMed]
72. Tervaert, T.W.C.; Mooyaart, A.L.; Amann, K.; Cohen, A.H.; TerenceCook, H.; Drachenberg, C.B.; Ferrario, F.; Fogo, A.B.; Haas, M.; De Heer, E.; et al. Pathologic classification of diabetic nephropathy. *J. Am. Soc. Nephrol.* **2010**, *21*, 556–563. [CrossRef]
73. Coughlan, M.T.; Nguyen, T.V.; Penfold, S.A.; Higgins, G.C.; Thallas-Bonke, V.; Tan, S.M.; Van Bergen, N.J.; Sourris, K.C.; Harcourt, B.E.; Thorburn, D.R.; et al. Mapping time-course mitochondrial adaptations in the kidney in experimental diabetes. *Clin. Sci.* **2016**, *130*, 711–720. [CrossRef] [PubMed]
74. Lee, W.C.; Chau, Y.Y.; Ng, H.Y.; Chen, C.H.; Wang, P.W.; Liou, C.W.; Lin, T.K.; Chen, J.B. Empagliflozin Protects HK-2 Cells from High Glucose-Mediated Injuries via a Mitochondrial Mechanism. *Cells* **2019**, *8*, 1085. [CrossRef]

75. Takagi, S.; Li, J.; Takagaki, Y.; Kitada, M.; Nitta, K.; Takasu, T.; Kanasaki, K.; Koya, D. Ipragliflozin improves mitochondrial abnormalities in renal tubules induced by a high-fat diet. *J. Diabetes Investig.* **2018**, *9*, 1025–1032. [CrossRef]
76. Szabo, C. Role of nitrosative stress in the pathogenesis of diabetic vascular dysfunction. *Br. J. Pharmacol.* **2009**, *156*, 713–727. [CrossRef] [PubMed]
77. Coughlan, M.T.; Sharma, K. Challenging the dogma of mitochondrial reactive oxygen species overproduction in diabetic kidney disease. *Kidney Int.* **2016**, *90*, 272–279. [CrossRef]
78. Cui, F.Q.; Tang, L.; Gao, Y.B.; Wang, Y.F.; Meng, Y.; Shen, C.; Shen, Z.L.; Liu, Z.Q.; Zhao, W.J.; Liu, W.J. Effect of Baoshenfang Formula on Podocyte Injury via Inhibiting the NOX-4/ROS/p38 Pathway in Diabetic Nephropathy. *J. Diabetes Res.* **2019**, *2019*, 2981705. [CrossRef] [PubMed]
79. Ilatovskaya, D.V.; Blass, G.; Palygin, O.; Levchenko, V.; Pavlov, T.S.; Grzybowski, M.N.; Winsor, K.; Shuyskiy, L.S.; Geurts, A.M.; Cowley, A.W.; et al. A NOX4/TRPC6 Pathway in Podocyte Calcium Regulation and Renal Damage in Diabetic Kidney Disease. *J. Am. Soc. Nephrol.* **2018**, *29*, 1917–1927. [CrossRef]
80. Liu, J.J.; Liu, S.; Gurung, R.L.; Ching, J.; Kovalik, J.P.; Tan, T.Y.; Lim, S.C. Urine Tricarboxylic Acid Cycle Metabolites Predict Progressive Chronic Kidney Disease in Type 2 Diabetes. *J. Clin. Endocrinol. Metab.* **2018**, *103*, 4357–4364. [CrossRef]
81. Jiang, H.; Shao, X.; Jia, S.; Qu, L.; Weng, C.; Shen, X.; Wang, Y.; Huang, H.; Wang, C.; Wang, Y.; et al. The mitochondria-targeted metabolic tubular injury in diabetic kidney disease. *Cell. Physiol. Biochem.* **2019**, *52*, 156–171. [CrossRef]
82. Cao, H.; Wu, J.; Luo, J.; Chen, X.; Yang, J.; Fang, L. Urinary mitochondrial DNA: A potential early biomarker of diabetic nephropathy. *Diabetes. Metab. Res. Rev.* **2019**, *35*. [CrossRef]
83. Al-Kafaji, G.; Aljadaan, A.; Kamal, A.; Bakhiet, M. Peripheral blood mitochondrial DNA copy number as a novel potential biomarker for diabetic nephropathy in type 2 diabetes patients. *Exp. Ther. Med.* **2018**, *16*, 1483–1492. [CrossRef] [PubMed]
84. Huang, Y.; Chi, J.; Wei, F.; Zhou, Y.; Cao, Y.; Wang, Y. Mitochondrial DNA: A New Predictor of Diabetic Kidney Disease. *Int. J. Endocrinol.* **2020**, *2020*, 3650937. [CrossRef] [PubMed]
85. Zhang, Q.; Raoof, M.; Chen, Y.; Sumi, Y.; Sursal, T.; Junger, W.; Brohi, K.; Itagaki, K.; Hauser, C.J. Circulating mitochondrial DAMPs cause inflammatory responses to injury. *Nature* **2010**, *464*, 104–107. [CrossRef]
86. Jin, L.; Yu, B.; Armando, I.; Han, F. Mitochondrial DNA-Mediated Inflammation in Acute Kidney Injury and Chronic Kidney Disease. *Oxid. Med. Cell. Longev.* **2021**, *2021*, 9985603. [CrossRef]
87. Chow, F.; Ozols, E.; Nikolic-Paterson, D.J.; Atkins, R.C.; Tesch, G.H. Macrophages in mouse type 2 diabetic nephropathy: Correlation with diabetic state and progressive renal injury. *Kidney Int.* **2004**, *65*, 116–128. [CrossRef] [PubMed]
88. Nguyen, D.; Ping, F.; Mu, W.; Hill, P.; Atkins, R.C.; Chadban, S.J. Macrophage accumulation in human progressive diabetic nephropathy. *Nephrology* **2006**, *11*, 226–231. [CrossRef]
89. Elmarakby, A.A.; Sullivan, J.C. Relationship between oxidative stress and inflammatory cytokines in diabetic nephropathy. *Cardiovasc. Ther.* **2012**, *30*, 49–59. [CrossRef]
90. Isaji, M. SGLT2 inhibitors: Molecular design and potential differences in effect. *Kidney Int.* **2011**, *79*, S14–S19. [CrossRef]
91. Bashier, A.; Khalifa, A.A.; Rashid, F.; Abdelgadir, E.I.; Al Qaysi, A.A.; Ali, R.; Eltinay, A.; Nafach, J.; Alsayyah, F.; Alawadi, F. Efficacy and Safety of SGLT2 Inhibitors in Reducing Glycated Hemoglobin and Weight in Emirati Patients With Type 2 Diabetes. *J. Clin. Med. Res.* **2017**, *9*, 499–507. [CrossRef] [PubMed]
92. Wanner, C.; Heerspink, H.J.L.; Zinman, B.; Inzucchi, S.E.; Koitka-Weber, A.; Mattheus, M.; Hantel, S.; Woerle, H.J.; Broedl, U.C.; Von Eynatten, M.; et al. Empagliflozin and kidney function decline in patients with type 2 diabetes: A slope analysis from the EMPA-REG OUTCOME trial. *J. Am. Soc. Nephrol.* **2018**, *29*, 2755–2769. [CrossRef] [PubMed]
93. Neal, B.; Perkovic, V.; Mahaffey, K.W.; de Zeeuw, D.; Fulcher, G.; Erondu, N.; Shaw, W.; Law, G.; Desai, M.; Matthews, D.R. Canagliflozin and Cardiovascular and Renal Events in Type 2 Diabetes. *N. Engl. J. Med.* **2017**, *377*, 644–657. [CrossRef]
94. Wiviott, S.D.; Raz, I.; Bonaca, M.P.; Mosenzon, O.; Kato, E.T.; Cahn, A.; Silverman, M.G.; Zelniker, T.A.; Kuder, J.F.; Murphy, S.A.; et al. Dapagliflozin and Cardiovascular Outcomes in Type 2 Diabetes. *N. Engl. J. Med.* **2018**, NEJMoa1812389. [CrossRef]
95. Perkovic, V.; Jardine, M.J.; Neal, B.; Bompoint, S.; Heerspink, H.J.L.; Charytan, D.M.; Edwards, R.; Agarwal, R.; Bakris, G.; Bull, S.; et al. Canagliflozin and Renal Outcomes in Type 2 Diabetes and Nephropathy. *N. Engl. J. Med.* **2019**, *380*, 2295–2306. [CrossRef]
96. Vergara, A.; Jacobs-Cachá, C.; Soler, M.J. Sodium-glucose cotransporter inhibitors: Beyond glycaemic control. *Clin. Kidney J.* **2019**, *12*, 322–325. [CrossRef] [PubMed]
97. Górriz, J.L.; Navarro-González, J.F.; Ortiz, A.; Vergara, A.; Nuñez, J.; Jacobs-Cachá, C.; Martínez-Castelao, A.; Soler, M.J. Sodium-glucose cotransporter 2 inhibition: Towards an indication to treat diabetic kidney disease. *Nephrol. Dial. Transplant.* **2020**, *35*, i13–i23. [CrossRef]
98. van Bommel, E.J.M.; Muskiet, M.H.A.; van Baar, M.J.B.; Tonneijck, L.; Smits, M.M.; Emanuel, A.L.; Bozovic, A.; Danser, A.H.J.; Geurts, F.; Hoorn, E.J.; et al. The renal hemodynamic effects of the SGLT2 inhibitor dapagliflozin are caused by post-glomerular vasodilatation rather than pre-glomerular vasoconstriction in metformin-treated patients with type 2 diabetes in the randomized, double-blind RED trial. *Kidney Int.* **2020**, *97*, 202–212. [CrossRef]
99. Lambadiari, V.; Thymis, J.; Kouretas, D.; Skaperda, Z.; Tekos, F.; Kousathana, F.; Kountouri, A.; Balampanis, K.; Parissis, J.; Andreadou, I.; et al. Effects of a 12-Month Treatment with Glucagon-like Peptide-1 Receptor Agonists, Sodium-Glucose Cotransporter-2 Inhibitors, and Their Combination on Oxidant and Antioxidant Biomarkers in Patients with Type 2 Diabetes. *Antioxidants* **2021**, *10*, 1379. [CrossRef]

100. Nabrdalik-Leśniak, D.; Nabrdalik, K.; Sedlaczek, K.; Główczyński, P.; Kwiendacz, H.; Sawczyn, T.; Hajzler, W.; Drożdż, K.; Hendel, M.; Irlik, K.; et al. Influence of SGLT2 Inhibitor Treatment on Urine Antioxidant Status in Type 2 Diabetic Patients: A Pilot Study. *Oxid. Med. Cell. Longev.* **2021**, *2021*, 5593589. [CrossRef]
101. Iannantuoni, F.; de Marañon, A.M.; Diaz-Morales, N.; Falcon, R.; Bañuls, C.; Abad-Jimenez, Z.; Victor, V.M.; Hernandez-Mijares, A.; Rovira-Llopis, S. The SGLT2 Inhibitor Empagliflozin Ameliorates the Inflammatory Profile in Type 2 Diabetic Patients and Promotes an Antioxidant Response in Leukocytes. *J. Clin. Med.* **2019**, *8*, 1814. [CrossRef]
102. Nandula, S.R.; Kundu, N.; Awal, H.B.; Brichacek, B.; Fakhri, M.; Aimalla, N.; Elzarki, A.; Amdur, R.L.; Sen, S. Role of Canagliflozin on function of CD34+ve endothelial progenitor cells (EPC) in patients with type 2 diabetes. *Cardiovasc. Diabetol.* **2021**, *20*, 44. [CrossRef]
103. Kamezaki, M.; Kusaba, T.; Komaki, K.; Fushimura, Y.; Watanabe, N.; Ikeda, K.; Kitani, T.; Yamashita, N.; Uehara, M.; Kirita, Y.; et al. Comprehensive renoprotective effects of ipragliflozin on early diabetic nephropathy in mice. *Sci. Rep.* **2018**, *8*, 4029. [CrossRef]
104. Sun, X.; Han, F.; Lu, Q.; Li, X.; Ren, D.; Zhang, J.; Han, Y.; Xiang, Y.K.; Li, J. Empagliflozin Ameliorates Obesity-Related Cardiac Dysfunction by Regulating Sestrin2-Mediated AMPK-mTOR Signaling and Redox Homeostasis in High-Fat Diet–Induced Obese Mice. *Diabetes* **2020**, *69*, 1292–1305. [CrossRef]
105. Rahadian, A.; Fukuda, D.; Salim, H.M.; Yagi, S.; Kusunose, K.; Yamada, H.; Soeki, T.; Sata, M. Canagliflozin Prevents Diabetes-Induced Vascular Dysfunction in ApoE-Deficient Mice. *J. Atheroscler. Thromb.* **2020**, *27*, 1141–1151. [CrossRef]
106. Shin, S.J.; Chung, S.; Kim, S.J.; Lee, E.-M.; Yoo, Y.-H.; Kim, J.-W.; Ahn, Y.-B.; Kim, E.-S.; Moon, S.-D.; Kim, M.-J.; et al. Effect of Sodium-Glucose Co-Transporter 2 Inhibitor, Dapagliflozin, on Renal Renin-Angiotensin System in an Animal Model of Type 2 Diabetes. *PLoS ONE* **2016**, *11*, e0165703. [CrossRef]
107. Ye, Y.; Bajaj, M.; Yang, H.-C.; Perez-Polo, J.R.; Birnbaum, Y. SGLT-2 Inhibition with Dapagliflozin Reduces the Activation of the Nlrp3/ASC Inflammasome and Attenuates the Development of Diabetic Cardiomyopathy in Mice with Type 2 Diabetes. Further Augmentation of the Effects with Saxagliptin, a DPP4 Inhibitor. *Cardiovasc. Drugs Ther.* **2017**, *31*, 119–132. [CrossRef]
108. Liang, Y.; Arakawa, K.; Ueta, K.; Matsushita, Y.; Kuriyama, C.; Martin, T.; Du, F.; Liu, Y.; Xu, J.; Conway, B.; et al. Effect of Canagliflozin on Renal Threshold for Glucose, Glycemia, and Body Weight in Normal and Diabetic Animal Models. *PLoS ONE* **2012**, *7*, e30555. [CrossRef] [PubMed]
109. Tahara, A.; Takasu, T. Prevention of progression of diabetic nephropathy by the SGLT2 inhibitor ipragliflozin in uninephrectomized type 2 diabetic mice. *Eur. J. Pharmacol.* **2018**, *830*, 68–75. [CrossRef] [PubMed]
110. Leng, W.; Ouyang, X.; Lei, X.; Wu, M.; Chen, L.; Wu, Q.; Deng, W.; Liang, Z. The SGLT-2 Inhibitor Dapagliflozin Has a Therapeutic Effect on Atherosclerosis in Diabetic ApoE−/− Mice. *Mediators Inflamm.* **2016**, *2016*, 6305735. [CrossRef] [PubMed]
111. Swe, M.T.; Thongnak, L.; Jaikumkao, K.; Pongchaidecha, A.; Chatsudthipong, V.; Lungkaphin, A. Dapagliflozin attenuates renal gluconeogenic enzyme expression in obese rats. *J. Endocrinol.* **2020**, *245*, 193–205. [CrossRef]
112. Li, X.; Römer, G.; Kerindongo, R.P.; Hermanides, J.; Albrecht, M.; Hollmann, M.W.; Zuurbier, C.J.; Preckel, B.; Weber, N.C. Sodium glucose co-transporter 2 inhibitors ameliorate endothelium barrier dysfunction induced by cyclic stretch through inhibition of reactive oxygen species. *Int. J. Mol. Sci.* **2021**, *22*, 6044. [CrossRef]
113. Uthman, L.; Homayr, A.; Juni, R.P.; Spin, E.L.; Kerindongo, R.; Boomsma, M.; Hollmanna Benedikt Preckel, M.W.; Koolwijk, P.; Van Hinsbergh, V.W.M.; Zuurbier, C.J.; et al. Empagliflozin and dapagliflozin reduce ROS generation and restore no bioavailability in tumor necrosis factor α-stimulated human coronary arterial endothelial cells. *Cell. Physiol. Biochem.* **2019**, *53*, 865–886. [CrossRef]
114. Zaibi, N.; Li, P.; Xu, S.Z. Protective effects of dapagliflozin against oxidative stress-induced cell injury in human proximal tubular cells. *PLoS ONE* **2021**, *16*, e0247234. [CrossRef] [PubMed]
115. Shirakawa, K.; Sano, M. Sodium-Glucose Co-Transporter 2 Inhibitors Correct Metabolic Maladaptation of Proximal Tubular Epithelial Cells in High-Glucose Conditions. *Int. J. Mol. Sci.* **2020**, *21*, 7676. [CrossRef]
116. Maeda, S.; Matsui, T.; Takeuchi, M.; Yamagishi, S.I. Sodium-glucose cotransporter 2-mediated oxidative stress augments advanced glycation end products-induced tubular cell apoptosis. *Diabetes. Metab. Res. Rev.* **2013**, *29*, 406–412. [CrossRef]
117. Maki, T.; Maeno, S.; Maeda, Y.; Yamato, M.; Sonoda, N.; Ogawa, Y.; Wakisaka, M.; Inoguchi, T. Amelioration of diabetic nephropathy by SGLT2 inhibitors independent of its glucose-lowering effect: A possible role of SGLT2 in mesangial cells. *Sci. Rep.* **2019**, *9*, 4703. [CrossRef] [PubMed]
118. Nangaku, M. More reasons to use SGLT2 inhibitors: EMPEROR-reduced and DAPA-CKD. *Kidney Int.* **2020**, *98*, 1387–1389. [CrossRef] [PubMed]
119. Packer, M.; Claggett, B.; Lefkowitz, M.P.; McMurray, J.J.V.; Rouleau, J.L.; Solomon, S.D.; Zile, M.R. Effect of neprilysin inhibition on renal function in patients with type 2 diabetes and chronic heart failure who are receiving target doses of inhibitors of the renin-angiotensin system: A secondary analysis of the PARADIGM-HF trial. *Lancet Diabetes Endocrinol.* **2018**, *6*, 547–554. [CrossRef]
120. Heerspink, H.J.L.; Stefánsson, B.V.; Correa-Rotter, R.; Chertow, G.M.; Greene, T.; Hou, F.-F.; Mann, J.F.E.; McMurray, J.J.V.; Lindberg, M.; Rossing, P.; et al. Dapagliflozin in Patients with Chronic Kidney Disease. *N. Engl. J. Med.* **2020**, *383*, 1436–1446. [CrossRef]
121. Nagoya, T.; Kamimura, K.; Goto, R.; Shinagawa-Kobayashi, Y.; Niwa, Y.; Kimura, A.; Sakai, N.; Ko, M.; Nishina, H.; Terai, S. Inhibition of sodium-glucose cotransporter 2 ameliorates renal injury in a novel medaka model of nonalcoholic steatohepatitis-related kidney disease. *FEBS Open Bio* **2019**, *9*, 2016–2024. [CrossRef]

122. Jin, J.; Jin, L.; Luo, K.; Lim, S.W.; Chung, B.H.; Yang, C.W. Effect of Empagliflozin on Tacrolimus-Induced Pancreas Islet Dysfunction and Renal Injury. *Am. J. Transplant.* **2017**, *17*, 2601–2616. [CrossRef] [PubMed]
123. Olgar, Y.; Tuncay, E.; Degirmenci, S.; Billur, D.; Dhingra, R.; Kirshenbaum, L.; Turan, B. Ageing-associated increase in SGLT2 disrupts mitochondrial/sarcoplasmic reticulum Ca 2+ homeostasis and promotes cardiac dysfunction. *J. Cell. Mol. Med.* **2020**, *24*, 8567–8578. [CrossRef]
124. Villani, L.A.; Smith, B.K.; Marcinko, K.; Ford, R.J.; Broadfield, L.A.; Green, A.E.; Houde, V.P.; Muti, P.; Tsakiridis, T.; Steinberg, G.R. The diabetes medication Canagliflozin reduces cancer cell proliferation by inhibiting mitochondrial complex-I supported respiration. *Mol. Metab.* **2016**, *5*, 1048–1056. [CrossRef] [PubMed]
125. Zhou, J.; Zhu, J.; Yu, S.J.; Ma, H.L.; Chen, J.; Ding, X.F.; Chen, G.; Liang, Y.; Zhang, Q. Sodium-glucose co-transporter-2 (SGLT-2) inhibition reduces glucose uptake to induce breast cancer cell growth arrest through AMPK/mTOR pathway. *Biomed. Pharmacother.* **2020**, *132*, 110821. [CrossRef] [PubMed]

biomolecules

MDPI

Article

Protective Effects of Purple Rice Husk against Diabetic Nephropathy by Modulating PGC-1α/SIRT3/SOD2 Signaling and Maintaining Mitochondrial Redox Equilibrium in Rats

Orawan Wongmekiat [1,*], Narissara Lailerd [2], Anongporn Kobroob [3] and Wachirasek Peerapanyasut [1,†]

1 Renal Physiology Unit, Department of Physiology, Faculty of Medicine, Chiang Mai University, Chiang Mai 50200, Thailand; wachirasek.pee@mahidol.ac.th
2 Nutrition and Exercise Unit, Department of Physiology, Faculty of Medicine, Chiang Mai University, Chiang Mai 50200, Thailand; narissara.lailerd@cmu.ac.th
3 Division of Physiology, School of Medical Science, University of Phayao, Phayao 56000, Thailand; anongporn.ko@up.ac.th
* Correspondence: orawan.wongmekiat@cmu.ac.th; Tel.: +66-53-935362
† Present address: Mahidol University, Nakhonsawan Campus, Nakhonsawan 60130, Thailand.

Abstract: Diabetic nephropathy (DN) is the primary cause of end-stage renal disease worldwide. Oxidative stress and mitochondrial dysfunction are central to its pathogenesis. Rice husk, the leftover from the milling process, is a good source of phytochemicals with antioxidant activity. This study evaluated the possible protection of purple rice husk extract (PRHE) against diabetic kidney injury. Type 2 diabetic rats were given vehicle, PRHE, metformin, and PRHE+metformin, respectively, while nondiabetic rats received vehicle. After 12 weeks, diabetic rats developed nephropathy as proven by metabolic alterations (increased blood glucose, insulin, HOMA-IR, triglycerides, cholesterol) and renal abnormalities (podocyte injury, microalbuminuria, increased serum creatinine, decreased creatinine clearance). Treatment with PRHE, metformin, or combination diminished these changes, improved mitochondrial function (decreased mitochondrial swelling, reactive oxygen species production, membrane potential changes), and reduced renal oxidative damage (decreased lipid peroxidation and increased antioxidants). Increased expression of PGC-1α, SIRT3, and SOD2 and decreased expression of Ac-SOD2 correlated with the beneficial outcomes. HPLC revealed protocatechuic acid and cyanidin-3-glucoside as the key components of PRHE. The findings indicate that PRHE effectively protects against the development of DN by retaining mitochondrial redox equilibrium via the regulation of PGC-1α-SIRT3-SOD2 signaling. This study creates an opportunity to develop this agricultural waste into a useful health product for diabetes.

Keywords: diabetes; kidney; mitochondria; *Oryza sativa*; oxidative stress; rice husk

Citation: Wongmekiat, O.; Lailerd, N.; Kobroob, A.; Peerapanyasut, W. Protective Effects of Purple Rice Husk against Diabetic Nephropathy by Modulating PGC-1α/SIRT3/SOD2 Signaling and Maintaining Mitochondrial Redox Equilibrium in Rats. *Biomolecules* **2021**, *11*, 1224. https://doi.org/10.3390/biom11081224

Academic Editors: Liang-Jun Yan and Juan Antonio Moreno Gutiérrez

Received: 23 May 2021
Accepted: 16 August 2021
Published: 17 August 2021

Publisher's Note: MDPI stays neutral with regard to jurisdictional claims in published maps and institutional affiliations.

1. Introduction

Diabetes mellitus is currently a serious public health problem that affects millions of individuals worldwide. Diabetic nephropathy (DN), also called diabetic kidney disease, is one of the most important long-term complications in terms of morbidity and mortality for individual patients with diabetes. It is also accepted as the leading cause of end-stage renal disease in many countries around the world [1]. The latest report by the International Diabetes Federation (IDF) shows that half a billion people worldwide are diabetic, and more than 80% of cases of end-stage renal disease are caused by diabetes [2]. Thailand is among the countries in Asia with a high prevalence of diabetes. Currently, diabetes in the Thai population is estimated at 4.2 million cases, or 7–8% of the population, of which about 44% are experiencing diabetic kidney disease [2,3]. Global records, including Thailand, indicate that the number of diabetic patients who require renal replacement therapy has progressively increased over the last two decades [4]. The issue not only produced a significant burden on economics but also caused psychosocial problems in terms of reduced quality of life and suffering of the patients. Importantly, data from the Thai National Health Examination Survey and the IDF atlas reveal that diabetes-related deaths are approximately 12–15% of total deaths in Thailand [2,5,6]. Finding strategies to prevent or slow down the onset and progression of kidney disease in people with diabetes remains an important and necessary task.

Rice is the seed of the grass species *Oryza sativa*. It is the agricultural commodity with the third highest worldwide production. Rice is the most widely consumed staple food in over half of the world's population, particularly in Asia and Africa [7]. Though rice can be cultivated worldwide, more than 90% is grown in Asia. Nowadays, the trend of health promotion with natural supplements is becoming increasingly popular. Rice, especially colored rice, is one of the popular choices, as it contains nutrients, vitamins, and phytochemicals. There is a variety of colored rice, including brown, red, and purple (or black) rice. Among these colored rice, purple rice has received the most considerable interest due to its anthocyanin content, which is higher by weight than that of other colored grains [8]. Studies have shown several health benefits of extracts from diverse parts of purple rice, i.e., rice grain and rice bran [9,10], and thus greatly increased the demand for this colored rice. Rice husks, a major organic source of waste of the rice milling process, are also accumulated. These wastes are often removed by burning, causing air pollution that directly affects climate change, quality of life, and human health. Therefore, turning these agricultural wastes into valuable stuffs is of interest and becomes challenging research. A study regarding rice husk, particularly from pigmented rice, indicated that rice husk is a good source of phytochemicals with antioxidant activity and suggested that it may have a great potential to turn into functional food or nutraceuticals to prevent diseases related to oxidative stress [11]. Recently, extract from purple rice husk has been shown to be a potent anticancer agent without any toxicity [12].

Mitochondria is well recognized as a major source of reactive oxygen species (ROS) production in the body, and thus, the maintenance of mitochondrial oxidative balance is essential. The peroxisome proliferator-activated receptor gamma coactivator 1-alpha-sirtuin 3-superoxide dismutase 2 (PGC-1α-SIRT3-SOD2) axis plays a significant role in the regulation of mitochondrial redox homeostasis [13,14]. PGC-1α is a nuclear-encoded transcriptional coactivator that regulates the expression of several nuclear-encoded mitochondrial proteins, including mitochondrial antioxidant and biogenesis genes [14]. SIRT3, a downstream target of PGC-1α, is a major mitochondrial deacetylase that directly deacetylates and activates numerous mitochondrial proteins, including the major mitochondrial antioxidant enzyme SOD2 [13–15]. SIRT3 is also being accepted as a global regulator playing a multifaceted role in the mitochondrial adaptive response to stress and is currently indicated as a new target for therapy aimed at improving end-organ damage and survival [14,15].

As reactive oxygen species (ROS) production, oxidative stress generation, and, particularly, mitochondrial dysfunction are central to the pathogenesis of diabetic nephropathy [1,16], it is assumed that purple rice husk extract (PRHE) may be able to prevent or reduce diabetes-induced renal deterioration. Herein, we examined the renoprotective potential of PRHE in a rat model of high-fat diet/streptozotocin-induced type 2 diabetes (T2DM) and explored whether the modulation of the PGC-1α-SIRT3-SOD2 axis contributes to protection by PRHE.

2. Materials and Methods

2.1. Plant Materials and Preparation of PRHE

Thai purple rice (*Oryza sativa* L. var. Indica) cv. Kum Doisaket was first identified by Associate Professor Dr. Dumnern Karladee, Faculty of Agriculture, Chiang Mai University, Thailand (Voucher Specimen No.: Rawiwan_001) and planted at Mae Hia Agricultural Research, Faculty of Agriculture, Chiang Mai University. After harvested, the purple rice was confirmed again by comparing it to the known specimen identity deposited at the Faculty of Pharmacy, Chiang Mai University (Herbarium No.: 023252). Then, the rice husk was separated from the rice paddy and soaked in 0.1% hydrochloric in absolute methanol for 48 h at room temperature. The procedure was repeated twice, and the entire solution was filtered through Whatman No. 1 filter paper. The supernatant was evaporated under reduced pressure and lyophilized to obtain PRHE, which was kept in tight-sealed dark containers and stored at −20 °C for further studies.

2.2. Phytochemical Analysis, Quantification of Bioactive Compounds, and Evaluation of Antioxidant Capacity of PRHE

PRHE was initially examined for its major phytochemical constituents. The total amounts of phenolic, flavonoid, and anthocyanin were measured by Folin–Ciocalteu, aluminum trichloride, and pH differential methods, respectively, according to procedures described previously [17,18]. The phenolic acids and anthocyanins in PRHE were further identified and quantified by high-performance liquid chromatography (HPLC) using conditions and external standards as previously published [6]. 2,2′-Azino-bis(3-ethylbenzothiazoline-6-sulphonic acid) (ABTS) radical cation decolorization and 2,2-diphenyl-1-picryl-hydrazyl (DPPH) radical scavenging capacity assays were performed to evaluate the antioxidant capacity of PRHE according to the standard method previously described [19].

2.3. Animals and Experimental Protocols

Male Wistar rats weighing 150–200 g (National Laboratory Animal Center, Mahidol University, Salaya, Thailand) were housed in a controlled environment (24 ± 1 °C, 55 ± 5% humidity, 12 h light/dark cycle) with free access to food and water. All procedures complied with the guidelines established by the National Research Council of Thailand and were approved by the Animal Care and Use Committee of the Faculty of Medicine, Chiang Mai University (Project Number: 41/2559).

After acclimatization, rats were assigned to receive a normal diet (ND, n = 6) or a high-fat diet (HFD, n = 24). The diet composition is shown in the Appendix A (Table A1). Diabetes was induced in the HFD-fed rats two weeks later by a single intraperitoneal injection of streptozotocin (35 mg/kg in 0.1 M sodium citrate buffer, pH 4.5) [20], while the ND-fed rats received an equal amount of sodium citrate buffer. Two weeks after injection, rats with fasting blood glucose ≥250 mg/dL without hypoinsulinemia were considered diabetes and were included in the study.

Diabetic rats were randomly divided into 4 groups (n = 6 each): diabetic control group received vehicle (DMV), diabetic positive control group treated orally with metformin 50 mg/kg/day (DMM), diabetic group treated orally with PRHE 300 mg/kg/day (DME), and diabetic group treated with a combination of metformin and PRHE (DMME). The dose of PRHE was based on our preliminary results showing its efficacy in improving renal function together with lowering blood glucose and insulin resistance appropriately (Appendix A, Table A2). All diabetic rats were maintained on HFD for a further 12 weeks, whereas those in the nondiabetic group received a normal diet and vehicle (NDV). The 24 h urine samples were collected at the end of the study using metabolic cages, and blood and kidney tissues were collected thereafter under thiopental anesthesia (60 mg/kg, i.p.). Parts of the kidneys were rapidly taken for mitochondrial and histopathological studies. The remainders were snap-frozen in liquid nitrogen and stored at −80 °C for further analysis.

2.4. Biochemical Assays

2.4.1. Determinations of Metabolic Indexes

Fasting blood glucose, cholesterol, and triglycerides were measured by enzymatic assay kits (ERBA Diagnostics Inc., Miami, FL, USA). Plasma insulin was quantified using ELISA kit (Millipore, Burlington, MA, USA). The Homeostasis Model Assessment for Insulin Resistance was calculated (HOMA-IR = fasting insulin level (ng/mL) × fasting glucose (mg/dL)/405.1) as described previously [21].

2.4.2. Determinations of Renal Functions

Serum creatinine, urine creatinine, and urine microalbumin were analyzed using the AU480 Chemistry Analyzer (Beckman Coulter, Inc., Brea, CA, USA). Glomerular filtration rate (GFR) was estimated by calculation of the creatinine clearance using a standard clearance formula (the ratio of creatinine in urine/serum and the volume of urine produced).

2.4.3. Determinations of Renal Oxidative Stress

The supernatant from kidney homogenate was used for oxidative stress assays. Malondialdehyde (MDA) was estimated using the TBARS assay kit (Cayman Chemical, Ann Arbor, MI, USA), glutathione (GSH) was assayed using the QuantiChrom™ Glutathione Assay Kit (Bioassay Systems, Hayward, CA, USA), and superoxide dismutase (SOD) and glutathione peroxidase (GPx) were determined using Calbiochem® Assay Kits (Merck Millipore, Darmstadt, Germany). All analyses were performed according to the manufacturer's instructions.

2.4.4. Determinations of Mitochondrial Functions

Kidney mitochondria were isolated by differential centrifugation, and the mitochondrial proteins obtained were used to study mitochondrial function according to the protocols previously published [22]. Mitochondrial ROS production was determined using a fluorogenic dye dichlorofluorescin diacetate (DCFDA). ROS level was assessed from the fluorescent product and expressed in arbitrary units of fluorescent intensity. An increasing intensity indicates an increase in mitochondrial ROS production. Mitochondrial membrane potential change was assessed by staining mitochondria with a lipophilic cationic fluorescence JC-1 dye. The fluorescence intensity of JC-1 aggregate (red fluorescence) and monomer (green fluorescence) was measured, and a decrease in the red/green fluorescence intensity ratio implies the loss of membrane potential. Mitochondrial swelling was determined through changes in the absorbance of the mitochondrial suspension at 540 nm, where a decrease in absorbance intensity indicates the swelling of mitochondria.

2.5. Histopathological Examinations

Kidney tissues were routinely processed for light and electron microscopic examinations using the protocols previously described [13]. For the light microscopic study, a paraffin section of 4 μm was stained with hematoxylin and eosin (H&E) and examined by an unbiased pathologist using a Leica DM750 photomicroscope (Leica Microsystems, Heerbrugg, Switzerland). For electron microscopy, renal cortical sections embedded in Epon resin were stained with uranyl acetate and lead citrate, then examined using a JEM-2200 FS transmission electron microscope (JEOL, Tokyo, Japan).

The formalin-fixed paraffin-embedded kidney tissues were also used for CD34 immunofluorescence staining. Kidney sections (4 μm) were deparaffinized in xylene then rehydrated through an ethanol series. Antigen retrieval with 10 mM sodium citrate buffer (pH 6.0) was performed using a hot water bath for 10 min. The sections were permeabilized in PBST containing 1% BSA and 0.4% Triton X-100 for 10 min, followed by blocking with 5% BSA for 30 min at room temperature. Next, the slides were incubated with anti-CD34 antibodies (US Biological Life Sciences, Salem, MA, USA) for 2 h at room temperature then overnight at 4 °C. Fluorescence was observed under an inverted fluorescence microscope (Nikon, Eclipse Ts2, Nikon Instruments Inc., Melville, NY, USA).

2.6. Western Blot Analysis

Renal cortical tissues were homogenized in lysis buffer containing a protease inhibitor cocktail (Sigma-Aldrich, St. Louis, MO, USA), and protein concentrations were measured using the Bio-Rad protein assay kit (Bio-Rad Laboratories, Hercules, CA, USA). The protein sample was subjected to 10% sodium dodecyl sulfate–polyacrylamide gel electrophoresis (SDS-PAGE) and transferred to a polyvinylidene fluoride (PVDF) membrane, where it was blocked and probed with primary antibodies against peroxisome proliferator-activated receptor gamma coactivator 1-alpha (PGC-1α), sirtuin 3 (SIRT3), acetylated superoxide dismutase 2 (Ac-SOD2), superoxide dismutase 2 (SOD2), and β-actin (Cell Signaling Technology, Danvers, MA, USA), followed by corresponding horseradish peroxidase-conjugated secondary antibodies (Millipore Sigma, Darmstadt, Germany). Protein band densities were detected by the ChemiDoc^{TM} Touch Imaging System (Bio-Rad Laboratories, Inc., Philadelphia, PA, USA) and quantified using ImageJ software (National Institute of Health, Bethesda, MD, USA).

2.7. Statistical Analyses

Results are presented as the mean ± SD. Values related to phytochemical analysis and the antioxidant capacity of PRHE were obtained from three independent experiments, where each experiment was repeated three times. Comparisons between groups were assessed using One-way ANOVA followed by Fisher's least-significant difference (SPSS 20.0, IBM Corporation, Armonk, NY, USA), and the significance level was set at $p < 0.05$.

3. Results

3.1. PRHE Possesses Antioxidant Capacity and Contains Protocatechuic Acid and Cyanidin-3-Glucoside as Major Phenolic-Based Compounds

Initial screening of phytochemical elements in PRHE revealed the presence of phenolics (98.66 ± 0.003 mg gallic acid equivalents/g extract), flavonoids (64.26 ± 0.002 mg quercetin equivalents/g extract), and anthocyanins (7.90 ± 0.030 mg cyanidin-3-glucoside equivalents/g extract). DPPH and ABTS radical scavenging assay demonstrated the antioxidant power of PRHE, being 12.42 ± 0.16 mg gallic acid equivalents/g extract and 152.96 ± 0.77 mM Trolox equivalents/g extract, respectively. HPLC further revealed a variety of phenolic acids (Figure 1) and anthocyanins (Figure 2), where protocatechuic acid (PCA) and cyanidin-3-glucoside (C3G) appeared to be the major phenolic acid and anthocyanin found in PRHE, respectively.

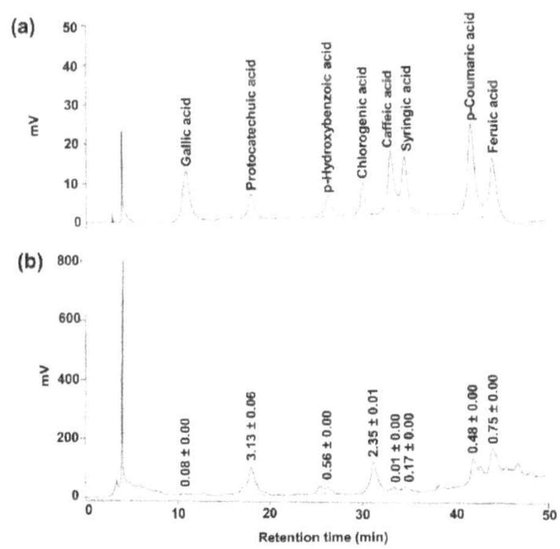

Figure 1. HPLC chromatogram at 280 nm showing major phenolic acids of purple rice husk extract. (**a**) Reference standards; (**b**) purple rice husk extract. The chromatographic peaks of the extract were confirmed by comparing their retention time and UV spectra with those of the reference standards. Numerical values denote the amount of individual compounds detected. Data (mean ± SD, n = 3) are expressed in mg/g of the extract.

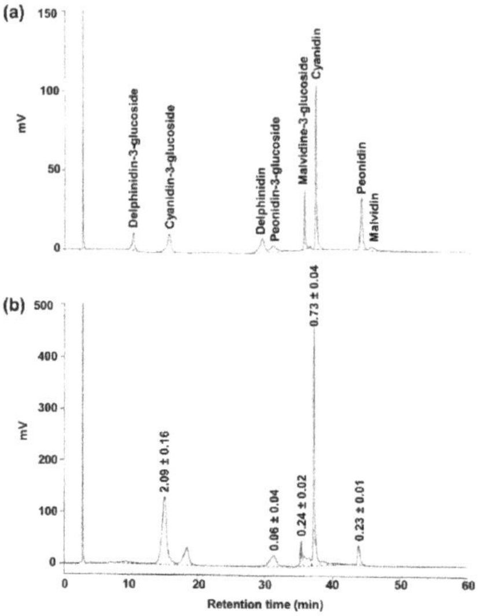

Figure 2. HPLC chromatogram at 520 nm showing major anthocyanins of purple rice husk extract. (**a**) Reference standards; (**b**) purple rice husk extract. The chromatographic peaks of the extract were confirmed by comparing their retention time and UV spectra with those of the reference standards. Numerical values denote the amount of individual compounds detected. Data (mean ± SD, $n = 3$) are expressed in mg/g of the extract.

3.2. PRHE Improves Diabetes-Induced Metabolic Alterations

The diabetic control group (DMV) showed significant body weight gain along with increased visceral fat, serum cholesterol, and triglycerides compared to the nondiabetic group (NDV), though the energy intake was comparable (Table 1). A significant increase in fasting blood glucose and insulin, including HOMA-IR, was also observed in the DMV group compared to the NDV group. Diabetes-induced metabolic alterations were significantly diminished in the PRHE-treated group (DME). Interestingly, treatment with PRHE was as effective as metformin alone (DMM) or metformin plus PRHE (DMME).

Table 1. Effects of purple rice husk extract (PRHE) and metformin on metabolic indexes.

Parameters	NDV	DMV	DMM	DME	DMME
Food intake (g/day)	20.83 ± 0.26 [a]	17.58 ± 0.53 [b]	17.57 ± 0.34 [b]	17.05 ± 0.38 [b]	17.64 ± 0.49 [b]
Energy intake (kcal/day)	77.90 ± 0.98 [a]	82.25 ± 2.46 [a]	82.22 ± 1.59 [a]	79.79 ± 1.78 [a]	82.56 ± 2.30 [a]
BW gain (%)	39.17 ± 1.81 [a]	76.22 ± 7.78 [b]	49.90 ± 3.05 [a]	46.61 ± 3.36 [a]	44.48 ± 4.75 [a]
VF/100 g BW	5.91 ± 0.62 [a]	13.33 ± 0.69 [b]	10.60 ± 0.40 [c]	9.26 ± 0.69 [c]	10.08 ± 0.56 [c]
Total cholesterol (mg/mL)	61.19 ± 1.77 [a]	107.60 ± 3.07 [b]	63.00 ± 2.17 [a]	61.92 ± 5.31 [a]	72.55 ± 8.60 [a]
Triglycerides (mg/mL)	52.25 ± 4.68 [a]	103.50 ± 4.47 [b]	62.51 ± 3.64 [a]	57.34 ± 2.24 [a]	65.82 ± 7.64 [a]
Glucose (mg/dL)	163.77 ± 7.14 [a]	386.47 ± 43.31 [b]	233.41 ± 19.18 [a]	223.82 ± 9.43 [a]	230.06 ± 29.79 [a]
Insulin (ng/mL)	1.92 ± 0.15 [a]	4.14 ± 0.90 [b]	1.90 ± 0.21 [a]	1.82 ± 0.30 [a]	1.94 ± 0.31 [a]
HOMA-IR	0.78 ± 0.08 a	3.79 ± 0.69 b	1.06 ± 0.09 a	1.01 ± 0.17 a	1.04 ± 0.14 a

Values are the mean ± SD ($n = 6$). NDV: vehicle-treated normal group; DMV: vehicle-treated diabetic group; DMM: metformin-treated diabetic group; DME: PRHE-treated diabetic group; DMME: metformin plus PRHE-treated diabetic group; BW: body weight; VF: visceral fat; HOMA-IR: homeostasis model assessment of insulin resistance. [a–c] Mean values with different letters in the same row are significantly different ($p < 0.05$).

3.3. PRHE Ameliorates Diabetes-Induced Renal Functional and Structural Impairments

Diabetic rats exhibited a significant increase in kidney weight/body weight ratio, serum creatinine, and microalbuminuria and a reduction in creatinine clearance (Table 2). These changes were significantly reduced, to a similar extent, when treated with PRHE, metformin, or both. In agreement with renal functions, kidney sections obtained from diabetic rats displayed perihilar focal segmental mesangial cell proliferation with mesangial area expansion and collagen deposits, together with a larger glomerular volume than normal rats (Figure 3, first panel). Immunofluorescence staining of CD34, a mesangial cell marker, also verified the presence of mesangial proliferation (Figure 3, second panel). Electron micrograph clearly showed glomerular basement membrane thickening with diffuse and extensive podocyte effacement (Figure 3, third panel), including the smaller size, abnormal shape, and decreased number of mitochondria within the proximal tubules (Figure 3, last panel). All these changes were significantly attenuated upon treatment with PRHE, metformin, or both.

Table 2. Effects of purple rice husk extract (PRHE) and metformin on renal function.

Parameters	NDV	DMV	DMM	DME	DMME
KW/BW (g/100 g BW)	0.43 ± 0.02 [a]	0.56 ± 0.01 [b]	0.44± 0.01 [a]	0.47 ± 0.03 [a]	0.48 ± 0.02 [a]
Serum creatinine (mg/dL)	0.41 ± 0.04 [a]	0.63 ± 0.03 [b]	0.45 ± 0.02 [a]	0.47 ± 0.02 [a]	0.47 ± 0.02 [a]
Creatinine clearance (mL/min/g KW)	0.95 ± 0.04 [a]	0.55 ± 0.03 [b]	0.97 ± 0.01 [a]	0.94 ± 0.03 [a]	0.98 ± 0.03 [a]
Urine microalbumin (mg/g creatinine)	22.75 ± 4.56 [a]	83.37 ± 16.62 [b]	40.40 ± 11.74 [a]	43.22 ± 16.09 [a]	41.12 ± 16.11 [a]

Values are the mean ± SD (n = 6). NDV: vehicle-treated normal group; DMV: vehicle-treated diabetic group; DMM: metformin-treated diabetic group; DME: PRHE-treated diabetic group; DMME: metformin plus PRHE-treated diabetic group; KW: kidney weight; BW: body weight. [a,b] Mean values with different letters in the same row are significantly different ($p < 0.05$).

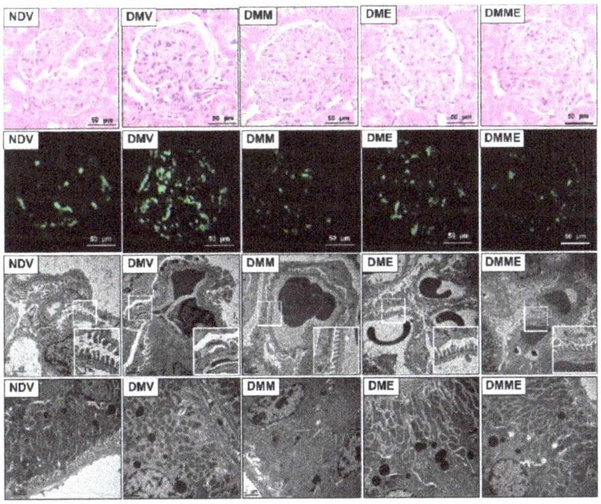

Figure 3. Photomicrographs of the kidney tissues following purple rice husk extract (PRHE) and metformin treatment. First panel shows kidney sections stained with hematoxylin and eosin (H&E, 40×). The second panel shows immunofluorescence staining of CD34 (magnification, 40×). The third and last panels show transmission electron micrographs of glomerulus and renal tubules, respectively (original magnification: 3000×). The boxed areas are magnified in the right lower panel. NDV: vehicle-treated normal group; DMV: vehicle-treated diabetic group; DMM: metformin-treated diabetic group; DME: PRHE-treated diabetic group; DMME: metformin plus PRHE-treated diabetic group.

3.4. PRHE Attenuates Diabetes-Induced Renal Oxidative Stress and Mitochondrial Dysfunction

A significant increase in MDA but a decrease in GSH, SOD, and GPx were detected in the kidney tissues of diabetic rats (Figure 4). Treatment with PRHE and metformin and the combination regimen showed comparative efficacy to significantly attenuate these alterations. Diabetic rats also demonstrated a significant increase in mitochondrial ROS production, a decrease in mitochondrial membrane potential, and a swelling of the mitochondria (Figure 5). Supplementation of PRHE to the diabetic rats significantly restored all the changes. Metformin alone or in combination with PRHE showed similar results to PRHE treatment alone.

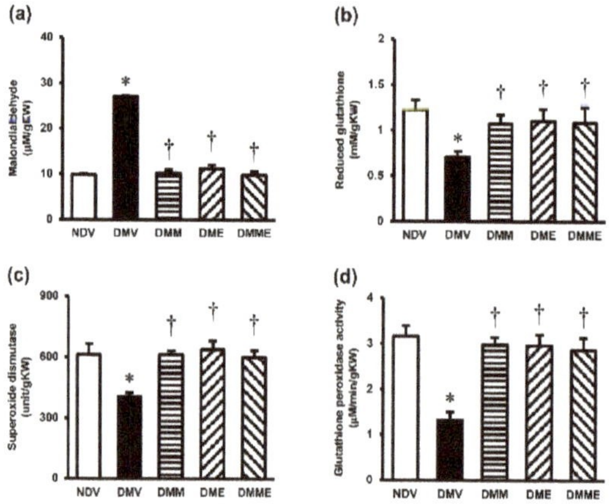

Figure 4. Effects of purple rice husk extract (PRHE) and metformin on renal oxidative stress indexes. (**a**) malondialdehyde; (**b**) reduced glutathione; (**c**) superoxide dismutase; (**d**) glutathione peroxidase. Values are the mean ± SD (n = 6). NDV: vehicle-treated normal group; DMV: vehicle-treated diabetic group; DMM: metformin-treated diabetic group; DME: PRHE-treated diabetic group; DMME: metformin plus PRHE-treated diabetic group; KW: kidney weight. * $p < 0.05$ vs. NDV, † $p < 0.05$ vs. DMV.

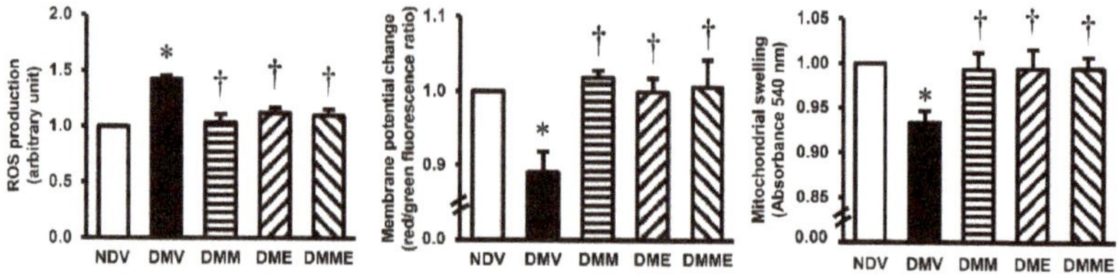

Figure 5. Effects of purple rice husk extract (PRHE) and metformin on kidney mitochondrial function. (**a**) ROS production; (**b**) membrane potential change; (**c**) mitochondrial swelling. Values are the mean ± SD (n = 6). NDV: vehicle-treated normal group; DMV: vehicle-treated diabetic group; DMM: metformin-treated diabetic group; DME: PRHE-treated diabetic group; DMME: metformin plus PRHE-treated diabetic group. * $p < 0.05$ vs. NDV, † $p < 0.05$ vs. DMV.

3.5. PRHE Modifies PGC-1α-SIRT3-Ac-SOD2-SOD2 Signaling Transduction

To determine the signaling pathway involved in the benefits of PRHE on diabetic kidney injury, the protein expression of PGC-1α, SIRT3, Ac-SOD2, and SOD2 was determined. In the DMV group, the expression levels of PGC-1α, SIRT3, and SOD2 were significantly decreased compared to the NDV group, while the expression of Ac-SOD2 was significantly enhanced (Figure 6). These alterations were significantly reversed upon treated diabetic rats with PRHE. Similar results were observed in metformin monotherapy and in combination with PRHE.

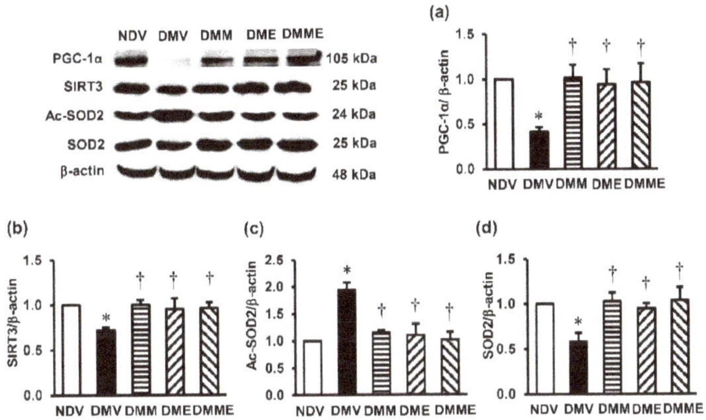

Figure 6. Effects of purple rice husk extract (PRHE) and metformin on renal cortical expression of (**a**) Peroxisome proliferator-activated receptor gamma coactivator 1-alpha (PGC-1α); (**b**) sirtuin 3 (SIRT3); (**c**) acetylated superoxide dismutase 2 (Ac-SOD2); (**d**) superoxide dismutase 2 (SOD2) and β-actin as a reference control. Values are the mean ± SD (n = 3). NDV: vehicle-treated normal group; DMV: vehicle-treated diabetic group; DMM: metformin-treated diabetic group; DME: PRHE-treated diabetic group; DMME: metformin plus PRHE-treated diabetic group. * $p < 0.05$ vs. NDV, † $p < 0.05$ vs. DMV.

4. Discussion

This study reveals the protection against nephropathy in T2DM by PRHE through antioxidant abilities and mitochondrial protection, possibly via its major phytochemicals, particularly PCA and C3G. Results suggest that modifications of PGC-1α/SIRT3/SOD2 signaling are linked to this protection.

A combination of a high-fat diet and streptozotocin injection was used to develop a rat model of T2DM in the present study. Feeding rats with a high-fat diet promotes the development of obesity and insulin resistance, while injection of streptozotocin selectively destroys pancreatic β-cells and, thus, impairs insulin secretion [20]. This model has been shown to replicate the natural history and metabolic characteristics of humans and is also suitable for pharmacological screening [20]. However, the reliability and appropriateness of the model depend on the number of pancreatic β-cells remaining, which is largely regulated by the dose of streptozotocin injection. A high dose of streptozotocin has been shown to critically damage pancreatic β-cell function, leading to insulin deficiency, which is considered to resemble T1DM rather than T2DM [23]. Regarding our model, hyperglycemia, hyperinsulinemia, insulin resistance (presented by increased HOMA-IR), and lipid abnormalities were all detected in the DMV group. These metabolic alterations are compatible with the well-recognized characteristics of T2DM, thus confirming the successful T2DM induction in the current study. We did not quantify the number of functioning pancreatic β-cells. However, the observations of hyperglycemia together with hyperinsulinemia in our diabetic rats implied that the number of intact pancreatic β-cells was sufficient to increase insulin secretion in response to high glucose, even though it could not overcome a progressive decline in insulin action caused by diabetes. Additionally, the degree of hyperglycemia and the dose of streptozotocin (35 mg/kg) used in our study corresponded to a previous report showing that streptozotocin at the dose of 30–40 mg/kg has a moderate effect on plasma glucose and on the islets of Langerhans at a cellular level [24]. Taken together, all evidence points towards the development of T2DM in the present study. Most importantly, our diabetic model also developed diabetic kidney injury, as indicated by accumulation of blood urea nitrogen and creatinine, reduced creatinine clearance, microalbuminuria, and histopathology that characterize diabetic nephropathy.

We found that diabetes-induced metabolic alterations were significantly diminished in the diabetic group given PRHE, suggesting the glucose-lowering ability of PRHE. To our knowledge, this is the first evidence showing antihyperglycemic potential in T2DM from the husk of purple rice, as a report is available only on its bran [25]. PCA and C3G, the main phytochemicals, may play a role in this action of PRHE. Previous publications showed the potential of PCA and C3G in reducing blood glucose, enhancing insulin sensitivity, and increasing glucose uptake, thereby improving glucose homeostasis in diabetes [26–28]. Studies also indicated that increased expression and translocation to the cell membrane of glucose transporters 1 (GLUT1) and 4 (GLUT4) contributed to the improvement of glucose tolerance by PCA and C3G [16–18]. The effectiveness on glycemic control by PCA and C3G was found to be comparable to metformin, a gold-standard antidiabetic drug, and was associated with the activation of AMPK [16]. Interestingly, our results also demonstrated a similar efficacy on metabolic improvement between PRHE, metformin, and PRHE+metformin. Metformin is an AMPK activator, and AMPK activation is recognized as the initial process for metabolic control by metformin [26]. It may be that PRHE (via PCA and C3G) stimulated AMPK and exerted its metabolic control through the same downstream signaling pathways as metformin; thus, no additional benefits were found in the combination therapy. Further study using cell culture experiments or the knockout mice model with an AMPK inhibitor or activator will be helpful to validate this possibility.

Our results showed that PRHE effectively protected against diabetic kidney injury. Because metabolic impairment was improved after PRHE treatment, it is likely that renal protection may result from glycemic control by PRHE. However, PRHE (particularly through PCA) may exert its benefits independent of glycemic control. This suggestion was based on a study using human mesangial cells exposed to high glucose in the presence or absence of PCA, which demonstrated that PCA was able to suppress mesangial cell proliferation in a dose-dependent manner by decreasing the levels of protein expression involved in mesangial expansion including type IV collagen, laminin, and fibronectin [29]. This in vitro evidence reinforces the potential direct action of PRHE at the kidney level in diabetes.

Studies have shown that oxidative stress plays an independent role in the development, progression, and severity of diabetic nephropathy [16,30]. ROS can damage renal cells by oxidizing membrane phospholipids, proteins, carbohydrates, and nucleic acids. ROS are also secondary messengers that activate many signaling cascades, leading to cell damage and deterioration of kidney function in diabetic kidney disease [16,30]. Mitochondria are well recognized as a major source of ROS production. There is evidence linking obesity and diabetes with oxidative stress and mitochondrial dysfunction [30]. The kidneys have high energy demand and are rich in mitochondria, thus mitochondrial dysfunction could contribute to renal oxidative cell injury and consequently renal functional and structural impairments. In line with this view, we detected mitochondrial damage in diabetic rats, as shown by increased ROS production, a dissipation of membrane potential, and a swollen and abnormal mitochondrial ultrastructure. Significant increases in the renal tissue levels of malondialdehyde (a lipid peroxidation product) and decreases in antioxidants (GSH, SOD, GPx) were also observed, indicating renal oxidative injury. These alterations were markedly diminished by PRHE, which is most likely mediated through the mitochondrial protective properties of PCA and C3G. PCA has been demonstrated to protect cardiac and brain mitochondrial dysfunction in streptozotocin-induced diabetic rats [31,32]. The protective role of C3G by modulating intracellular signaling and maintaining mitochondrial function, rather than acting solely as an antioxidant, has also been reported in cardiac ischemia-reperfusion injury [33]. Studies using HK-2 cells further demonstrated that C3G inhibited high glucose-induced ROS production, prevented mitochondrial membrane potential loss, and protected against mitochondrial dysfunction-mediated cell death [34]. Taken together, it is proposed that treatment with PRHE preserves mitochondrial integrity, decreases oxidative stress, and, finally, leads to the protection of diabetic kidney injury. It is interesting to mention here that there was no greater production of mitochondrial ROS (including the kidney tissue levels of MDA) in the metformin-treated diabetic group, though blood glucose level was slightly higher than normal. Beyond its well-known blood glucose regulatory action, metformin has been confirmed to possess antioxidant properties [35]. This action of metformin is mediated through activation of AMPK secondary to its inhibition on the mitochondrial respiratory chain complex 1, the major site of ROS generation in mitochondria [35].

SIRT3, a major mitochondrial deacetylase, was recently highlighted as a novel regulator of mitochondrial function and redox homeostasis [13]. SIRT3 plays a key role in deacetylating and modifying the enzymatic activities of several mitochondrial proteins, including SOD2 [36]. SOD2 is the main antioxidant enzyme in the mitochondria that is responsible for scavenging the superoxide anion, a byproduct of the mitochondrial electron transport chain. This function allows SOD2 to clear mitochondrial ROS, maintain mitochondrial oxidative equilibrium, and confer protection against mitochondrial injury and death [37]. Evidence has suggested that the deacetylation of SOD2 by SIRT3 is necessary for the activation of SOD2, and this deacetylase activity of SIRT3 is further modulated by PGC-1α [13].

Herein, we detected a significant reduction in the expression of PGC-1α and SIRT3 in parallel with the increased Ac-SOD2 and decreased SOD2 expression in diabetic rats. Importantly, PRHE supplementation to the diabetes group was able to normalize these changes. This suggests that the protection against mitochondrial dysfunction and redox imbalance in diabetic nephropathy by PRHE may be a result of the modification of PGC-1α-SIRT3-SOD2 signaling transduction. Consistent with our results, an increased expression ratio of Ac-SOD2/SOD2 followed by increased mitochondrial oxidative stress and mitochondrial dysfunction were observed in Zucker diabetic fatty rats upon reduction of SIRT3 activity [38]. A study of SIRT3 knockout mice demonstrated similar results [36,39]. A recent publication also showed the protection against diabetic kidney injury in BTBR ob/ob mice by honokiol, a polyphenolic compound isolated from magnolia bark, through the maintenance of mitochondrial stability by activating SOD2 and restoring PGC-1α expression [40].

In this study, increased PGC-1α, SIRT3, and SOD2 expressions and decreased Ac-SOD2 expression in diabetic rats treated with a combination of metformin and PRHE were found to be very similar compared to the single therapy with metformin or PRHE. A previous study showed the upregulation of SIRT3 expression followed by a decrease in mitochondrial ROS formation after metformin treatment [41]. There is also a report that activation of PGC-1α is one of the therapeutic mechanisms of metformin in diabetes [42]. As PGC-1α and its subsequent downstream signals appear to be the same target for both metformin and PRHE, this may underlie the lack of additional therapeutic effects in diabetes treated with a combination regimen. However, this issue deserves future exploration.

5. Conclusions

This study provides convincing evidence indicating the promising role of PRHE in preventing the development and progression of diabetic nephropathy. The potential of PRHE is apparently associated with its ability to retain mitochondrial integrity and redox equilibrium within the kidney through the activation of the PGC-1α-SIRT3-SOD2 signaling pathway. The outcomes substantiate the worth of this agricultural waste and highlight the opportunity to develop purple rice husk as a dietary supplement or health product, which will be of great value in developing countries with limited resources and a high incidence of diabetes mellitus.

Author Contributions: Conceptualization, O.W. and N.L.; methodology, O.W., N.L., A.K. and W.P.; validation, O.W., A.K. and W.P.; formal analysis, O.W. and A.K.; investigation, O.W., N.L., A.K. and W.P.; data curation, O.W., N.L., A.K. and W.P.; writing manuscript, O.W.; project administration,

O.W.; funding acquisition, O.W. All authors have read and agreed to the published version of the manuscript.

Funding: This research was funded by Chiang Mai University (Grant Number: ผด07/2561).

Institutional Review Board Statement: The study was conducted according to the guidelines of the Declaration of Helsinki and approved by the Institutional Animal Care and Use Committee at the Faculty of Medicine, Chiang Mai University (Project Number: 41/2559, approved on 3 October 2019).

Informed Consent Statement: Not applicable.

Data Availability Statement: The data presented in this study are available on request from the corresponding author.

Acknowledgments: Special thanks to the Faculty of Medicine, Chiang Mai University, for providing infrastructural support to carry out this research. We also extend sincere appreciation to Jannarong Intakhad and Parichat Tojing for their technical help during the experiment.

Conflicts of Interest: The authors declare no conflict of interest.

Appendix A

Table A1. Nutritional compositions of the experimental diets.

Compositions	Normal Diet		High-Fat Diet	
	g/100 g Diet	% Energy	g/100 g Diet	% Energy
Carbohydrates	56.21	60.11	26.38	22.54
Fat	4.55	10.95	27.89	53.63
Protein	27.06	28.94	28.81	23.93
Vitamin and mineral	6.54	-	9.92	-
Fiber	3.43	-	4.32	-
Total energy (%)		100		100
Caloric value (kcal/g)		3.74		4.68

Table A2. Effects of different concentrations of purple rice husk extract (PRHE) on metabolic indexes and renal function (a preliminary study).

Parameters	NDV	DMV	DME 150	DME 300	DME 600
Food intake (g/day)	20.54 ± 0.17	16.87 ± 0.43 *	17.56 ± 0.76 *	17.22 ± 0.56 *	17.00 ± 0.37 *
Energy intake (kcal/day)	76.82 ± 0.62	78.92 ± 1.99 *	82.20 ± 3.55	80.59 ± 2.63	79.57 ± 1.01
BW gain (%)	41.60 ± 1.45	87.49 ± 4.92 *	56.63 ± 3.87 *,†	46.14 ± 5.12 †	37.57 ± 4.41 †
Glucose (mg/dL)	161.18 ± 10.98	433.99 ± 48.50 *	283.00 ± 15.51*,†	238.24 ± 3.72 *,†	260.37 ± 22.51 *,†
Insulin (ng/mL)	1.98 ± 0.12	4.57 ± 1.31 *	3.88 ± 0.35 *,#	1.92 ± 0.40 †	2.74 ± 0.36 †
HOMA-IR	0.80 ± 0.10	4.53 ± 0.79 *	2.69 ± 0.20 *,†,#	1.12 ± 0.22 †	1.77 ± 0.35 †
Serum creatinine (mg/dL)	0.45 ± 0.03	0.63 ± 0.02 *	0.55 ± 0.10	0.48 ± 0.03 †	0.52 ± 0.14
Creatinine clearance (ml/min/g KW)	0.97 ± 0.05	0.54 ± 0.03 *	0.63 ± 0.16 *	0.91 ± 0.04 †	0.85 ± 0.15 †

Values are mean ± SD ($n = 4$). NDV: vehicle-treated normal group; DMV: vehicle-treated diabetic group; DME (150, 300, 600): PRHE-treated diabetic group at 150, 300, 600 mg/kg, respectively; BW: body weight; HOMA-IR: homeostasis model assessment of insulin resistance; KW: kidney weight. * $p < 0.05$ vs. NDV, † $p < 0.05$ vs. DMV, # $p < 0.05$ vs. DME 300.

References

1. Wu, M.; Li, S.; Yu, X.; Chen, W.; Ma, H.; Shao, C.; Zhang, Y.; Zhang, A.; Huang, S.; Jia, Z. Mitochondrial activity contributes to impaired renal metabolic homeostasis and renal pathology in STZ-induced diabetic mice. *Am. J. Physiol. Renal Physiol.* **2019**, *317*, F593–F605. [CrossRef] [PubMed]
2. International Diabetes Federation. *IDF Diabetes Atlas*, 9th ed.; Brussels, Belgium, 2019; Available online: https://www.diabetesatlas.org (accessed on 10 June 2021).
3. Saritsiri, S.; Mongkolchati, A.; Suksaroj, T. Prevalence of chronic kidney disease and related factors among diabetic patients in primary care, Bangkok, Thailand. *J. Public Health Dev.* **2021**, *19*, 1–8.

4. Yu, S.M.; Bonventre, J.V. Acute kidney injury and progression of diabetic kidney disease. *Adv. Chronic Kidney Dis.* **2018**, *25*, 166–180. [CrossRef]
5. Aekplakorn, W.; Chariyalertsak, S.; Kessomboon, P.; Sangthong, R.; Inthawong, R.; Putwatana, P.; Taneepanichskul, S. Thai National Health Examination Survey IV Study Group. Prevalence and management of diabetes and metabolic risk factors in Thai adults: The Thai National Health Examination Survey IV, 2009. *Diabetes Care.* **2011**, *34*, 1980–1985. [CrossRef] [PubMed]
6. Aekplakorn, W.; Chariyalertsak, S.; Kessomboon, P.; Assanangkornchai, S.; Taneepanichskul, S.; Putwatana, P. Prevalence of Diabetes and Relationship with Socioeconomic Status in the Thai Population: National Health Examination Survey, 2004–2014. *J. Diabetes Res.* **2018**, *2018*, 1654530. [CrossRef]
7. Liu, W.; Liu, J.; Triplett, L.; Leach, J.E.; Wang, G.L. Novel insights into rice innate immunity against bacterial and fungal pathogens. *Annu. Rev. Phytopathol.* **2014**, *52*, 213–241. [CrossRef]
8. Abdel-Aal, E.S.M.; Young, J.C.; Rabalski, I. Anthocyanin composition in black, blue, pink, purple, and red cereal grains. *J. Agric. Food Chem.* **2006**, *54*, 4696–4704. [CrossRef]
9. Deng, G.F.; Xu, X.R.; Zhang, Y.; Li, D.; Gan, R.Y.; Li, H.B. Phenolic compounds and bioactivities of pigmented rice. *Crit. Rev. Food Sci. Nutr.* **2013**, *53*, 296–306. [CrossRef]
10. Sohail, M.; Rakha, A.; Butt, M.S.; Iqbal, M.J.; Rashid, S. Rice bran nutraceutics: A comprehensive review. *Crit. Rev. Food Sci. Nutr.* **2017**, *57*, 3771–3780. [CrossRef]
11. Butsat, S.; Siriamornpun, S. Antioxidant capacities and phenolic compounds of the husk, bran and endosperm of Thai rice. *Food Chem.* **2010**, *119*, 606–613. [CrossRef]
12. Chariyakornkul, A.; Punvittayagul, C.; Taya, S.; Wongpoomchai, W. Inhibitory effect of purple rice husk extract on AFB1-induced micronucleus formation in rat liver through modulation of xenobiotic metabolizing enzymes. *BMC Complement. Altern. Med.* **2019**, *19*, 237. [CrossRef]
13. Peerapanyasut, W.; Kobroob, A.; Palee, S.; Chattipakorn, N.; Wongmekiat, O. Activation of sirtuin 3 and maintenance of mitochondrial integrity by N-acetylcysteine protects against bisphenol A-induced kidney and liver toxicity in rats. *Int. J. Mol. Sci.* **2019**, *11*, 267. [CrossRef]
14. Zhang, J.; Xiang, H.; Liu, J.; Chen, Y.; He, R.R.; Liu, B. Mitochondrial sirtuin 3: New emerging biological function and therapeutic target. *Theranostics* **2020**, *10*, 8315–8342. [CrossRef]
15. Benigni, A.; Perico, L.; Macconi, D. Mitochondrial Dynamics is linked to longevity and protects from end-organ Injury: The emerging role of sirtuin 3. *Antioxid. Redox Signal.* **2016**, *25*, 185–199. [CrossRef]
16. Sagoo, M.K.; Gnudi, L. Diabetic nephropathy: Is there a role for oxidative stress? *Free Radic. Biol. Med.* **2018**, *116*, 50–63. [CrossRef]
17. Lin, J.Y.; Tang, C.Y. Determination of total phenolic and flavonoid content in selected fruits and vegetables as well as their stimulatory effects on mouse splenocyte proliferation. *Food Chem.* **2007**, *101*, 140–147. [CrossRef]
18. Sutharut, J.; Sudarat, J. Total anthocyanin content and antioxidant activity of germinated colored rice. *Int. Food Res. J.* **2012**, *19*, 215–221.
19. Umnajkitikorn, K.; Faiyue, B.; Saengnil, K. Enhancing antioxidant properties of germinated Thai rice (*Oryza sativa* L.) cv. Kum Doi Saket with salinity. *J. Rice Res.* **2013**, *1*, 103. [CrossRef]
20. Srinivasan, K.; Viswanad, B.; Asrat, L.; Kaul, C.L.; Ramarao, P. Combination of high-fat diet-fed and low-dose streptozotocin-treated rat: A model for type 2 diabetes and pharmacological screening. *Pharmacol. Res.* **2005**, *52*, 313–320. [CrossRef]
21. Matthews, D.R.; Hosker, J.P.; Rudenski, A.S.; Naylor, B.A.; Treacher, D.F.; Turner, R.C. Homeostasis model assessment: Insulin resistance and beta-cell function from fasting plasma glucose and insulin concentrations in man. *Diabetologia* **1985**, *28*, 412–419. [CrossRef]
22. Kobroob, A.; Peerapanyasut, W.; Chattipakorn, N.; Wongmekiat, O. Damaging effects of bisphenol A on the kidney and the protection by melatonin: Emerging evidences from in vivo and in vitro studies. *Oxid. Med. Cell. Longev.* **2018**, *2018*, 3082438. [CrossRef]
23. Suman, R.K.; Ray Mohanty, I.; Borde, M.K.; Maheshwari, U.; Deshmukh, Y.A. Development of an experimental model of diabetes co-existing with metabolic syndrome in rats. *Adv. Pharmacol. Sci.* **2016**, *2016*, 9463476. [CrossRef]
24. Mythili, M.D.; Vyas, R.; Akila, G.; Gunasekaran, S. Effect of streptozotocin on the ultrastructure of rat pancreatic islets. *Microsc. Res. Tech.* **2004**, *63*, 274–281. [CrossRef]
25. Boue, S.M.; Daigle, K.W.; Chen, M.H.; Cao, H.; Heiman, M.L. Antidiabetic potential of purple and red rice (*Oryza sativa* L.) bran extracts. *J. Agric. Food Chem.* **2016**, *64*, 5345–5353. [CrossRef]
26. Talagavadi, V.; Rapisarda, P.; Galvano, F.; Pelicci, P.; Giorgio, M. Cyanidin-3-O-β-glucoside and protocatechuic acid activate AMPK/mTOR/S6K pathway and improve glucose homeostasis in mice. *J. Func. Foods.* **2016**, *21*, 338–348. [CrossRef]
27. Scazzocchio, B.; Filesi, C.; Gaudio, I.D.; Archivio, M.D.; Santangelo, C.; Iacovelli, A.; Galvano, F.; Pluchinotta, F.R.; Giovannini, C.; Masella, R. Protocatechuic acid activates key components of insulin signaling pathway mimicking insulin activity. *Mol. Nutr. Food Res.* **2015**, *59*, 1472–1481. [CrossRef]
28. Sasaki, R.; Nishimura, N.; Hoshino, H.; Isa, Y.; Kadowaki, M.; Ichi, T.; Tanaka, A.; Nishiumi, S.; Fukuda, I.; Ashida, H.; et al. Cyanidin 3-glucoside ameliorates hyperglycemia and insulin sensitivity due to down regulation of retinol binding protein 4 expression in diabetic mice. *Biochem. Pharmacol.* **2007**, *74*, 1619–1627. [CrossRef]
29. Ma, Y.; Chen, F.; Yang, S.; Chen, B.; Shi, J. Protocatechuic acid ameliorates high glucose-induced extracellular matrix accumulation in diabetic nephropathy. *Biomed. Pharmacother.* **2018**, *98*, 18–22. [CrossRef]

30. Wei, P.Z.; Szeto, C.C. Mitochondrial dysfunction in diabetic kidney disease. *Clin. Chim. Acta* **2019**, *496*, 108–116. [CrossRef]
31. Semaming, Y.; Kumfu, S.; Pannangpetch, P.; Chattipakorn, S.C.; Chattipakorn, N. Protocatechuic acid exerts a cardioprotective effect in type 1 diabetic rats. *J. Endocrinol.* **2014**, *223*, 13–23. [CrossRef]
32. Semaming, Y.; Sripetchwandee, J.; Sa-Nguanmoo, P.; Pintana, H.; Pannangpetch, P.; Chattipakorn, N.; Chattipakorn, S.C. Protocatechuic acid protects brain mitochondrial function in streptozotocin-induced diabetic rats. *Appl. Physiol. Nutr. Metab.* **2015**, *40*, 1078–1081. [CrossRef]
33. Liobikas, J.; Skemiene, K.; Trumbeckaite, S.; Borutaite, V. Anthocyanins in cardioprotection: A path through mitochondria. *Pharmacol. Res.* **2016**, *113*, 808–815. [CrossRef]
34. Wei, J.; Wu, H.; Zhang, H.; Li, F.; Chen, S.; Hou, B.; Shi, Y.; Zhao, L.; Duan, H. Anthocyanins inhibit high glucose-induced renal tubular cell apoptosis caused by oxidative stress in db/db mice. *Int. J. Mol. Med.* **2018**, *41*, 1608–1618. [CrossRef]
35. Foretz, M.; Guigas, B.; Bertrand, L.; Pollak, M.; Viollet, B. Metformin: From mechanisms of action to therapies. *Cell Metab.* **2014**, *20*, 953–966. [CrossRef]
36. Kitada, M.; Ogura, Y.; Monno, I.; Koya, D. Sirtuins and type 2 diabetes: Role in inflammation, oxidative stress, and mitochondrial function. *Front. Endocrinol.* **2019**, *10*, 187. [CrossRef] [PubMed]
37. He, J.; Liu, X.; Su, C.; Wu, F.; Sun, J.; Zhang, J.; Yang, X.; Zhang, C.; Zhou, Z.; Zhang, X.; et al. Inhibition of mitochondrial oxidative damage improves reendothelialization capacity of endothelial progenitor cells via SIRT3 (Sirtuin3)-enhanced SOD2 (Superoxide dismutase 2) deacetylation in hypertension. *Arterioscler. Thromb. Vasc. Biol.* **2019**, *39*, 1682–1698. [CrossRef] [PubMed]
38. Ogura, Y.; Kitada, M.; Monno, I.; Kanasaki, K.; Watanabe, A.; Koya, D. Renal mitochondrial oxidative stress is enhanced by the reduction of Sirt3 activity in Zucker diabetic fatty rats. *Redox Rep.* **2018**, *23*, 153–159. [CrossRef]
39. Jing, E.; Emanuelli, B.; Hirschey, M.D.; Boucher, J.; Lee, K.Y.; Lombard, D.; Verdin, E.M.; Kahn, C.R. Sirtuin-3 (Sirt3) regulates skeletal muscle metabolism and insulin signaling via altered mitochondrial oxidation and reactive oxygen species production. *Proc. Natl. Acad. Sci. USA* **2011**, *108*, 14608–14613. [CrossRef]
40. Locatelli, M.; Zoja, C.; Zanchi, C.; Corna, D.; Villa, S.; Bolognini, S.; Novelli, R.; Perico, L.; Remuzzi, G.; Benigni, A.; et al. Manipulating Sirtuin 3 pathway ameliorates renal damage in experimental diabetes. *Sci. Rep.* **2020**, *10*, 8418. [CrossRef]
41. Wang, C.; Yang, Y.; Zhang, Y.; Liu, J.; Zhenjun Yao, Z.; Zhang, C. Protective effects of metformin against osteoarthritis through upregulation of SIRT3-mediated PINK1/Parkin-dependent mitophagy in primary chondrocytes. *Biosci. Trends.* **2018**, *12*, 605–612. [CrossRef] [PubMed]
42. Driver, C.; Bamitale, K.D.S.; Kazi, A.; Olla, M.; Nyane, N.A.; Owira, P.M.O. Cardioprotective effects of metformin. *J. Cardiovasc. Pharmacol.* **2018**, *72*, 121–127. [CrossRef] [PubMed]

biomolecules

MDPI

Article

Elevated Levels of Renalase, the β-NAD(P)H Isomerase, Can Be Used as Risk Factors of Major Adverse Cardiovascular Events and All-Cause Death in Patients with Chronic Kidney Disease

Wojciech Knop [1], Natalia Maria Serwin [2,*], Elżbieta Cecerska-Heryć [2], Bartłomiej Grygorcewicz [2], Barbara Dołęgowska [2], Aleksandra Gomółka [1], Magda Wiśniewska [1] and Kazimierz Ciechanowski [1]

[1] Clinical Department of Nephrology, Transplantology, and Internal Medicine, Pomeranian Medical University in Szczecin, 70-111 Szczecin, Poland; wojciechknop@gmail.com (W.K.); aemgomolka@gmail.com (A.G.); mwisniewska35@gmail.com (M.W.); kazcie@pum.edu.pl (K.C.)

[2] Department of Laboratory Medicine, Pomeranian Medical University in Szczecin, 70-111 Szczecin, Poland; blanka23@pum.edu.pl (E.C.-H.); bartlomiej.grygorcewicz@pum.edu.pl (B.G.); barbara.dolegowska@pum.edu.pl (B.D.)

* Correspondence: natser@pum.edu.pl

Abstract: Background: Renalase is an enzyme and a cytokine involved in cell survival. Since its discovery, associations between it and both cardiovascular and kidney disease have been noted. Recognizing this, we conducted a study in which we followed patients with chronic kidney disease. Material and methods: The study involved 90 CKD patients with varying stages of the disease and 30 healthy controls. Renalase was measured with an ELISA kit, and patients were followed-up after a median of 18 months. During the follow-up, we asked about the occurrence of MACE, all-cause mortality and the need for dialysis initiation. Results: In CKD subgroups, RNSL correlated with all-cause death only in the HD group (Rs = 0.49, $p < 0.01$). In the whole CKD population, we found a positive correlation of RNSL concentration and both MACE occurrence (Rs = 0.38, $p < 0.001$) and all-cause death (Rs = 0.34, $p < 0.005$). There was a significant increase in MACE occurrence probability in patients with elevated renalase levels (>25 µg/mL). Conclusions: Elevated renalase levels can be used as a risk factor of MACE in patients with CKD, but its long-term utility needs further research. High renalase levels are a risk factor of death among CKD patients. In HD patients, all deaths were observed among patients with >30 µg/mL; this level could be used as a "red flag" marker in future studies.

Keywords: renalase; chronic kidney disease; major adverse cardiovascular outcomes

Citation: Knop, W.; Serwin, N.M.; Cecerska-Heryć, E.; Grygorcewicz, B.; Dołęgowska, B.; Gomółka, A.; Wiśniewska, M.; Ciechanowski, K. Elevated Levels of Renalase, the β-NAD(P)H Isomerase, Can Be Used as Risk Factors of Major Adverse Cardiovascular Events and All-Cause Death in Patients with Chronic Kidney Disease. *Biomolecules* **2021**, *11*, 1514. https://doi.org/10.3390/biom11101514

Academic Editor: Liang-Jun Yan

Received: 28 July 2021
Accepted: 9 October 2021
Published: 14 October 2021

Publisher's Note: MDPI stays neutral with regard to jurisdictional claims in published maps and institutional affiliations.

1. Introduction

Renalase (RNLS) is a small flavoprotein produced mainly by the kidney. However, the latest investigations show that RNLS might be an "organolase", as the *RNLS* gene is expressed in many other cells and tissues, including the nervous system, endocrinal and digestive tract organs, lungs or heart in humans and some other mammals [1]. RNLS shows both intracellular and extracellular activity. Intracellular RNLS acts as an enzyme that, in the presence of a FAD cofactor, oxidizes 2- and 6- DHNAD(P) to β-NAD(P)+, its biologically active form. This action prevents toxicity resulting from the inhibition of many β-NAD(P)+ dependent enzymes and reactions. Additionally, extracellular renalase and RP-200 and RP-220 peptides, which are fragments of the protein, activate some signaling pathways, including Akt and MAP kinases, therefore promoting cell survival. This activity is mediated by the binding of renalase to its recently discovered receptor–plasma membrane Ca^{2+}-ATPase-4b (PMCA4b) [2].

Despite reported discrepancies in observed serum renalase levels in humans, most analyses indicate that serum RNLS levels significantly increase in people with chronic kidney disease (CKD). As in the case of many other markers, this relationship would demonstrate the usefulness of the RNLS concentration assessment in the diagnosis and prognosis of kidney diseases and accompanying disorders. Seeing that CVD is the most significant contributor to mortality of CKD patients, we sought to answer whether RNLS could be a predictive factor for CVD in CKD.

2. Materials and Methods

2.1. Study Design

One hundred twenty people (aged 40–79) were enrolled into the study. We divided 90 CKD patients into subgroups, each consisting of 30 participants: CKD stage III (CKD III), CKD stage IV (CKD IV), CKD stage 5D who were hemodialyzed (HD) and 30 adults (15 women and 15 men) with no signs of chronic kidney disease (control). Both groups were signed up for the study during treatment at a Nephrology Ward and its Admissions Department, Outpatient Clinic and Dialysis Centre.

2.2. Material

To obtain blood serum, blood was drawn into test tubes (S-Monovette, Sarstedt, Germany) and left for 30 min at room temperature. The tubes were then centrifuged (10 min, $1000\times g$, room temperature). The obtained material was then transferred to separate tubes and frozen at −80 °C until renalase levels were measured. eGFR, HDL, LDL and TG levels were assessed in an analytical laboratory as soon as possible after sample collection.

2.3. Methods

eGFR was calculated using the CKD-EPI Equation (1), where Scr is serum creatinine (mg/dL), κ is 0.7 for females and 0.9 for males, α is −0.329 for females and −0.411 for males, min indicates the minimum of Scr/κ or 1 and max indicates the maximum of Scr/κ or 1) [3]. Renalase levels were measured in blood serum using an ELISA kit (Cloud-Clone Corp, Houston, TX, USA). Biochemical parameters in serum were evaluated using routine laboratory methods.

$$eGFR = 141 \times \min\left(\frac{Scr}{\kappa},1\right)^{\alpha} \times \max\left(\frac{Scr}{\kappa},1\right)^{-1.209} \times 0.993^{\text{age}} \times 1.018\,[\text{if female}] \times 1.159\,[\text{if black}] \quad (1)$$

Cardiovascular risk (CV) was calculated using the ASCVD (atherosclerotic cardiovascular risk) algorithm, which includes: age, sex, race, systolic and diastolic blood pressure, total cholesterol and HDL/LDL fractions, diabetes, smoking and hypertension [4]. Participants with acute kidney injury or a history of kidney transplantation were excluded from the study. All patients have given written informed consent.

Follow-up was administered by means of a questionnaire during an outpatient clinic visit, dialysis treatment or a telephone call. Mean follow-up time was 17.3 months. Two patients were lost to follow-up—one patient in control and one in the CKD IV group. In the follow-up, we asked about the first major adverse cardiovascular event (MACE). We defined MACE as an occurrence of either ischemic stroke, acute coronary syndrome, hospitalization due to heart failure or death attributed to a cardiac event. We also looked for all-cause mortality and a need to start renal replacement therapy (RRT).

2.4. Statistical Analysis

Obtained data were evaluated using Statistica 13.0 software (StatSoft, Tulsa OK, USA). Since almost all parameters had distributions different than normal, non-parametric tests were used. Differences between two groups (control and CKD) were evaluated using the Mann–Whitney U test. Differences between analyzed parameters in subgroups (control, CKD III, CKD IV, HD) were assessed using Kruskal–Wallis ANOVA with post-hoc Dunn's test. Correlations were evaluated using Spearman's Rank Correlation Coefficient. The analysis of the occurrence of MACE and overall survival was performed using the Kaplan–Meier method. The significance level was set at α = 0.05.

3. Results

3.1. Data Obtained during Recruitment

The study population's demographic and descriptive data are shown in Tables 1 and 2 (divided as control and CKD stage-based subgroups) and Table 3 (divided as control and whole CKD group).

Table 1. Descriptive data of the subgroups together with statistical evaluation (Kruskal-Wallis ANOVA). Data are shown as mean ± SD and median (lower quartile–upper quartile).

Parameter	Control n = 30		CKD III n = 30		CKD IV n = 30		HD n = 30		p-Value
	Mean ± SD	Median (LQ-UQ)	Mean ± SD	Median (LQ-UQ)	Mean ± SD	Median (LQ-UQ)	Mean ± SD	Median (LQ-UQ)	
Age (Years)	57 ± 11 [1,2]	56 (45–67)	68 ± 8 [1]	69 (62–74)	66 ± 11 [2]	67 (60–75)	64 ± 11	65 (60–74)	0.0028
Systolic BP (mmHg)	131 ± 21 [1]	125 (120–140)	134 ± 12	132 (130–140)	140 ± 18 [1]	140 (130–160)	140 ± 21	140 (130–150)	0.0216
Diastolic BP (mmHg)	80 ± 11	80 (70–85)	79 ± 9	80 (75–85)	80 ± 11	80 (75–90)	77 ± 12	80 (7–80)	0.44
eGFR	85 ± 13	85 (72–91)	41 ± 8	41 (34–48)	22 ± 5	21 (17–26)	7 ± 3	7 (5–9)	<0.001
Body Weight (kg)	83 ± 20	80 (70–94)	84 ± 15 [1]	84 (75–92)	82 ± 16	80 (71–92)	72 ± 16 [1]	72 (64–83)	0.0286
Height (cm)	171 ± 8	170 (164–176)	169 ± 8	169 (160–176)	167 ± 9	167 (160–175)	168 ± 11	165 (158–178)	0.50
LDL (mg/dL)	122 ± 35	124 (97–148)	119 ± 47	128 (79–156)	119 ± 52	106 (82–151)	110 ± 46	105 (70–136)	0.55
HDL (mg/dL)	55 ± 16	52 (45–61)	50 ± 17	48 (39–61)	52 ± 27	41 (37–65)	49 ± 11	48 (42–54)	0.37
TC (mg/dL)	174 ± 48	187 (130–208)	187 ± 47	188 (151–232)	191 ± 64	84 (144–247)	179 ± 57	162 (136–203)	0.66
TG (mg/dL)	167 ± 91	161 (105–214)	168 ± 102	135 (96–208)	202 ± 26	171 (116–246)	159 ± 81	145 (103–196)	0.60
Renalase (µg /mL)	21.8 ± 9.2 [1]	23.9 (19.4–25.9)	20.2 ± 3.1 [2,4]	21.0 (18–21.5)	24.9 ± 4.1 [3,4]	24.6 (2.0–26.8)	35.6 ± 13.5 [1,2,3]	31.79 (28.9–34.9)	$p < 0.01$
CV Risk Prediction Score (%)	11.3 ± 12.1 [1,2]	6.1 (1.4–19.8)	24.0 ± 14.6 [1]	24.0 (11.3–31.5)	25.0 ± 17.2 [2]	22.7 (9.8–39.9)	20 ± 17	17 (6–31)	0.0012

Abbreviations: BP—blood pressure; eGFR—estimated glomerular filtration rate; LDL—low-density cholesterol; HDL—high-density cholesterol; TC—total cholesterol; TG—triglycerides; CV—Cardiovascular. [1,2,3,4]—the result of the post-hoc analysis showing differences between the indicated groups.

Table 2. Qualitative risk factors occurrence in analyzed subgroups (p-values derived from Kruskal–Wallis ANOVA analysis).

Parameter	Control n = 30	CKD III n = 30	CKD IV n = 30	HD n = 30	p-Value
Hypertension n = 95	12 [1,2,3]	28 [1]	29 [2]	26 [3]	<0.05
Diabetes n = 37	7	10	11	9	>0.05
Smoking n = 21	7	6	5	3	>0.05
No Residual Diuresis n = 6	0	0	0	6	<0.05

Arterial hypertension was significantly more common in all CKD subgroups than in control. There was no statistical difference between diagnosed diabetes and active smoking in all groups. [1,2,3]—the result of the post-hoc analysis showing differences between the indicated groups.

Table 3. Demographic and descriptive data and statistical evaluation between control and the whole CKD group (Mann–Whitney U test).

Parameter	Control		CKD		*p*-Value
	Mean ± SD	Median (LQ–UQ)	Mean ± SD	Median (LQ-UQ)	
Age (Years)	57 ± 11	56 (45–67)	66 ± 7	67 (60–74)	<0.001
Systolic BP (mmHg)	131 ± 21	125 (120–140)	135 ± 18	132 (120–140)	0.01
Diastolic BP (mmHg)	80 ± 11	80 (70–85)	80 ± 10	80 (70–90)	0.94
eGFR	85 ± 13	85 (72–91)	49 ± 28	41 (26–72)	<0.001
Body Weight (kg)	83 ± 20	80 (70–94)	83 ± 17	80 (72–92)	0.74
Height (cm)	171 ± 8	170 (164–176)	169 ± 9	170 (161–176)	0.20
LDL (mg/dL)	122 ± 35	124 (97–148)	120 ± 44	123 (84–151)	0.34
HDL (mg/dL)	55 ± 16	52 (45–61)	52 ± 20	50 (40–63)	0.07
TC (mg/dL)	174 ± 48	187 (130–208)	184 ± 53	184 (144–217)	0.63
TG (mg/dL)	167 ± 91	161 (105–214)	179 ± 107	161 106–214)	0.98
Renalase (μg/mL)	21.8 ± 9.2	23.9 (19.4–25.9)	22.3 ± 6.3	22.5 (19.9–25.1)	0.07
CV risk (%)	11.3 ± 12.1	6.1 (1.4–19.8)	22.8 ± 16.3	20.8 (9.4–31.5)	<0.001

Abbreviations: BP—blood pressure; eGFR—estimated glomerular filtration rate; LDL—low-density cholesterol; HDL—high-density cholesterol; TC—total cholesterol; TG—triglycerides; CV—Cardiovascular.

When analyzing the control group and CKD-based subgroups, renalase levels were significantly higher in HD than in any other group and the CKD IV group compared to the CKD III group. There was no significant difference in renalase levels between the control group and CKD III and IV groups.

Renalase levels in subgroups are shown graphically below in Figure 1.

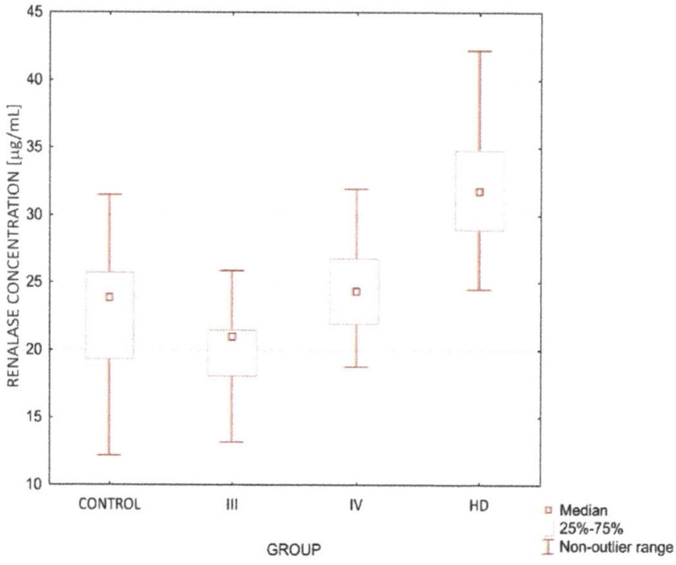

Figure 1. Renalase levels in the studied groups.

In the control group, renalase correlated negatively with body weight (Rs = −0.39, $p < 0.05$). There was also a borderline correlation with systolic (Rs = −0.31, $p = 0.08$) and diastolic (Rs = −0.32, $p = 0.09$) blood pressure.

In the CKD III group, renalase did not correlate with any of the analyzed parameters, but in the stage IV CKD group, it correlated again with eGFR (Rs = −0.48, $p < 0.01$) and body weight (Rs = −0.44, $p < 0.05$).

In the HD group, renalase correlated negatively with eGFR (Rs = −0.55, $p < 0.05$) and positively with residual diuresis (Rs = −0.36, $p < 0.05$).

Comparing adults with no kidney disease and the whole CKD group, there was no significant difference in renalase levels, but this difference was at the borderline significance level ($p = 0.07$).

In the whole CKD group, we found a strong negative correlation between renalase levels and eGFR (Rs = −0.83, $p < 0.001$), as shown in Figure 2.

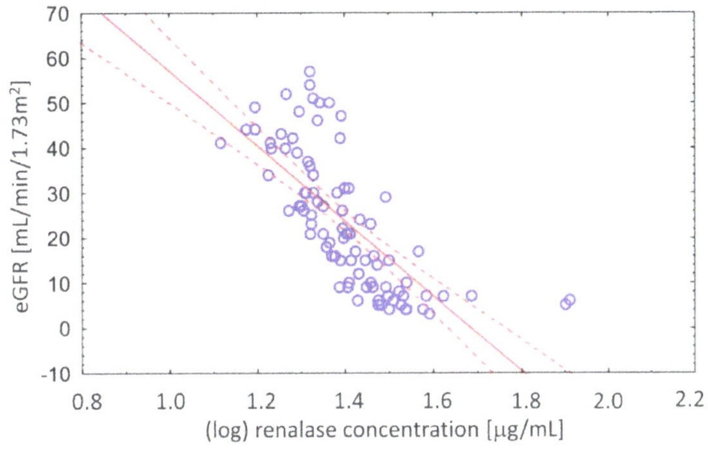

Figure 2. The negative, strong correlation between renalase and eGFR in the CKD group.

When analyzing data from all participants together (control and CKD patients), renalase correlates with diastolic blood pressure (Rs = −0.18, $p < 0.05$), eGFR (Rs = −0.59, $p < 0.001$) and body weight (Rs = −0.32, $p < 0.001$).

3.2. Follow-Up Data

During the follow-up period, we recorded 22 MACEs—2 in control, 2 in CKD III, 5 in CKD IV, and 13 in the HD group. There were 11 deaths; none in control and CKD III, four in CKD IV and seven in HD. We observed five HD initiations. There was no RRT other than hemodialysis initiated. Data are shown in Tables 4 and 5.

Table 4. Qualitative data on recorded endpoints; data are shown as number of individuals (*p*-values derived from Kruskal–Wallis ANOVA analysis).

Parameter	CONTROL n = 29	CKD III n = 30	CKD IV n = 29	HD n = 30	*p*-Value
MACE n = 22	2	2	5	13	0.0006
Deaths n = 11	0	0	4	7	0.0035
Dialysis Initiation n = 5	0	0	5	N/A	N/A

Abbreviations: MACE—major adverse cardiovascular events, CKD—chronic kidney disease.

Table 5. Quantitative data on recorded endpoints with statistical evaluation (Kruskal-Wallis ANOVA). Data are shown as mean ± SD and median (lower quartile—upper quartile).

Parameter	CONTROL n = 29		CKD III n = 30		CKD IV n = 29		HD n = 30		*p*-Value
	Mean ± SD	Median (LQ-UQ)	Mean ± SD	Median (LQ-UQ)	Mean ± SD	Median (LQ-UQ)	Mean ± SD	Median (LQ-UQ)	
Time to MACE (Months)	6 ± 5	6 (3–10)	9 ± 3	9 (7–12)	14 ± 2 [1]	14 (13–15)	7 ± 4 [1]	8 (4–9)	0.028
Time to Death (Months)	N/A	N/A	N/A	N/A	15 ± 4.5	14.5 (14–16)	11 ± 4	12 (9–14)	>0.05
Time to HD Initiation (Months)	N/A	N/A	N/A	N/A	10 ± 4.5	12 (11–12)	N/A	N/A	N/A

Abbreviations: MACE—major adverse cardiovascular events; CKD—chronic kidney disease; HD—hemodialysis. [1]—the result of the post-hoc analysis showing differences between the indicated groups.

Even though statistical analysis showed significant differences in both MACE occurrence and all-cause mortality in all subgroups, the post-hoc analysis did not show specific significance between particular groups. All RRT initiations were observed in the CKD IV group.

Observed time to first MACE was significantly shorter in the HD group than in the CKD IV group. There was no statistical difference between time to all-cause death between groups in which deaths were observed. The median time to RRT initiation was 12 months.

In CKD subgroups, RNLS correlated with all-cause death occurrence in the HD group (Rs = 0.49, $p < 0.01$), but no other significant correlation was found.

When analyzing the whole CKD patient population, we found a positive correlation of RNLS concentration and MACE occurrence (Rs = 0.38, $p < 0.001$) and a negative correlation with time to first MACE occurrence (Rs = −0.52, $p < 0.05$). The correlation of RNLS and MACE occurrence was stronger than that of the algorithm derived, calculated risk (Rs = 0.25, $p < 0.05$). All-cause death also correlated positively with renalase concertation (Rs = 0.34, $p < 0.005$).

Data pooled from all participants (both with and without CKD) concerning endpoints have shown a correlation between RNLS and MACE occurrence (Rs = 0.33, $p < 0.001$), time to MACE (Rs = −0.50, $p < 0.05$) and all-cause death (Rs = 0.31, $p < 0.001$). This conclusion takes into account that a small number of endpoints were observed in the control group.

3.3. Analysis of Survival and MACE Occurrence

Based on obtained data, specifically, renalase levels and correlations between renalase and occurrence of MACE and all-cause death in the CKD group, we first divided the CKD group into two subgroups with RNLS levels higher and lower than 25 µg/mL.

MACE occurred in 20 patients with CKD: 11 with elevated renalase levels (>25 µg/mL) and 9 with lower renalase levels (<25 µg/mL). A Mantel–Cox test $p < 0.01$ showed a significant difference in MACE occurrence between analyzed groups, the odds of which were significantly higher in the group with elevated renalase levels, as shown in Figure 3.

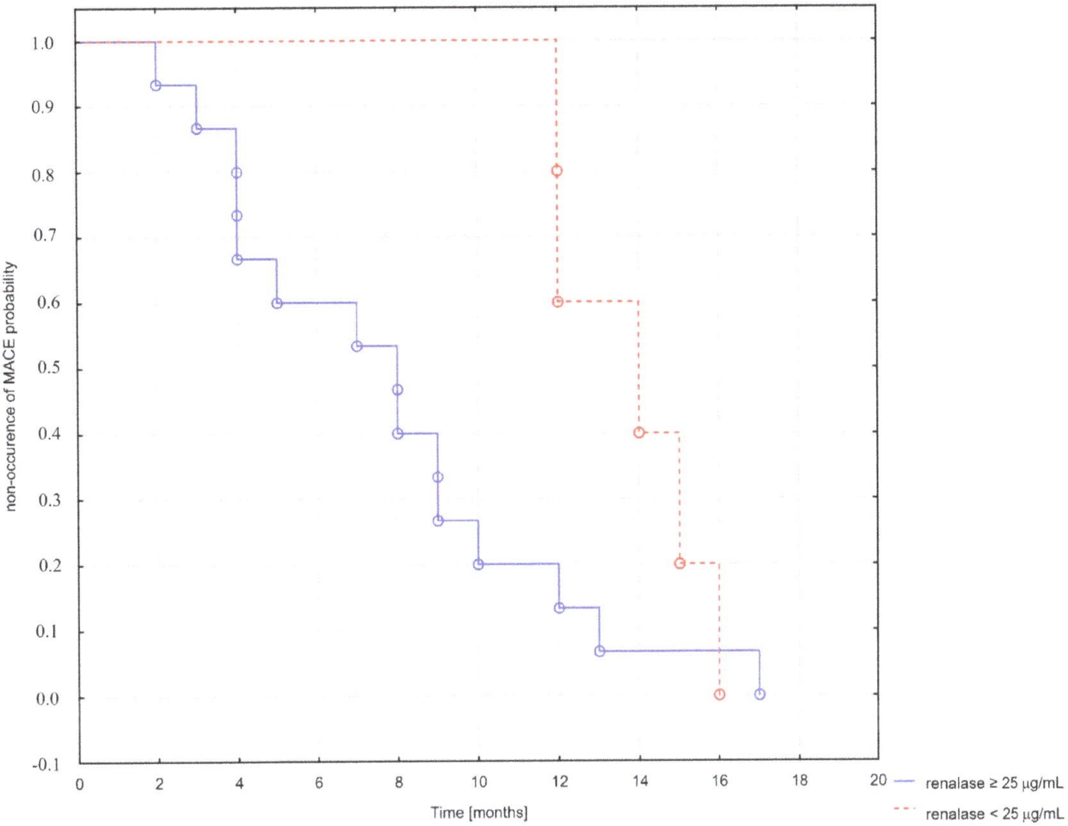

Figure 3. Kaplan Meier curve for MACE in groups with low and elevated renalase levels.

There was no difference in the occurrence of death between subgroups divided basing on the 25 µg/mL threshold.

When a higher threshold was assumed, and patients with CKD were divided into groups with levels of RNLS higher or lower than 30 µg/mL, not only did the higher probability of MACE remain significant, but a higher probability of all-cause death was also observed (but without statistical significance in log-rank test: $p = 0.09$), as presented in Figure 4.

In the HD group, in which a significant correlation between RNSL levels and all-cause death was observed, a Kaplan–Meier analysis of survival was not possible as the deaths were observed only in patients with renalase concentration higher than 30 µg/mL. At the same time, there was no significant difference in MACE occurrence in this subgroup.

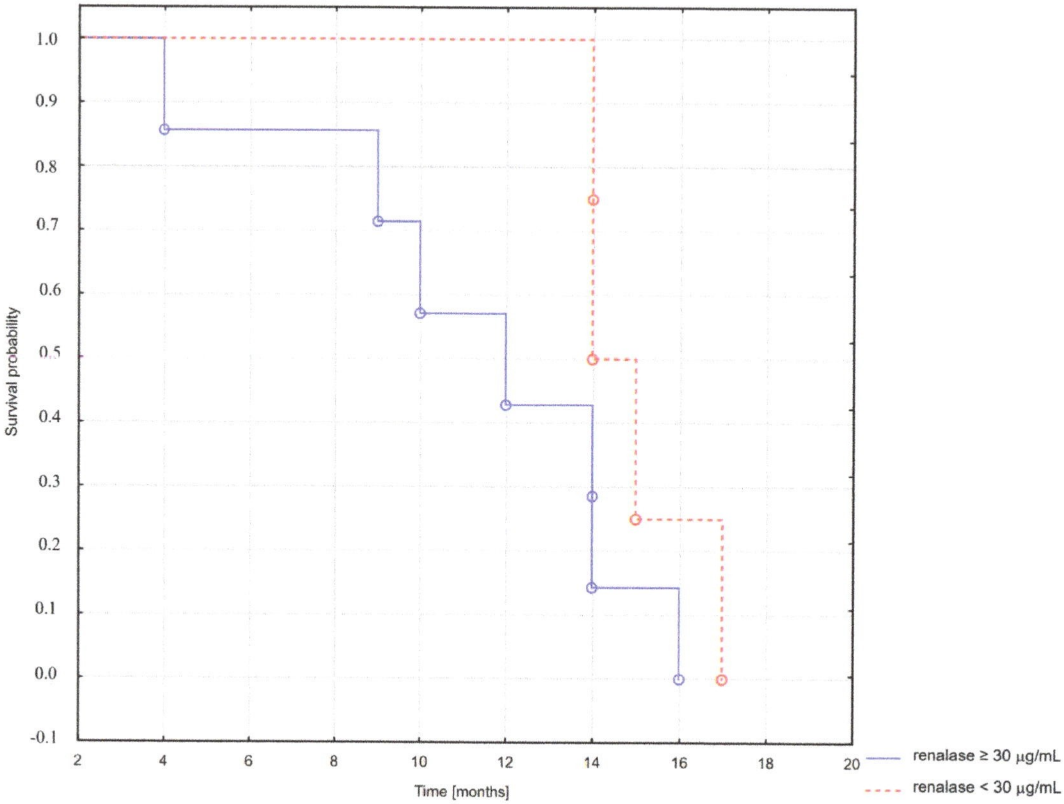

Figure 4. Kaplan–Meier curve for all-cause death in groups with low and high renalase levels.

4. Discussion

CKD significantly contributes globally to both morbidity and mortality. Exact numbers are unknown, but screening programs in high-income countries have shown that more than 10% of adults have markers of kidney disease. The global burden of CKD in 2017 was estimated at 6975 million cases. Even though primary causes of CKD vary in different populations, the most common causes are hypertension and diabetes [5]. In Poland, an estimated 4.2 million people have CKD, and about 6500 yearly develop ESRD and require RRT [6]. The lives of patients afflicted with CKD, whether they deal with early-stage disease or end-stage renal disease (ESRD), are associated with high morbidity and increased healthcare utilization [7].

Patients with CKD have an established, higher CVD-related morbidity and mortality, and multiple analyses of studies have shown that the presence of CKD is independently associated with cardiovascular events, especially in groups with preexisting known risk factors [8–10]. The exponential increase in risk for cardiovascular mortality risk at low eGFR was shown in a large meta-analysis comprising ten cohorts with 266,975 patients. Hazard ratios at eGFRs of 60, 45 and 15 mL/min per 1.73 m^2 were 1.03, 1.38 and 3.11, respectively, compared to an eGFR of 95, after adjustment for albuminuria and cardiovascular risk factors [11]. A total of 7.6% of all CVD deaths (1.4 million globally in 2017) could be attributed to impaired kidney function [12].

CVD burden is also qualitatively different in CKD patients than those without renal pathology. Aside from atherosclerosis-related diseases (coronary artery disease, ischemic stroke and peripheral artery disease), other cardiac events become more and more prevalent as kidney function declines. Non-atherosclerotic CVD, arrhythmias, sudden cardiac deaths, arterial calcifications, valve calcifications and haemorrhagic strokes may be caused by uraemia, CKD-MBD, LVH or be dialysis-related events [13]. A growing trend of non-ischemic CVD, to which CKD undoubtedly contributes, has been observed in developed countries [14].

Standard risk factors and assessment tools underperform in CKD patients, as their predictive risk is much lower than observed CVD events [15]. Moreover, we do not have particularly well-designed or widely used recalibration tools, which should at least account for eGFR and albuminuria [10]. Standard and widespread CVD risk assessment tools name CKD as an additional factor to be taken into consideration but do not clearly stratify the eventual risk. CVD risk equations can be of great importance to both patients and clinicians during decision-making. Even the classic symptoms of acute myocardial infarction are reported far less often by CKD patients than people without kidney pathology [13]. Risk-aware and evidence-based medical interventions are needed as CKD prevalence continues to grow. Early action can delay disease progression and mitigate the fact that patients with ESRD consume a disproportionate share of health care resources.

Renalase has been linked to CKD and CVD in numerous studies. In a seminal paper describing renalase discovery, Xu et al. postulated a link to the cardiovascular system [16]; unfortunately, the enzymatic function of catecholamine degradation primarily proposed in the study, directly linking renalase to CVD, was disputed [17–21] and eventually refuted [22]. Different enzymatic activity was observed and described; renalase was found to be an oxidase/isomerase, using molecular oxygen to convert α-NAD(P)H into β-NAD(P)+, with hydrogen peroxide as a reaction byproduct [23], which also turned out to be incorrect [24]. Eventually, renalase has come to be described as an isomerase catalyzing rapid oxidation of 6DHNAD(P) and 2DHNAD(P) (which are inhibitory to metabolism dehydrogenases) to their active form—β-NAD(P)+ [22].

Additionally, since 2014, renalase's other function has been observed—that of a cytokine. It was found that a part of renalase protein – RP 220 (described previously [25], containing amino acids 220–239) as well as its variations RP-224 (amino acid 224–233) and RP-H220 (histidine-tagged RP 2020) – improved HK-2 cell line survival when treated with cisplatin, independently of amine oxidase activity by activation of the AKT and MAPK signaling pathways [26]. A receptor for renalase was found a year later. In 2015, PMCA4b (plasma membrane ATPase) was discovered to bind with renalase and activate the aforementioned signaling pathways [27], thus rekindling the search for the causative interplay of renalase and CVD, as PMCA4b is connected to both cardiac hypertrophy [28] and hypertension [29]. Aside from catalytic and cytokine mechanisms, yet another metabolic role came to light through further research in 2019: data collected in models of liver injury indicated that renalase might activate SIRT1 by elevating NAD+ levels [30].

Renalase affects redox balance in all functions and is beneficial in states of ischemia and/or reperfusion injury.

Its role in kidney disease has been shown in animal model interventional studies. Contrast-induced nephropathy in rat models was mitigated by supplying exogenous renalase by reducing oxidative stress, among other influences [31]. Renalase has also been found to reduce cisplatin-induced acute kidney injury by decreasing mitochondrial fission and reactive oxygen production [30].

Heart disease is also linked to redox state [32]. As such, studies have shown associations between higher renalase levels and unstable angina pectoris [33], lower ejection fraction in heart failure patients [34,35] and coronary microvascular dysfunction diagnosed with Rb-82 PET/CT imaging [36]. When performed in patients with stable angina, Percutaneous Coronary Intervention decreased serum renalase levels within a few days [37]. Extensive reviews of other observed relationships can be found elsewhere [38–40].

Even though we do not fully understand renalase's mechanism of action, as paradigm shifts seem to occur every few years, renalase displays a wide range of correlations and effects in CVD and both chronic kidney disease and acute kidney injury, which have been found in observational and interventional studies.

Other studies have also tried to assess renalase levels as a predictive factor. A study by Gluba-Brzózka et al. [41] positively correlated renalase with worsening CKD stages, as did markers (both biochemical and functional) of cardiovascular risk, but no observational follow-up was mentioned. Chang et al. [42] conducted a study in which patients (all male) were observed up to 10 years. A statistically significant correlation was found between renalase levels at baseline and both 1- and 5-year all-cause mortality. No correlation was found at 10 years. Using data from a Korean K-STAR cohort, Na et al. [43] observed higher all-cause mortality but not more frequent MACE in CKD patients with higher renalase levels. Another group studied patients recruited during scheduled PCI. In the group characterized by higher renalase levels, more endpoints (comprising myocardial infarction, stroke or death) were recorded during follow-up (median follow-up time lasted 4.1 years) [37]. In yet another recent study by Cerqueira et al., 40 pre-dialysis CKD patients were followed up for a mean period of 65 months. RNLS was associated with CKD progression, hospitalizations and all-cause mortality, but not with MACE occurrence [44]. Our study results are in line with these findings. Renalase levels found in studies mentioned in this paragraph are summarized in Table 6.

Table 6. Renalase levels in various groups in studies by other authors.

Publication	Group	Level	Group	Level
Markers of Increased Cardiovascular Risk in Patients with Chronic Kidney Disease [41]	Control Group n = 45	mean: 251.0 ± 157.0 ng/mL	CKD all stages n = 132	mean: 316.1 ± 155.3
Identification of Two Forms of Human Plasma Renalase, and Their Association with All-Cause Mortality [42]	Normal renal function n = 10	mean: 20.39 ± 7.70 µg/mL	All patients n = 267	mean: 18.8 ± 8.5 µg/mL
Circulating renalase predicts all-cause mortality and renal outcomes in patients with advanced chronic kidney disease [43]	Control group n = 16	mean: 28.2 ± 5.1 µg/mL	CKD patients n = 383	mean: 75.8 ± 34.8 µg/mL
Serum Renalase Levels Are Predicted by Brain-Derived Neurotrophic Factor and Associated with Cardiovascular Events and Mortality after Percutaneous Coronary Intervention [37]	Before percutaneous coronary intervention n = 152	mean: 47.5 ± 17.3 ng/mL	After percutaneous coronary intervention n = 152	mean: 35.9 ± 11.3 ng/mL
Circulating Renalase as Predictor of Renal and Cardiovascular Outcomes in Pre-Dialysis CKD Patients: A 5-Year Prospective Cohort Study [44]	CKD stages 1 and 2 n = 17	mean: 42.03 µg/mL	CKD stages 4 and 5 n = 14	mean: 83.53 µg/mL

Abbreviations: CKD—chronic kidney disease.

5. Conclusions

Loss of kidney function is accompanied by an accumulation of renalase in blood serum.

- Elevated renalase levels (>25 μL/mL) are a risk factor of MACE in patients with chronic kidney disease, but its long-term utility needs further research.
- High renalase levels (>30 μg/mL) can be a risk factor of death among CKD patients.
- In HD patients, all deaths were observed among patients with a renalase serum concentration greater than 30 μg/mL. Further research could show a possible "threshold" level, prompting an evaluation of patients to either intensify treatment or start discussing palliative measures.

Author Contributions: Conceptualization, W.K.; methodology, W.K. and N.M.S.; statistical analysis, B.G.; investigation, E.C.-H.; resources, A.G.; writing—original draft preparation, W.K.; writing—review and editing, B.D. and K.C.; supervision, M.W.; funding acquisition, K.C. All authors have read and agreed to the published version of the manuscript.

Funding: This research received no external funding.

Institutional Review Board Statement: The study was conducted according to the guidelines of the Declaration of Helsinki and approved by the Ethics Committee of Pomeranian Medical University in Szczecin (No. KB-0012/07/17 from 16 January 2017).

Informed Consent Statement: Informed consent was obtained from all subjects involved in the study.

Conflicts of Interest: The authors declare no conflict of interest. The funders had no role in the design of the study, in the collection, analyses, or interpretation of data, in the writing of the manuscript or in the decision to publish the results.

References

1. Uhlén, M.; Fagerberg, L.; Hallström, B.M. Tissue Expression of RNLS—Summary—The Human Protein Atlas. Available online: https://www.proteinatlas.org/ENSG00000184719-RNLS/tissue (accessed on 29 November 2020).
2. Wiśniewska, M.; Serwin, N.; Dziedziejko, V.; Marchelek-Myśliwiec, M.; Dołęgowska, B.; Domański, L.; Ciechanowski, K.; Safranow, K.; Pawlik, A. Chronic Kidney Disease Is Associated with Increased Levels of Renalase in Serum and Decreased in Erythrocytes. *Pol. Arch. Intern. Med.* **2019**, *129*, 790–797. [CrossRef] [PubMed]
3. Levey, A.S.; Inker, L.A.; Coresh, J. GFR Estimation: From Physiology to Public Health. *Am. J. Kidney Dis.* **2014**, *63*, 820–834. [CrossRef]
4. Goff, D.C.; Lloyd-Jones, D.M.; Bennett, G.; Coady, S.; D'Agostino, R.B.; Gibbons, R.; Greenland, P.; Lackland, D.T.; Levy, D.; O'Donnell, C.J.; et al. 2013 ACC/AHA Guideline on the Assessment of Cardiovascular Risk: A Report of the American College of Cardiology/American Heart Association Task Force on Practice Guidelines. *Circulation* **2014**, *129*, 49–73. [CrossRef]
5. Bikbov, B.; Purcell, C.A.; Levey, A.S.; Smith, M.; Abdoli, A.; Abebe, M.; Adebayo, O.M.; Afarideh, M.; Agarwal, S.K.; Agudelo-Botero, M.; et al. Global, Regional, and National Burden of Chronic Kidney Disease, 1990–2017: A Systematic Analysis for the Global Burden of Disease Study 2017. *Lancet* **2020**, *395*, 709–733. [CrossRef]
6. Kalinowska, A.; Kowalczyk, M.; Pruszko, C.; Dostęp, P.T. Do Świadczeń Nefrologicznych w Polsce. Raport 2019. Available online: http://www.korektorzdrowia.pl/wp-content/uploads/raport_dostep_do_swiadczen_nefrologicznych.pdf (accessed on 24 July 2021).
7. Go, A.S.; Chertow, G.M.; Fan, D.; McCulloch, C.E.; Hsu, C. Chronic Kidney Disease and the Risks of Death, Cardiovascular Events, and Hospitalization. *N. Engl. J. Med.* **2004**, *351*, 1296–1305. [CrossRef]
8. Fox, C.S.; Matsushita, K.; Woodward, M.; Bilo, H.J.G.; Chalmers, J.; Lambers Heerspink, H.J.; Lee, B.J.; Perkins, R.M.; Rossing, P.; Sairenchi, T.; et al. Associations of Kidney Disease Measures with Mortality and End-Stage Renal Disease in Individuals with and without Diabetes: A Meta-Analysis. *Lancet* **2012**, *380*, 1662–1673. [CrossRef]
9. Mahmoodi, B.K.; Matsushita, K.; Woodward, M.; Blankestijn, P.J.; Cirillo, M.; Ohkubo, T.; Rossing, P.; Sarnak, M.J.; Stengel, B.; Yamagishi, K.; et al. Associations of Kidney Disease Measures with Mortality and End-Stage Renal Disease in Individuals with and without Hypertension: A Meta-Analysis. *Lancet* **2012**, *380*, 1649–1661. [CrossRef]
10. Matsushita, K.; Coresh, J.; Sang, Y.; Chalmers, J.; Fox, C.; Guallar, E.; Jafar, T.; Jassal, S.K.; Landman, G.W.D.; Muntner, P.; et al. Estimated Glomerular Filtration Rate and Albuminuria for Prediction of Cardiovascular Outcomes: A Collaborative Meta-Analysis of Individual Participant Data. *Lancet Diabetes Endocrinol.* **2015**, *3*, 514–525. [CrossRef]

11. Van Der Velde, M.; Matsushita, K.; Coresh, J.; Astor, B.C.; Woodward, M.; Levey, A.; De Jong, P.; Gansevoort, R.T.; El-Nahas, M.; Eckardt, K.U.; et al. Lower Estimated Glomerular Filtration Rate and Higher Albuminuria Are Associated with All-Cause and Cardiovascular Mortality. A Collaborative Meta-Analysis of High-Risk Population Cohorts. *Kidney Int.* **2011**, *79*, 1341–1352. [CrossRef]
12. Carney, E.F. The Impact of Chronic Kidney. *Nat. Rev. Nephrol.* **2020**, *16*, 251. [CrossRef] [PubMed]
13. Sarnak, M.J.; Amann, K.; Bangalore, S.; Cavalcante, J.L.; Charytan, D.M.; Craig, J.C.; Gill, J.S.; Hlatky, M.A.; Jardine, A.G.; Landmesser, U.; et al. Chronic Kidney Disease and Coronary Artery Disease: JACC State-of-the-Art Review. *J. Am. Coll. Cardiol.* **2019**, *74*, 1823–1838. [CrossRef]
14. Shah, N.S.; Molsberry, R.; Rana, J.S.; Sidney, S.; Capewell, S.; O'Flaherty, M.; Carnethon, M.; Lloyd-Jones, D.M.; Khan, S.S. Heterogeneous Trends in Burden of Heart Disease Mortality by Subtypes in the United States, 1999-2018: Observational Analysis of Vital Statistics. *BMJ-BRIT Med. J.* **2020**, *370*. [CrossRef] [PubMed]
15. Weiner, D.E.; Tighiouart, H.; Elsayed, E.F.; Griffith, J.L.; Salem, D.N.; Levey, A.S.; Sarnak, M.J. The Framingham Predictive Instrument in Chronic Kidney Disease. *J. Am. Coll. Cardiol.* **2007**, *50*, 217–224. [CrossRef] [PubMed]
16. Xu, J.; Crowley, S.; Desir, G.V.; Xu, J.; Li, G.; Wang, P.; Velazquez, H.; Yao, X.; Li, Y. Renalase Is a Novel, Soluble Monoamine Oxidase That Regulates Cardiac Function and Blood Pressure Find the Latest Version: Renalase Is a Novel, Soluble Monoamine Oxidase That Regulates Cardiac Function and Blood Pressure. *J. Clin. Investig.* **2005**, *115*, 1275–1280. [CrossRef]
17. Pandini, V.; Ciriello, F.; Tedeschi, G.; Rossoni, G.; Zanetti, G.; Aliverti, A. Synthesis of Human Renalase1 in Escherichia Coli and Its Purification as a FAD-Containing Holoprotein. *Protein Expr. Purif.* **2010**, *72*, 244–253. [CrossRef] [PubMed]
18. Eikelis, N.; Hennebry, S.C.; Lambert, G.W.; Schlaich, M.P. Does Renalase Degrade Catecholamines ? *Kidney Int.* **2011**, *79*, 1380. [CrossRef]
19. Weir, M.R.; Eikelis, N.; Hennebry, S.C.; Lambert, G.W.; Schlaich, M.P. Does Renalase Degrade Catecholamines ? The Author Replies. *Kidney Int.* **2011**, *79*, 1380–1381. [CrossRef]
20. Desir, G. Novel Insights into the Physiology of Renalase and Its Role in Hypertension and Heart Disease. *Pediatr. Nephrol.* **2012**, *27*, 719–725. [CrossRef]
21. Medvedev, A.E.; Veselovsky, A.V.; Fedchenko, V.I. Renalase, a New Secretory Enzyme Responsible for Selective Degradation of Catecholamines: Achievements and Unsolved Problems. *Biochemistry* **2010**, *75*, 951–958. [CrossRef] [PubMed]
22. Beaupre, B.A.; Hoag, M.R.; Moran, G.R. Renalase Does Not Catalyze the Oxidation of Catecholamines. *Arch. Biochem. Biophys.* **2015**, *579*, 62–66. [CrossRef] [PubMed]
23. Beaupre, B.A.; Carmichael, B.R.; Hoag, M.R.; Shah, D.D.; Moran, G.R. Renalase Is an α-NAD(P)H Oxidase/Anomerase. *J. Am. Chem. Soc.* **2013**, *135*, 13980–13987. [CrossRef]
24. Moran, G.R. The Catalytic Function of Renalase: A Decade of Phantoms. *Biochim. Biophys. Acta Proteins Proteom.* **2016**, *1864*, 177–186. [CrossRef]
25. Milani, M.; Ciriello, F.; Baroni, S.; Pandini, V.; Canevari, G.; Bolognesi, M.; Aliverti, A. FAD-Binding Site and NADP Reactivity in Human Renalase: A New Enzyme Involved in Blood Pressure Regulation. *J. Mol. Biol.* **2011**, *411*, 463–473. [CrossRef]
26. Wang, L.; Velazquez, H.; Moeckel, G.; Chang, J.; Ham, A.; Lee, H.T.; Safirstein, R.; Desir, G.V. Renalase Prevents AKI Independent of Amine Oxidase Activity. *J. Am. Soc. Nephrol.* **2014**, *25*, 1226–1235. [CrossRef] [PubMed]
27. Wang, L.; Velazquez, H.; Chang, J.; Safirstein, R.; Desir, G.V. Identification of a Receptor for Extracellular Renalase. *PLoS ONE* **2015**, *10*, e0122932. [CrossRef]
28. Mohamed, T.M.A.; Abou-Leisa, R.; Stafford, N.; Maqsood, A.; Zi, M.; Prehar, S.; Baudoin-Stanley, F.; Wang, X.; Neyses, L.; Cartwright, E.J.; et al. The Plasma Membrane Calcium ATPase 4 Signalling in Cardiac Fibroblasts Mediates Cardiomyocyte Hypertrophy. *Nat. Commun.* **2016**, *7*, 11074. [CrossRef] [PubMed]
29. Little, R.; Cartwright, E.J.; Neyses, L.; Austin, C. Plasma Membrane Calcium ATPases (PMCAs) as Potential Targets for the Treatment of Essential Hypertension. *Pharmacol. Ther.* **2016**, *159*, 23–34. [CrossRef]
30. Zhang, T.; Gu, J.; Guo, J.; Chen, K.; Li, H.; Wang, J. Renalase Attenuates Mouse Fatty Liver Ischemia/Reperfusion Injury through Mitigating Oxidative Stress and Mitochondrial Damage via Activating SIRT1. *Oxid. Med. Cell. Longev.* **2019**, *2019*, 1–21. [CrossRef] [PubMed]
31. Zhao, B.; Zhao, Q.; Li, J.; Xing, T.; Wang, F.; Wang, N. Renalase Protects against Contrast-Induced Nephropathy in Sprague-Dawley Rats. *PLoS ONE* **2015**, *10*, e0116583. [CrossRef]
32. Daiber, A.; Hahad, O.; Andreadou, I.; Steven, S.; Daub, S.; Münzel, T. Redox-Related Biomarkers in Human Cardiovascular Disease—Classical Footprints and Beyond. *Redox Biol.* **2021**, *42*, 101875. [CrossRef]
33. Izadpanah, P.; Asadian, F.; Jangjou, A. Association of Serum Renalase Levels and Renalase Rs10887800 Polymorphism with Unstable Angina Pectoris Patients Having Metabolic Syndrome. *Diabetes Metab. Syndr. Obes. Targets Ther.* **2020**, *13*, 3249–3259. [CrossRef] [PubMed]
34. Stojanovic, D.; Mitic, V.; Stojanovic, M.; Petrovic, D.; Ignjatovic, A.; Stefanovic, N.; Cvetkovic, T.; Kocic, G.; Bojanic, V.; Deljanin Ilic, M. The Partnership between Renalase and Ejection Fraction as a Risk Factor for Increased Cardiac Remodeling Biomarkers in Chronic Heart Failure Patients. *Curr. Med. Res. Opin.* **2020**, *36*, 909–919. [CrossRef] [PubMed]
35. Stojanovic, D.; Mitic, V.; Petrovic, D.; Stojanovic, M.; Ignjatovic, A.; Stefanovic, N.; Cvetkovic, T.; Bojanic, V.; Kocic, G.; Ilic, M.D. Association of Plasma Renalase and Left Ventricle Mass Index in Heart Failure Patients Stratified to the Category of the Ejection Fraction: A Pilot Study. *Dis. Markers* **2019**, *2019*, 1–9. [CrossRef] [PubMed]

36. Safdar, B.; Guo, X.; Johnson, C.; D'Onofrio, G.; Dziura, J.; Sinusas, A.J.; Testani, J.; Rao, V.; Desir, G. Elevated Renalase Levels in Patients with Acute Coronary Microvascular Dysfunction—A Possible Biomarker for Ischemia. *Int. J. Cardiol.* **2019**, *279*, 155–161. [CrossRef] [PubMed]
37. Lee, I.-T.; Sheu, W. Serum Renalase Levels Are Predicted by Brain-Derived Neurotrophic Factor and Associated with Cardiovascular Events and Mortality after Percutaneous Coronary Intervention. *J. Clin. Med.* **2018**, *7*, 437. [CrossRef]
38. Czubilińska-Łada, J.; Gliwińska, A.; Badeński, A.; Szczepańska, M. Associations between Renalase Concentration and the Occurrence of Selected Diseases. *Endokrynol. Pol.* **2020**, *71*, 334–342. [CrossRef]
39. Li, Y.; Wu, W.; Liu, W.; Zhou, M. Roles and Mechanisms of Renalase in Cardiovascular Disease: A Promising Therapeutic Target. *Biomed. Pharmacother.* **2020**, *131*, 110712. [CrossRef] [PubMed]
40. Wiśniewska, M.; Serwin, N.M.; Gomółka, A.; Knop, W.; Heryć, R.; Cecerska-Heryć, E.; Skwirczyńska, E.; Dołęgowska, B. Renalaza—Działanie, Aspekty Kliniczne i Potencjał Terapeutyczny. *Forum Nefrol.* **2020**, *13*, 59–68.
41. Gluba-Brzózka, A.; Michalska-Kasiczak, M.; Franczyk-Skóra, B.; Nocuń, M.; Banach, M.; Rysz, J. Markers of Increased Cardiovascular Risk in Patients with Chronic Kidney Disease. *Lipids Health Dis.* **2014**, *13*, 135. [CrossRef]
42. Chang, J.; Guo, X.; Rao, V.; Gromisch, E.S.; Chung, S.; Kluger, H.M.; Cha, C.; Gorelick, F.; Testani, J.; Safirstein, R.; et al. Identification of Two Forms of Human Plasma Renalase, and Their Association with All-Cause Mortality. *Kidney Int. Rep.* **2020**, *5*, 362–368. [CrossRef]
43. Na, K.Y.; Baek, S.H.; Cha, R.-H.; Kang, S.W.; Park, C.W.; Cha, D.R.; Kim, S.G.; Yoon, S.A.; Kim, S.; Han, S.-Y.; et al. Circulating Renalase Predicts All-Cause Mortality and Renal Outcomes in Patients with Advanced Chronic Kidney Disease. *Korean J. Intern. Med.* **2019**, *34*, 858–866. [CrossRef]
44. Cerqueira, A.; Quelhas-Santos, J.; Ferreira, I.; Sampaio, S.; Relvas, M.; Marques, N.; Dias, C.C.; Pestana, M. Circulating Renalase as Predictor of Renal and Cardiovascular Outcomes in Pre-Dialysis CKD Patients: A 5-Year Prospective Cohort Study. *Life* **2021**, *11*, 210. [CrossRef] [PubMed]

biomolecules

MDPI

Article

Preservation of Mitochondrial Coupling and Renal Function by Controlled Oxygenated Rewarming of Porcine Kidney Grafts

Hristo Zlatev, Charlotte von Horn and Thomas Minor *

Surgical Research Department, Clinic for General, Visceral and Transplantation Surgery, University Hospital Essen, Hufelandstr. 55, D-45147 Essen, Germany; chirfor@uk-essen.de (H.Z.); charlotte.von-horn@uk-essen.de (C.v.H.)
* Correspondence: thomas.minor@uk-essen.de; Tel.: +49-201-723-2713; Fax: +49-201-723-6859

Abstract: Background: Warm reperfusion after previous cold storage has been shown to have a negative impact on mitochondrial function of organ grafts. Here, we wanted to investigate whether a more controlled warming up of the cold graft by ex vivo machine perfusion with gradually elevated temperature from cold to normothermia (including comparison of two warming up protocols) prior to implantation would be effective in preventing mitochondrial dysfunction upon reperfusion. Methods: All experiments were conducted on porcine kidneys retrieved 15 min after cardiac arrest. After 18 h of cold storage in HTK solution (CS, n = 6), kidneys (n = 6) were subjected to 2 h of reconditioning machine perfusion starting with a hypothermic period followed by a gradual increase in perfusion temperature up to 35 °C (controlled oxygenated rewarming—COR). For a second group (n = 6), the slow warming up was begun instantly after connecting the graft onto the machine (iCOR). Functional recovery of all grafts was then observed upon normothermic reperfusion in vitro. At the conclusion of the experiments, tissue specimens were taken for immediate isolation and analysis of renal mitochondria. Results: COR resulted in a significantly and more than 3-fold increased glomerular filtration rate upon reperfusion, along with a significant higher tubular sodium reabsorption and lesser loss of glucose in comparison to the controls. Enzyme release (AST) was also massively reduced during the reperfusion period. Specific analysis at the mitochondrial level revealed significantly better coupling efficiency and spare respiratory capacity in the COR group compared to the cold storage group. Interestingly, additional experiments revealed that the omission of a hypothermic perfusion period did not deteriorate any of the results after COR, provided that the instant temperature increase from 10 to 35 °C was effectuated in the same controlled manner. Conclusion: Controlled rewarming after extended cold preservation effectively improves mitochondrial recovery upon reperfusion and early functional outcome of kidney grafts.

Keywords: controlled oxygenated rewarming; mitochondrial uncoupling; rewarming injury; temperature paradox

Citation: Zlatev, H.; von Horn, C.; Minor, T. Preservation of Mitochondrial Coupling and Renal Function by Controlled Oxygenated Rewarming of Porcine Kidney Grafts. *Biomolecules* **2021**, *11*, 1880. https://doi.org/10.3390/biom11121880

Academic Editor: Liang-Jun Yan

Received: 10 November 2021
Accepted: 9 December 2021
Published: 14 December 2021

1. Introduction

Organ damage resulting from preservation or reperfusion still represents a major issue in transplantation medicine. Extended preservation times or graft donation after cardiocirculatory standstill in the donor are often conflicted with a reduced recovery after transplantation affecting early graft function, as well as long-term survival.

Although the underlying mechanistic pathways of preservation injury are not yet fully understood, altered mitochondrial integrity upon reperfusion has been deciphered as a major culprit for graft dysfunction after transplantation [1,2]. During cold ischemic preservation, opening of the mitochondrial transition pore (MTP) could be observed along with mitochondrial swelling [3,4]. However, mitochondrial respiratory capacity has been shown to remain stable for up to 24 h of cold storage and only deteriorates after unusual prolongation of cold ischemia or upon subsequent warm reperfusion [5].

Likewise, cytosolic release of cytochrome c and sequential induction of apoptosis are only observed after subsequent rewarming [4,6,7]. In line with these observations, main parts of organ preservation injury substantiate not during cold storage but upon and in consequence of abrupt rewarming [8,9]. Hence, significant alleviation of functional impairment of kidney grafts after transplantation could be achieved by modifying the rise in temperature during reoxygenation using a controlled oxygenated rewarming (COR) protocol subsequent to cold storage by thermoregulated machine perfusion [10,11]. The use of an incremental temperature rise instead of an abrupt return to normothermia improved post-ischemic recovery of discarded human donor livers [12] and resulted in superior results after experimental [10] and clinical [13] kidney transplantation. COR was also found to be operative in significantly reducing the initiation of the mitochondrial apoptotic pathway upon reperfusion along, with a better preservation of mitochondrial content of NAD+ [14]. Likewise, COR improved global recovery of tissue energetics [9] and largely preserved oxygen utilization efficiency at the whole organ level that was otherwise found significantly disturbed [11].

Controlled oxygenated rewarming during machine perfusion classically comprises three phases. A first period of oxygenated cold perfusion is undertaken to fuel residual aerobic metabolism to ameliorate subcellular homeostasis while avoiding major temperature shifts [15,16]. Then follows a thermal transition phase of slow and adapted increase in the perfusion temperature [9] and a final steady state period of perfusion at the final temperature for a variable time span [7].

Although the positive effects of a brief hypothermic perfusion period alone have been thoroughly established in that the cellular aerobic energetic homeostasis could be improved in the cold and that the tissue is better prepared for the following abrupt warm reperfusion [15–17], the relevance of an extended cold perfusion phase versus an immediate start of the gentle warming up protocol in the setting of COR has not yet been addressed.

Thus, the aim of the present study was to scrutinize the relative impacts of a hypothermic equilibration phase in the setting of a controlled rewarming protocol with a special focus on mitochondrial function and coupling status.

2. Materials and Methods

All experiments were performed in accordance with the federal law regarding the protection of animals. The principles of laboratory animal care (NIH publication no. 85-23, revised 1985) were followed.

Kidneys were removed from dead German Landrace pigs weighing between 25 and 30 kg. Then, 15 min after cardio-circulatory standstill, the renal artery was cannulated and the kidneys were flushed by 100 cm gravity with 100 mL of HTK solution (Köhler Chemie, Bensheim, Germany) on the back-table at 4 °C. No heparin was given at any time. After 18 h of static cold preservation in HTK solution, the grafts were randomly assigned to one of the following groups (n = 6, resp.):

1. In the control group, kidneys were exposed to 18 h of static cold storage at 4 °C without additional treatment (CS);
2. In the second group, controlled oxygenated rewarming (COR) was performed after 18 h of CS by end-ischemic machine perfusion, as described earlier [11,18]. Perfusion with Aqix-RS-I (Life Science Group, Bedford, UK) was started at a temperature of 8 °C with consecutive rewarming of the perfusate from 8 °C to 35 °C during the first 90 min. The rise in temperature was accompanied by an adapted increase of the perfusion pressure from 30 to 75 mmHg. The last 30 min of perfusion was kept constant at 35 °C (cf. Figure 1);
3. A third group was investigated, whereby the slow warming up was begun instantly after connecting the graft onto the machine, i.e., omitting the hypothermic equilibration phase (cf. Figure 1).

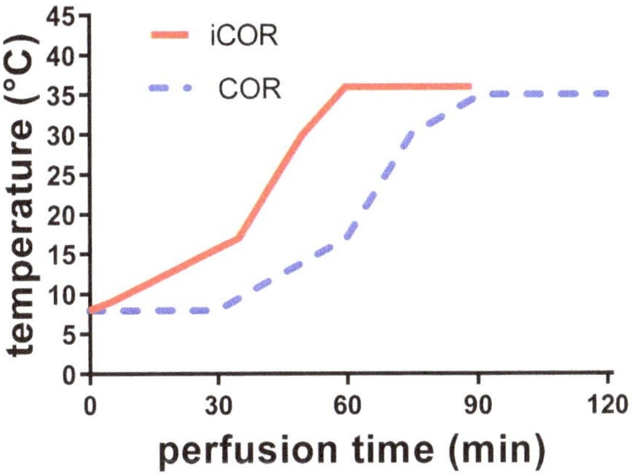

Figure 1. Representation of the thermal kinetic during controlled oxygenated rewarming by machine perfusion of porcine kidney grafts. Gentle rewarming of kidneys is either performed after an initial 30 min hypothermic equilibration period (COR, hatched line) or by starting the rewarming protocol immediately after connection to the machine (iCOR, continuous line).

Perfusion with Aqix-RS-I (Life Science Group, Bedford, UK) was started at a temperature of 8 °C but instant controlled oxygenated rewarming (iCOR) was started increasing temperature up to 35 °C during the first 60 min in a hyperbolic pattern comparable to the middle period in group 2. The rise in temperature was accompanied by an adapted increase of the perfusion pressure from 30 to 75 mmHg. The last 30 min of perfusion was kept constant at 35 °C.

2.1. Reperfusion Model

Prior to reperfusion all grafts were flushed with 100 mL of cold saline solution and exposed to no flow conditions at room temperature for 20 min in order to imitate the time of surgical engraftment.

The functional recovery of the grafts was tested using an established in vitro model, as previously described [19], which was modified regarding the replacement of fluid loss by urine production during ongoing reperfusion [20].

In brief, kidneys were put into a moist chamber and perfused at 37 °C with 1000 mL Krebs–Henseleit buffer to which were added 2.2% bovine serum albumin (PAN-Biotech, Aidenbach, Germany) and 20 mL of concentrated amino acid solution (RPMI 1640 Amino Acids Solution, 50×, PAN-Biotech, Aidenbach, Germany).

Perfusate was oxygenated with a mixture of 95% oxygen and 5% carbon dioxide by a hollow fiber oxygenator (Hilite LT 1000, Medos, Stolberg, Germany) and supplemented with 0.05 g/L of creatinine to allow for calculation of renal clearances. Cannulation of the ureter was performed with PE tubing for urine collection throughout the reperfusion period.

Urine produced during reperfusion was collected and reinfused after filtration (13 μm) at 100 mL intervals to the reservoir in order to prevent alterations in the composition of the perfusate over time, which would be encountered when replacing urine loss by adding balanced salt solution [21]. The volume of urine production was measured for each individual 30 min interval and representative aliquots from each fraction were pooled for the respective intervals for later analysis of metabolites.

Kidney perfusion pressure was set at 90 mmHg and automatically maintained by a servo-controlled roller pump connected to a pressure sensor placed in the inflow line immediately prior to the renal artery.

At the end of the perfusion, kidneys were removed from the perfusion chamber and samples were taken for biochemical analysis and for isolation of mitochondria.

2.2. Mitochondria Isolation and Analysis

The mitochondria were isolated as described in [22]. Mitochondria isolation buffer (MIB) was prepared with final concentrations of 70 mM sucrose, 210 mM mannitol, 53 mM HEPES, and 1 mM EGTA, (pH adjusted to 7.4). Mitochondrial assay buffer (MAS) was composed of 70 mM sucrose, 220 mM mannitol, 10 mM KH_2PO_4, 2 mM $MgCl_2$, 2 mM HEPES, 1 mM EGTA, and 0.2% *w/v* fatty-acid-free BSA (pH 7.4).

The mitochondrial oxygen consumption rate (OCR) and extracellular acidification rate (ECAR) were assessed using a Seahorse XFe24 Analyzer (Agient Technologies, Santa Clara, CA, US) and Seahorse XFe24 FluxPack (Agilent Technologies, Santa Clara, CA, US) as described by Rogers et al. (11).

Mitochondria were incubated with substrate (2 mM rotenone and 0.5 M succinate), following by sequential addition of 1M adenosine 5′-diphosphate potassium salt (ADP), 5 mg/mL oligomycin (ATP synthase inhibitor), 10 mM carbonyl cyanide-4-(trifluoromethoxy) phenylhydrazone (FCCP, mitochondrial uncoupler), and 40 mM antimycin A (complex 3 inhibitor). All substances were purchased from Sigma-Aldrich, Germany. The results were analyzed with Wave Seahorse Software (Agilent Technologies, Santa Clara, CA, USA).

2.3. Analytical Procedures

Analytical routines were performed as described previously [20]. The activities of aspartate aminotransferase (AST) and concentrations of creatinine were determined in a routine fashion by reflectance photometry on a Reflotron Plus point of care unit (Roche Diagnostics, Mannheim, Germany).

Clearances were calculated for the respective intervals as urinary creatinine x urine flow/perfusate creatinine.

The albumin concentration in urine was measured in a routine fashion at the Laboratory Center of the University Hospital and the amount of protein normalized against the corresponding concentrations of creatinine as the urinary albumin-to-creatinine ratio (mg/mg).

Oxygen partial pressure and perfusate concentrations of sodium were measured in a pH-blood gas analyzer (ABL 815flex acid-base laboratory, Radiometer, Copenhagen).

Oxygen consumption (VO_2) was calculated from the differences between arterial and venous sites and expressed as mL min^{-1} g^{-1} according to the trans-renal flow and kidney mass.

The efficiency of renal O_2 utilization was approximated by the ratio of total kidney transport of Na (TNa), accounting for the vast majority of energy consuming processes in the kidney [23], and VO_2, with TNa being equal to filtered Na minus excreted Na:

$$\text{TNa} = (\text{GFR} \times \text{Perfusate Na}) - (\text{urinary Na} \times \text{urine flow})$$

Fractional excretion of sodium (FE Na^+) was calculated according to:

$$\text{FE Na}^+ = \text{Na}^+_{\ (\text{urine})} \times \text{Creatinine}_{(\text{perfusate})} / \text{Na}^+_{\ (\text{perfusate})} \times \text{Creatinine}_{\ (\text{urine})} \times 100$$

Measurement of neutrophil-gelatinase-associated lipocalin (NGAL) was performed with a commercialized ELISA kit (USCN life science, Wuhan, China) according to the instructions of the manufacturer on a fluorescence microplate reader (Tecan, Grailsheim, Germany).

2.4. Histology

At the conclusion of the experiment, tissue samples were collected for later histological examination. Specimens were cut into 3 mm blocks, fixed by immersion in 4% buffered formalin, and embedded in paraffin. Tissue slides were prepared on a SM 2000R microtome (Leica Instruments, Nußloch, Germany). Light microscopy (20× magnification) of periodic acid Schiff-stained sections were used to demonstrate changes in morphology.

2.5. Statistical Analysis

Kidneys were randomized with n = 6 kidneys per group. Results are expressed as means ± standard deviation. Stochastic significance of differences was assessed using one way analysis of variance and Dunn's multiple comparisons test, if not otherwise indicated.

Data were analyzed with GraphPad Prism version 8.0.0 (GraphPad Software, San Diego, CA, USA, www.graphpad.com). Significance was defined as $p < 0.05$.

3. Results

The glomerular filtration function was evaluated by measurement of the clearance of creatinine during warm reperfusion (cf. Figure 2). Controlled oxygenated rewarming (COR) was followed by a significant enhancement of glomerular filtration rate when compared to cold-stored controls. Of note, omitting the hypothermic equilibration phase in the instant rewarming (iCOR) group did not result in any adverse effect in comparison to the COR group.

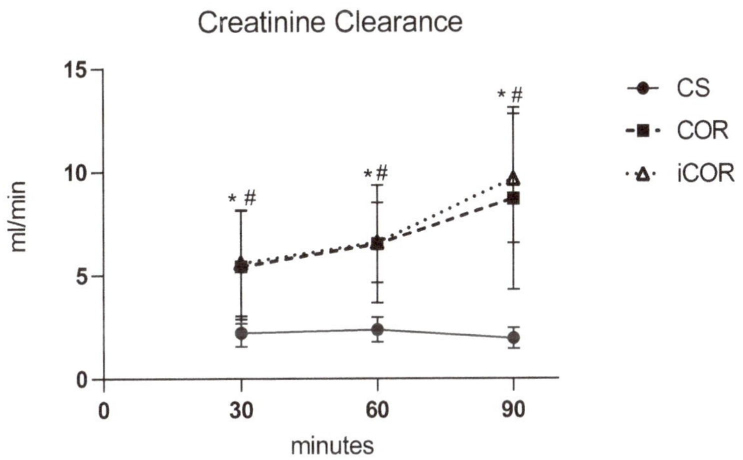

Figure 2. Clearance of creatinine during reperfusion in vitro after 20 h cold storage (CS) or after 2 h of subsequent controlled oxygenated rewarming with (COR) or without (iCOR) an initial hypothermic perfusion period (*: $p < 0.05$ COR vs. CS; #: $p < 0.05$ iCOR vs. CS).

Likewise, urinary leakage of albumin (expressed as the albumin/creatinine ratio) was markedly increased to 0.40 ± 0.14 in cold-stored kidneys but significantly ($p < 0.05$) reduced after COR (0.13 ± 0.05), as well as after iCOR treatment (0.12 ± 0.03) prior to reperfusion.

Tubular cell integrity was also found to be protected by both of the rewarming protocols. The concentration of neutrophil-gelatinase-associated lipocalin (NGAL) in the circulating perfusate was used as an indicator of tubular cell stress. It amounted to 6.8 ± 0.6 ng/mL in the cold storage group but was significantly ($p < 0.05$) reduced to 3.4 ± 0.6 and 3.9 ± 1.1 ng/mL in the COR and iCOR groups, respectively.

Differences were even more pronounced regarding the enzyme leakage of AST into the perfusate, which amounted to 659 ± 328 U/L, 140 ± 73 U/L * and 182 ± 249 U/L * after CS, COR, and iCOR, resp. ($p < 0.05$).

Tubular cell function was followed by measuring the fractional excretion of sodium from the renal ultrafiltrate (cf. Figure 3). Large amounts of the filtrated sodium could not be re-absorbed and were, thus, excreted with the urine in the cold storage group, although the excretion rate slightly improved during ongoing reperfusion. In contrast, significantly better values were observed after controlled oxygenated rewarming, and this benefit was independent from the inclusion of a hypothermic starting period during the rewarming protocol.

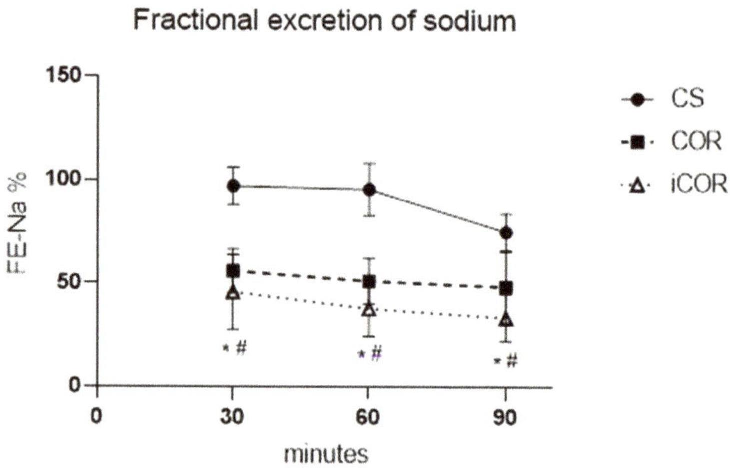

Figure 3. Course of fractional excretion of sodium (FENa) during isolated reperfusion in vitro after 20 h cold storage (CS) or after 2 h of subsequent controlled oxygenated rewarming with (COR) or without (iCOR) an initial hypothermic perfusion period (*: $p < 0.05$ COR vs. CS; #: $p < 0.05$ iCOR vs. CS).

The efficiency of oxygen utilization by the renal tissue is approximated by the ratio between total sodium transport (TNa), accounting for the vast majority of renal energy consumption, and corresponding oxygen consumption (VO_2). Cold-stored kidneys showed a markedly reduced TNa/VO_2 ratio upon reperfusion, indicating a massively disturbed aerobic efficiency (Figure 4A). However, this impairment of oxygen utilization efficiency was significantly ameliorated by using one of the controlled rewarming protocols prior to reperfusion. Again, the protective effects of both rewarming protocols were virtually identical.

Specific mitochondrial function upon reperfusion was evaluated by respiratory assays with freshly isolated mitochondria from renal tissue at the end of reperfusion. The coupling assay reflects the oxidative phosphorylation efficiency of mitochondria.

In line with the data on oxygen utilization efficiency at the whole organ level, mitochondrial coupling efficiency was notably compromised in the CS group (Figure 4B), whereas controlled rewarming using either the COR or the iCOR protocol resulted in a significantly better preserved oxidative electro-chemical coupling rate at the mitochondrial level.

Spare respiratory capacity (SRC) characterizes mitochondrial reserve to meet an increasing energy demand in response to stress conditions. Mitochondria of only cold-stored kidneys showed nearly no functional reserve upon warm reperfusion, whereas mitochondria of COR or iCOR treated kidneys yielded considerably higher SRC values.

Significant differences between the two rewarming protocols regarding mitochondrial function could not be observed in this setting (Figure 4C).

Light microscopy performed on tissue samples obtained after conclusion of the experiment did not disclose any glomerular damage, with only mild alterations in either group. However, some alterations of normal structural appearance were observed, mainly comprising tubular cell vacuolization, which was notably less prominent after COR or iCOR than in the cold storage group. Sporadic signs of necrosis were also observed only after cold storage (Figure 5).

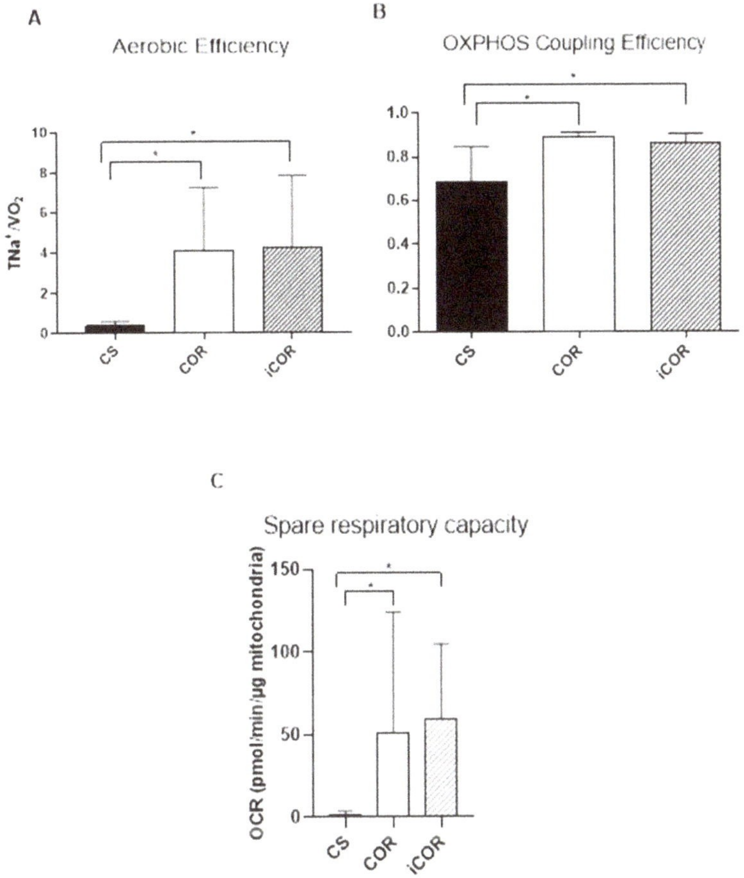

Figure 4. Parameters of mitochondrial function upon isolated reperfusion in vitro after 20 h cold storage (CS) or after 2 h of subsequent controlled oxygenated rewarming with (COR) or without (iCOR) an initial hypothermic perfusion period: (**A**) renal aerobic efficiency reflected by the ratio of total tubular sodium transport (TNa) and oxygen consumption (VO2); (**B**) OXPHOS coupling efficiency and (**C**) spare respiratory capacity of isolated mitochondria at the end of the 90 min reperfusion period (*: $p < 0.05$ vs. CS).

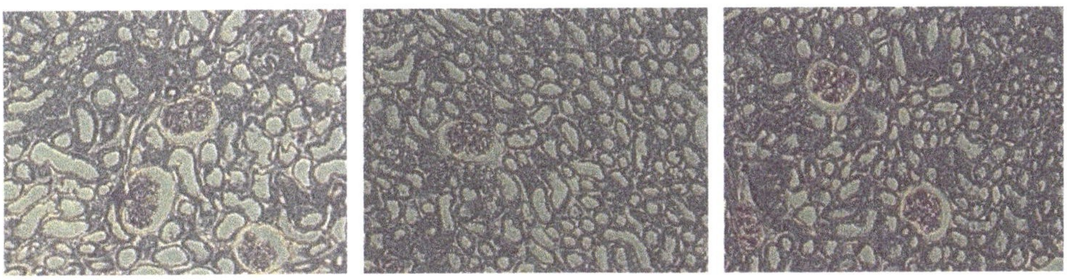

Figure 5. Representative sections of kidney tissue after reperfusion subsequent to preservation by either cold storage with controlled oxygenated rewarming (COR), instant controlled oxygenated rewarming (iCOR), or by cold storage without controlled rewarming (CS). PAS staining, 20× original magnification. Note the reduced tubular cell vacuolization and absence of necrosis in the groups COR and iCOR.

4. Discussion

Gentle elevation of tissue temperature after extended periods of hypothermia during cold preservation has been shown to favor a significantly more thorough restitution of mitochondrial function. Concomitantly, the whole-graft outcome during early reperfusion could be notably improved.

The latter observation is in line with previous reports, indicating controlled oxygenated rewarming to mitigate rewarming injury upon reperfusion in vitro [11], in vivo [10], and in clinical kidney transplantation [13].

Our present study, however, went further. We disclosed a notable reduction in the ratio of total sodium absorption and oxygen consumption (TNa/VO2) after cold storage and reperfusion that was significantly ameliorated by way of gentle warming up of the grafts prior to reperfusion. Thus, oxygen expenditure is only incompletely met by useful endergonic metabolism. At first sight, this seems to be attributable to an increasing uncoupling phenomenon at the mitochondrial level triggered upon warm reperfusion after cold preservation and significantly attenuated upon only slow elevation of temperature.

Another possible explanation for the reduced TNa/VO2 ratio might relate to oxygen-consuming cellular repair processes or resurrection of brush border microvilli [24]. Here, we provide actual direct evidence for mitochondrial uncoupling during the early reperfusion period after cold preservation. In addition, spare respiratory capacity, i.e., the amount of extra ATP that can be produced by oxidative phosphorylation in case of a sudden increase in energy demand, is largely depressed after cold storage. Hence, mitochondrial dysfunction rather than deviant oxygen utilization is responsible for the reduced TNa/VO2 ratio.

Interestingly, controlled oxygenated rewarming is operative in preserving mitochondrial performance upon reperfusion after hypothermic preservation, and this effect was not dependent on a longer hypothermic perfusion phase at the start of the rewarming protocol. This is in ostensible contrast to previous work from others, as well as from us, which consistently shows that a short oxygenated hypothermic perfusion prior to transplantation effectively improves the graft's resilience to reperfusion injury [15,16,25]. In contrast, hypothermic machine perfusion partly restores energetic and metabolic tissue homeostasis and helps to mitigate ischemia or reperfusion injury. However, even satisfactory aerobic conditions in the cold cannot prevent the adverse effects of an abrupt thermal transition to normothermia and a dramatic impairment of the mitochondrial respiratory control ratio [8].

Previous comparative studies have shown that although brief hypothermic machine perfusion improved functional outcome of cold-stored kidney [14] and liver grafts [9], notably better protection was provided by additional controlled oxygenated rewarming. Apparently, even energized tissue remains susceptible to the temperature paradox phenomenon that incurs along with the abrupt rewarming. On the other hand, slowing down of the rewarming process alone appears to be sufficient to allow for a graduated resumption of subcellular, especially mitochondrial homeostasis adapted to physiological demands [7].

In clinical practice, the process of connecting the graft to the machine and starting the ex vivo perfusion is more easily done in the cold, when hypothermia of the perfusate and graft allows for meticulous cannulation of the vessels without danger of normothermic tissue ischemia in the machine. Sometimes, however, limitation of the total perfusion time on the machine may be desirable, e.g., if the recipient is readily prepared or the total preservation time should be kept at a minimum. In these cases, it will, thus, be a suitable measure to start the rewarming protocol in the cold but to initiate the graduated rise in temperature immediately after organ connection has been effectuated and spare up to 30 min of perfusion in the cold.

The well-known limitations of all in vitro models also apply to our study. In contrast to cell culture experiments, which do not allow for functional evaluation of the whole organ, our model strictly mimics the clinical ex vivo graft perfusion and provides functional data on early reperfusion outcome at the whole organ level. Although no long term evaluation of graft function has been possible in the present study, nor could interferences of whole blood with the post-ischemic tissue be accounted for, close similarities of functional data after COR could be found between the present study and earlier investigations in vivo [10]. Apart from this, the primary goal of the present study, i.e., to decipher the mechanistic role of early mitochondrial dysfunction, triggered upon abrupt warm reperfusion after cold storage and its prevention by controlled oxygenated rewarming could be sensibly addressed by the used model.

In conclusion, our data provide evidence for mitochondrial uncoupling upon abrupt warm reperfusion after extended cold preservation of renal rafts, which can effectively be alleviated by controlling the rewarming process.

Author Contributions: H.Z. performed research, analyzed and interpreted data, and wrote the paper. C.v.H. analyzed data and revised the paper. T.M. designed research, performed research, and wrote the paper. All authors have read and agreed to the published version of the manuscript.

Funding: This research received no external funding.

Institutional Review Board Statement: All experiments were performed in accordance with the federal law regarding the protection of animals. The principles of laboratory animal care (NIH publication no. 85-23, revised 1985) were followed.

Informed Consent Statement: Not applicable.

Data Availability Statement: Data are available from the authors upon reasonable request.

Conflicts of Interest: The authors declare no conflict of interest.

References

1. Schlegel, A.; Muller, X.; Mueller, M.; Stepanova, A.; Kron, P.; de Rougemont, O.; Muiesan, P.; Clavien, P.A.; Galkin, A.; Meierhofer, D.; et al. Hypothermic oxygenated perfusion protects from mitochondrial injury before liver transplantation. *EBioMedicine* **2020**, *60*, 103014. [CrossRef] [PubMed]
2. Saeb-Parsy, K.; Martin, J.L.; Summers, D.M.; Watson, C.J.; Krieg, T.; Murphy, M.P. Mitochondria as Therapeutic Targets in Transplantation. *Trends Mol. Med.* **2021**, *27*, 185–198. [CrossRef] [PubMed]
3. Saris, N.E.; Eriksson, K.O. Mitochondrial dysfunction in ischaemia-reperfusion. *Acta Anaesthesiol. Scand.* **1995**, *107*, 171–176. [CrossRef]
4. Salahudeen, A.K. Cold ischemic injury of transplanted kidneys: New insights from experimental studies. *Am. J. Physiol. Ren. Physiol.* **2004**, *287*, F181–F187. [CrossRef] [PubMed]
5. Sammut, I.A.; Burton, K.; Balogun, E.; Sarathchandra, P.; Brooks, K.J.; Bates, T.E.; Green, C.J. Time-Dependent Impairment of Mitochondrial Function After Storage and Transplantation of Rabbit Kidneys1. *Transplantation* **2000**, *69*, 1265–1275. [CrossRef] [PubMed]
6. Duval, M.; Plin, C.; Elimadi, A.; Vallerand, D.; Tillement, J.P.; Morin, D.; Haddad, P.S. Implication of mitochondrial dysfunction and cell death in cold preservation–warm reperfusion-induced hepatocyte injury. *Can. J. Physiol. Pharmacol.* **2006**, *84*, 547–554. [CrossRef] [PubMed]
7. Minor, T.; von Horn, C. Rewarming Injury after Cold Preservation. *Int. J. Mol. Sci.* **2019**, *20*, 2059. [CrossRef]
8. Leducq, N.; Delmas-Beauvieux, M.C.; Bourdel-Marchasson, I.; Dufour, S.; Gallis, J.L.; Canioni, P.; Diolez, P. Mitochondrial permeability transition during hypothermic to normothermic reperfusion in rat liver demonstrated by the protective effect of cyclosporin A. *Biochem. J.* **1998**, *336 Pt 2*, 501–506. [CrossRef]
9. Minor, T.; Efferz, P.; Fox, M.; Wohlschlaeger, J.; Lüer, B. Controlled oxygenated rewarming of cold stored liver grafts by thermally graduated machine perfusion prior to reperfusion. *Am. J. Transplant.* **2013**, *13*, 1450–1460. [CrossRef]
10. Von Horn, C.; Zlatev, H.; Kaths, M.; Paul, A.; Minor, T. Controlled Oxygenated Rewarming Compensates for Cold Storage-induced Dysfunction in Kidney Grafts. *Transplantation* **2021**. [CrossRef]
11. Von Horn, C.; Minor, T. Improved approach for normothermic machine perfusion of cold stored kidney grafts. *Am. J. Transl. Res.* **2018**, *10*, 1921–1929.
12. Boteon, Y.L.; Laing, R.W.; Schlegel, A.; Wallace, L.; Smith, A.; Attard, J.; Bhogal, R.H.; Neil, D.A.; Hübscher, S.; Perera, M.T.P.; et al. Combined Hypothermic and Normothermic Machine Perfusion Improves Functional Recovery of Extended Criteria Donor Livers. *Liver Transplant.* **2018**, *24*, 1699–1715. [CrossRef]
13. Zlatev, H.; von Horn, C.; Kaths, M.; Paul, A.; Minor, T. Clinical Use of Controlled Oxygenated Rewarming of Kidney Grafts Prior to Transplantation by Ex Vivo Machine Perfusion: A Pilot Study. *Eur. J. Clin. Investig.* **2021**, *34*, e13691. [CrossRef]
14. Schopp, I.; Reissberg, E.; Lüer, B.; Efferz, P.; Minor, T. Controlled Rewarming after Hypothermia: Adding a New Principle to Renal Preservation. *Clin. Transl. Sci.* **2015**, *8*, 475–478. [CrossRef]
15. Stegemann, J.; Minor, T. Energy charge restoration, mitochondrial protection and reversal of preservation induced liver injury by hypothermic oxygenation prior to reperfusion. *Cryobiology* **2009**, *58*, 331–336. [CrossRef]

16. Kron, P.; Schlegel, A.; de Rougemont, O.; Oberkofler, C.E.; Clavien, P.A.; Dutkowski, P. Short, Cool, and Well Oxygenated—HOPE for Kidney Transplantation in a Rodent Model. *Ann. Surg.* **2016**, *264*, 815–822. [CrossRef]
17. Minor, T.; Paul, A. Hypothermic reconditioning in organ transplantation. *Curr. Opin. Organ. Transplant.* **2013**, *18*, 161–167. [CrossRef]
18. Minor, T.; von Horn, C.; Gallinat, A.; Kaths, M.; Kribben, A.; Treckmann, J.; Paul, A. First-in-man controlled rewarming and normothermic perfusion with cell-free solution of a kidney prior to transplantation. *Am. J. Transplant.* **2020**, *20*, 1192–1195. [CrossRef] [PubMed]
19. Von Horn, C.; Minor, T. Isolated kidney perfusion: The influence of pulsatile flow. *Scand. J. Clin. Lab. Investig.* **2018**, *78*, 131–135. [CrossRef] [PubMed]
20. Minor, T.; von Horn, C.; Paul, A. Role of erythrocytes in short-term rewarming kidney perfusion after cold storage. *Artif. Organs* **2019**, *43*, 584–592. [CrossRef] [PubMed]
21. Nizet, A. The isolated perfused kidney: Possibilities, limitations and results. *Kidney Int.* **1975**, *7*, 1–11. [CrossRef]
22. Rogers, G.W.; Brand, M.D.; Petrosyan, S.; Ashok, D.; Elorza, A.A.; Ferrick, D.A.; Murphy, A.N. High Throughput Microplate Respiratory Measurements Using Minimal Quantities of Isolated Mitochondria. *PLoS ONE* **2011**, *6*, e21746. [CrossRef] [PubMed]
23. Pei, L.; Solis, G.; Nguyen, M.T.; Kamat, N.; Magenheimer, L.; Zhuo, M.; Li, J.; Curry, J.; McDonough, A.A.; Fields, T.A.; et al. Paracellular epithelial sodium transport maximizes energy efficiency in the kidney. *J. Clin. Investig.* **2016**, *126*, 2509–2518. [CrossRef] [PubMed]
24. Herminghuysen, D.; Welbourne, C.J.; Welbourne, T.C. Renal sodium reabsorption, oxygen consumption, and gamma-glutamyltransferase excretion in the postischemic rat kidney. *Am. J. Physiol.* **1985**, *248*, F804–F809. [CrossRef]
25. Gallinat, A.; Paul, A.; Efferz, P.; Lüer, B.; Kaiser, G.; Wohlschlaeger, J.; Treckmann, J.; Minor, T. Hypothermic Reconditioning of Porcine Kidney Grafts by Short-Term Preimplantation Machine Perfusion. *Transplantation* **2012**, *93*, 787–793. [CrossRef] [PubMed]

biomolecules

MDPI

Article

Suppression of Inflammation-Associated Kidney Damage Post-Transplant Using the New PrC-210 Free Radical Scavenger in Rats

Torsten R. Goesch [1], Nancy A. Wilson [2], Weifeng Zeng [2], Bret M. Verhoven [2], Weixiong Zhong [2], Maya M. Coumbe Gitter [3] and William E. Fahl [1,3,*]

1 Obvia Pharmaceuticals Ltd., Madison, WI 53719, USA; torsten@goesch.com
2 Department of Surgery, Division of Organ Transplant, University of Wisconsin-Madison, Madison, WI 53706, USA; nawilson@medicine.wisc.edu (N.A.W.); wzeng28@wisc.edu (W.Z.); verhoven@surgery.wisc.edu (B.M.V.); wzhong3@wisc.edu (W.Z.)
3 Department of Oncology, Wisconsin Institutes for Medical Research, University of Wisconsin-Madison, Madison, WI 53706, USA; coumbe@wisc.edu
* Correspondence: fahl.bill@gmail.com

Abstract: Allograft kidney transplantation, which triggers host cellular- and antibody-mediated rejection of the kidney, is a major contributor to kidney damage during transplant. Here, we asked whether PrC-210 would suppress damage seen in allograft kidney transplant. Brown Norway (BN) rat kidneys were perfused in situ (UW Solution) with or without added 30 mM PrC-210, and then immediately transplanted into Lewis (LEW) rats. 20 h later, the transplanted BN kidneys and LEW rat plasma were analyzed. Kidney histology, and kidney/serum levels of several inflammation-associated cytokines, were measured to assess mismatch-related kidney pathology, and PrC-210 protective efficacy. Twenty hours after the allograft transplants: (i) significant histologic kidney tubule damage and mononuclear inflammatory cell infiltration were seen in allograft kidneys; (ii) kidney function metrics (creatinine and BUN) were significantly elevated; (iii) significant changes in key cytokines, i.e., TIMP-1, TNF-alpha and MIP-3A/CCL20, and kidney activated caspase levels were seen. In PrC-210-treated kidneys and recipient rats, (i) kidney histologic damage (Banff Scores) and mononuclear infiltration were reduced to untreated background levels; (ii) creatinine and BUN were significantly reduced; and (iii) activated caspase and cytokine changes were significantly reduced, some to background. In conclusion, the results suggest that PrC-210 could provide broadly applicable organ protection for many allograft transplantation conditions; it could protect transplanted kidneys during and after all stages of the transplantation process—from organ donation, through transportation, re-implantation and the post-operative inflammation—to minimize acute and chronic rejection.

Keywords: kidney allograft; kidney rejection; ischemia

Citation: Goesch, T.R.; Wilson, N.A.; Zeng, W.; Verhoven, B.M.; Zhong, W.; Coumbe Gitter, M.M.; Fahl, W.E. Suppression of Inflammation-Associated Kidney Damage Post-Transplant Using the New PrC-210 Free Radical Scavenger in Rats. *Biomolecules* **2021**, *11*, 1054. https://doi.org/10.3390/biom11071054

Academic Editor: Liang-Jun Yan

Received: 30 June 2021
Accepted: 12 July 2021
Published: 19 July 2021

1. Introduction

End-stage renal failure causes greater than 1.2 million deaths annually worldwide [1]. Kidney transplantation is the preferred treatment for patients with end-stage renal disease. Over 90,000 kidney transplants are performed each year worldwide.

The transplant process, itself, induces significant cellular and organ injury to the kidney, which reduces long-term survival of the organ. The three primary insults to a kidney during an allograft transplant are (i) reactive oxygen and nitrogen species (ROS and RNS)-induced damage during cold ischemia ('cold-storage') [2], (ii) ROS-induced damage upon implant ('re-perfusion injury') [3], and (iii) post-allograft-transplant inflammation, which triggers the innate immune response and antibody-mediated rejection (ABMR) [4].

Neutrophils and macrophages migrate into the damaged transplant within 6 h of reperfusion and stimulate chemokine synthesis in resident dendritic cells that then activate

T lymphocytes and recruit adaptive immune cells. Once these immune cells infiltrate the proximal tubule epithelial cells, they produce myeloperoxidase in neutrophils and nicotinamide adenine dinucleotide phosphate (NADPH) oxidase in macrophages, both of which contribute to local free radical production. These inflammatory processes lead to an activation of the complement pathway and further cell remodeling and lysis in the kidney allograft [5].

ABMR can occur as a result of either, or both, preformed alloantibody against the graft or through the *de novo* development of donor-specific antibody (dnDSA) [5–7]. The acute (min/days), transitioning to chronic (days/weeks), inflammatory response within the allograft kidney, with continuous production of ROS and inflammatory cytokines, can establish a severe, self-perpetuating response that causes kidney organ failure.

To better understand cellular and molecular pathways involved in the pathogenesis of kidney allograft inflammation and rejection, we developed and characterized a rat model that replicates most of the clinical criteria of innate immune response, ABMR and kidney organ loss [8]. This model has been used to evaluate a number of novel post-allograft transplant strategies.

The two currently acknowledged approaches for reducing the acute and long-term immune response against the kidney allograft are: (i) to increase the chance of finding a cross-matched donor, and (ii) to remove preexisting antibodies against the kidney allograft using desensitization protocols [9,10].

In the work described in this manuscript, we asked whether a third approach to suppress acute and longer-term inflammation severity would be beneficial. We administered the immediate-acting, free radical scavenger, PrC-210, both to the implanted allograft kidney and to the recipient rat, to determine whether inflammation-associated ROS damage could be suppressed. Though the concept of suppressing inflammation-associated ROS in kidney transplant is not new, the use here of the new, immediate-acting PrC-210 ROS scavenger is. Both immediate and chronic scavenging and inactivation of inflammation-generating, and generated, free radicals within the newly transplanted allograft kidney would significantly enhance the existing strategies to suppress allograft rejection and would provide another pathway to reduce post-transplant kidney cell damage, and with it, suppress Delayed Graft Function to improve survival of the kidney allograft.

PrC-210 is a new small-molecule, aminothiol, free radical scavenger [11]; it has no measurable nausea/emesis nor hypotension side effects [12]. Unlike traditional antioxidants that act *indirectly* over hours to days via NrF-2 to activate expression of protective genes [13], PrC-210 *directly* scavenges ROS to confer 100% protection in seconds [11]. PrC-210 was the most potent of the 13 commonly studied "antioxidants" screened in an assay that scored the ability of molecules to prevent x-ray-induced damage to naked DNA; the majority of the tested "antioxidants" showed no protection [14,15]. In a related assay, addition of PrC-210 30 s before a 60 s pulse of •OH to naked DNA provided complete protection against the •OH insult that induced >95% DNA damage in unprotected controls [16]. In two previous rodent kidney transplant studies [16,17], PrC-210 was shown to suppress ROS-induced kidney damage induced during (i) 30 h cold storage [17] and (ii) reperfusion injury upon implant [16] to *background* levels, thus removing two substantial sources of injury to the transplanted kidneys. The PrC-210 molecule has also been shown to suppress free radical-induced injury in several other organ settings [15,18]. Thus, we hypothesized that PrC-210 should also be able to protect an allograft against oxidative stress that is generated by BOTH (i) cellular- and (ii) antibody-mediated rejection processes that produce free radicals as a byproduct.

To explore this hypothesis, we developed a new rat model which avoided the induction of major ischemic and reperfusion events, and administered PrC-210 both pre- and post-implantation. Brown rat kidneys were flushed with UW solution containing PrC-210 and immediately transplanted into syngeneic Lewis rat recipients. Cold ischemic time was virtually eliminated. Immediately following implant, and for 8 h following kidney implant, recipient rats received systemic PrC-210 injections at doses that would enable continuous free radical-scavenging within the transplanted kidney. Transplanted kidneys and blood plasma were then harvested 20 h following transplant to enable measurement of both PrC-210-conferred (i) suppression of inflammatory byproducts and (ii) kidney protection.

2. Materials and Methods

2.1. Animals

Adult (200–250 g) male Lewis and BN rats were purchased from Envigo (Indianapolis, IN, USA) and housed in the animal care facility at the University of Wisconsin in Madison, WI, USA. All procedures were performed in accordance with the Animal Care and Use Policies at the University of Wisconsin. Animal health maintenance, including animal deaths, room temperature, 12 h light/dark cycles, and cage cleaning, among other sanitation duties, were performed daily by animal care staff. Food and water were available ad libitum. This research was prospectively approved by School of Medicine and Public Health Institutional Animal Care and Use Committee at the University of Wisconsin (Animal Protocol #M005204). All groups contained 4–6 animals.

2.2. Materials

Synthesis of the PrC-210 HCl aminothiol, a preclinical molecule, was described separately [19,20]. PrC-210 HCl crystals (3-(methylamino)-2-(methylaminomethyl)propane-1-thiol) are stored under a nitrogen atmosphere at −20 °C, and even with routine thawing, use, and re-storage, crystalline PrC-210 is completely stable for greater than 4 years by mass spectrometry analysis. Other chemical reagents were obtained from Sigma Aldrich (St. Louis, MO, USA). UW Organ Preservation Solution was purchased from Bridge to Life, Columbia, SC, USA.

2.3. Surgical and Experimental Procedure

The transplant procedure used in these experiments is shown in Figure 1. In the BN donor rat, after double ligation of the aorta, ligation of the right renal artery and vein, and surgical section of the left renal vein, the left rat kidney was perfused in situ using 5 mL of room temperature UW Solution (over a 15 s period). The perfusate was either UW Solution alone (for the "0 h" and the "20 h No Treatment" groups), or UW Solution to which crystalline PrC-210, to achieve 30 mM [17], had been added, dissolved immediately, and then pH adjusted to the starting UW Solution pH of 7.4 by adding 0.0619 μL 5N NaOH per μmol of PrC-210 HCL salt (FW: 220). The half-life of PrC-210 thiol (active form) is approximately 3.5 h in physiologic pH solutions such as UW Solution and human blood [14]. Following in situ perfusion, the left BN kidney was surgically removed and then sutured by blunt anastomosis of vessels and ureter into the vacated left kidney site of the LEW recipient rat. The right LEW kidney was ligated and removed immediately before. Five minutes after surgical closure of the LEW rat, the rat received a systemic PrC-210 dose (121 ug PrC-210 HCl per gm body weight, which equals 0.24 X Maximum Tolerated Dose) by intraperitoneal injection. As shown in the Figure 1 schematic, the rat also received intraperitoneal injection doses of PrC-210 (0.24 MTD) at +4 h and +8 h following the transplant. Rats were euthanized at +20 h following transplant, and kidneys and plasma samples were collected for analysis. There were a minimum of five rats in each treatment group.

2.4. Serum BUN and Creatinine Measurements

BUN and creatinine were measured in serum samples using the Catalyst One Analyzer Technology (IDEXX Laboratories, Westbrook, ME, USA).

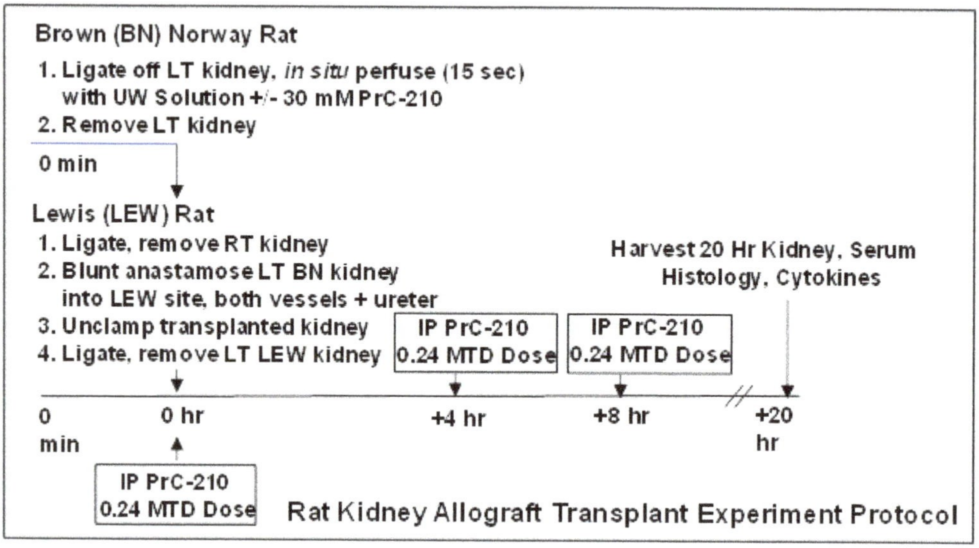

Figure 1. Experimental schematic showing surgery and PrC-210 administration times for the BN rat kidney transplants into LEW rats. LT (**left**), RT (**right**), MTD (maximum tolerated dose), and IP (intraperitoneal).

2.5. Enzyme-Linked Immunosorbent Assays

Assays were performed as described in the ELISA kit protocols (Rat TIMP-1, Cat# RTM-100; Rat MIP3-A, Cat# DY540; Rat TNF-alpha, Cat# RTA00; R&D Systems, Minneapolis, MN, USA). Briefly, dilutions of rat plasma were added to precoated plates, incubated for 2 h at 37 °C. Biotin-conjugated antibody specific for the assayed protein was then added and incubated for 1 h at 37 °C, washed, avidin-conjugated horseradish peroxidase was added, followed by washing and TMB substrate addition. The reaction was incubated for 10–30 min, stopped with sulfuric acid and read at 450 nm.

2.6. Proteome Profiler Rat Cytokine Array

Assays were performed essentially as described in the product protocol. Briefly, plasma was incubated with nitrocellulose membranes spotted with capture and control antibodies. After incubation, the membranes were washed and incubated with streptavidin HRP. In a deviation from protocol, we utilized SuperSignal West Femto (ThermoFisher, Madison, WI, USA; Cat# 34094) for chemiluminescent detection, as it gave a stronger signal. Blots were visualized on a FotoDyne gel doc system.

2.7. Histology

Formalin-fixed (10% formalin), paraffin-embedded, kidneys were cut into 5 um sections. Slides were deparaffinized, rehydrated from xylene through a graded ethanol series to water and subsequently treated as described below. Slides were scanned using a 20× objective in an Aperio Digital Pathology Slide Scanner. All H&E slides were reviewed by Dr. Weixiong Zhong, MD, PhD, transplant pathologist, and scored for ptc, glomerulitis (g), vasculitis (v)/intimal arteritis, interstitial inflammation (i) and C4d staining, according to Banff 2009 [21].

Separately, slides were assigned a blinded number, and non-overlapping digital images of renal tubules were taken at the interface between the medulla and the cortex from each H/E slide. Care was taken to not include large vessel lumens and glomeruli. Automated quantification of red and blue pixels in each 10× kidney image was performed using a custom macro written in ImageJ software (https://imagej.nih.gov/ij/index.html

accessed on 13 April 2021). Red pixels reflected proximal tubular thickness including brush border. Nuclei were quantified in the blue channel. The ratio of blue nuclear pixels to red tubules provided an Inflammatory Infiltration Score for the white blood cell infiltration in the post-transplant kidneys. Scores were averaged and plotted using Graphpad Prism.

2.8. Activated Caspase Enzyme Activity

Activated caspase 3 and 7 activity in kidney homogenate supernates was determined using the Apo-ONE fluorescent substrate (Promega, Madison, WI, USA) [16]. Briefly, thawed kidneys were mixed with an 8-fold excess of lysis buffer containing 50 mM Na HEPES, pH 7.4, 100 mM NaCl, 1 mM EDTA, 10 mM DTT, 10% glycerol and homogenized at 4 °C for 30 s with an Omni tissue homogenizer. The kidney homogenate was centrifuged at 4 °C (16,000× *g*) in an Eppendorf microfuge for 20 min. The supernates were immediately assayed for caspase activity, and protein content by the Bradford method using bovine serum albumin as the standard. The activated caspase assay was performed as follows: 5 µL supernate (~40 µg of supernate protein) was diluted to a total volume of 50 µL with the above lysis buffer, was mixed with 50 µL of the undiluted Apo-ONE substrate in the well of a black, opaque, 96 well plate to initiate the 60 min reaction. Plates were shaken at 200 RPM at 37 °C for 60 min. The DEVD caspase substrate peptide cleavage was measured using a BMG Clariostar fluorescent plate reader at an excitation wavelength of 499 nm and an emission wavelength of 521 nm. A caspase standard was included in each experiment.

2.9. Rat Kidney Mitochondria

The purified mitochondrial fraction was prepared from homogenized rat kidneys by a standard centrifugation technique [22]. The purified mitochondria were suspended in 0.15 M Tris HCl buffer, pH 7.4.

To determine whether the addition of exogenous PrC-210 suppresses ROS-induced fragmentation of mitochondrial DNA [22], in a 25 µL reaction volume (in a PCR tube), we added: 10 µL purified mitochondria, 5 µL PrC-210 dilution or water (PrC-210 was added 10 min before the Fe^{++} + ADP +H_2O_2 •OH generator), and 10 µL containing $FeCl_2$ (2.5 mM; FW:127), adenosine 5′-diphosphate sodium salt (10 mM; FW: 427) and H_2O_2 (0.003% final concentration). After 20 min at 37 °C, 10 µL of the reaction was mixed with 5 µL of 6× gel loading dye containing 0.3% SDS; tubes sat in 60 °C water for 1 min, 10 ul was then loaded into a well of a 1% agarose TAE gel, and after 60 min at 60 volts, gels were stained and photographed. A minimum of three replicates were done for each assay point to enable statistical comparison.

2.10. Statistical Analysis

Data are expressed as the means +/− STDs. Student's t-tests were used to determine statistical difference and p values using GraphPad Prism 7.03 software. p-values less than 0.05 were considered significant.

3. Results

3.1. PrC-210 Suppression of Kidney Allograft Pathology

At 20 h post-transplant, the histology of the transplanted BN kidneys (Figure 2A–D) treated with UW Solution alone clearly showed increased Banff scores for tubulitis and peritubular capillaritis (red and yellow arrows); the summed pathology scores are shown in panel E. BN kidneys perfused with PrC-210-containing UW Solution and receiving post-implant intraperitoneal systemic doses of PrC-210 (Figure 2D) showed clearly suppressed inflammatory pathology to the kidney and suppressed Banff scores for tubulitis and peritubular capillaritis equal to Banff scores of the "0 h" control BN group (Figure 2E).

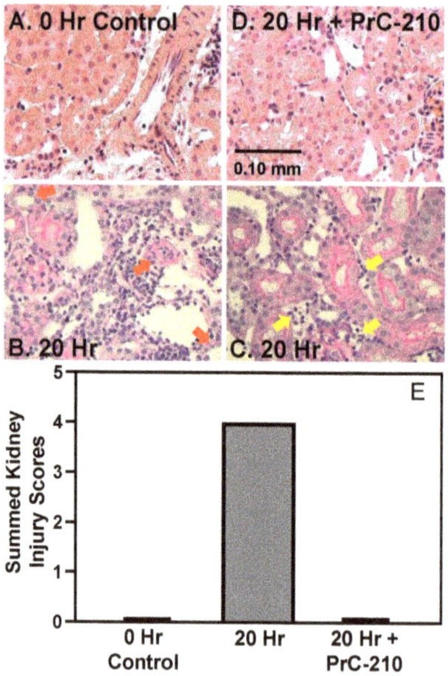

Figure 2. Histology of (**A**) BN kidney upon removal from BN rat. (**B–D**) BN kidneys 20 h following transplant into LEW rats. The "20 h" control kidneys (**B**,**C**) were flushed with room temperature UW Solution prior to transplant, and +PrC-210 kidneys (**D**) were flushed with room temperature UW Solution containing 30 mM PrC-210 prior to transplant, and recipient LEW rat also received three intraperitoneal systemic PrC-210 doses in the eight hours following transplant. Panel (**B**) red arrows highlight tubulitis pathology, and Panel C yellow arrows highlight peritubular capillaritis pathology. Panel (**E**) shows summed tubulitis and peritubular capillaritis severity scores for three kidneys in each of the indicated groups.

The "20 h" histology of the BN kidneys only flushed with UW Solution showed a significantly reduced thickness of the renal tubule brush border epithelium (Figure 3B) in comparison to the "0 h" control BN kidneys just removed from a BN rat (Figure 3A). The histology of BN kidneys receiving PrC-210 via both the perfusing UW Solution and intraperitoneal injections, clearly showed a preserved integrity of the renal tubular brush border (Figure 3C).

The same blinded histology sections were examined for mononuclear white cell infiltration. BN kidneys at 20 h only flushed with UW Solution showed a significant increase in the number of blue nuclei (scored as blue pixels), which reflects the infiltration of mononuclear white blood cells. In contrast, mononuclear infiltration in the BN kidneys flushed and treated with PrC-210 was significantly suppressed versus the untreated BN kidneys ($p = 0.011$) and the Inflammatory Infiltration Score was statistically the same as the 0 h control group (Figure 3D).

Serum creatinine and serum BUN were also measured to assess function in the BN allograft kidneys 20 h after transplant (Figure 4).

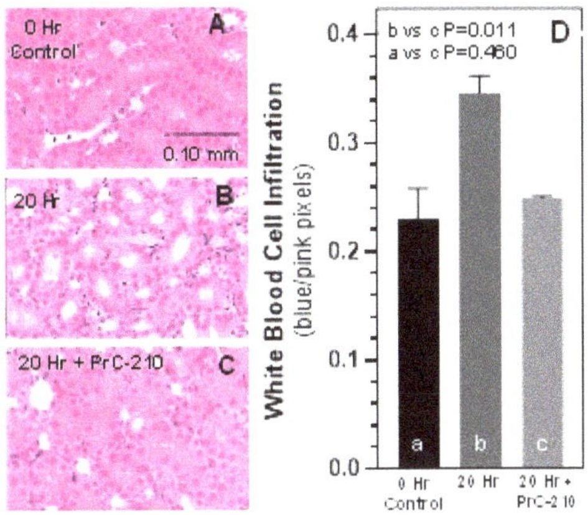

Figure 3. (**A–C**) Kidney histology 20 h following transplant. Operator randomly collected sample images of renal tubules from each of the indicated kidney groups in panels (**A–C**). Then, using an ImageJ macro (see Methods), H/E images (e.g., Figure 3 panels (**A–C**)) were analyzed, and above-threshold pink or blue pixels were selected and enumerated, and their ratio was then calculated and plotted to provide an estimate of kidney white blood cell infiltration (Panel (**D**)). In Panel (**D**), a,b,c designations are used to enable statistical comparisons between groups.

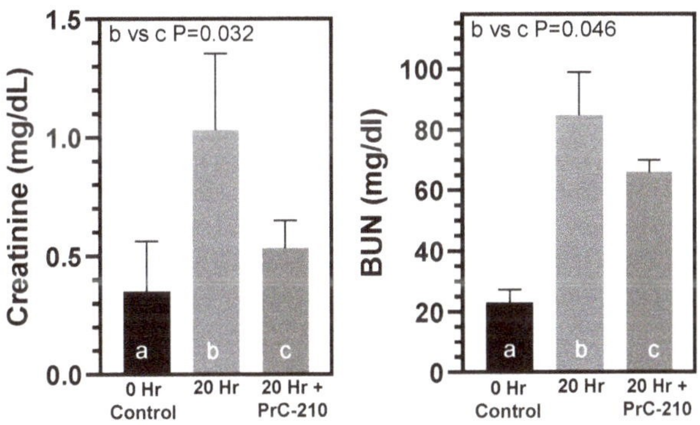

Figure 4. Effects of PrC-210 administration on kidney function 20 h after transplant of the BN kidney into a LEW rat. Rats either received no PrC-210 ("20 h") or three IP PrC-210 injections in the 20 h period following transplant ("20 h + PrC-210"), before recipient rat was euthanized and serum was collected. Serum BUN and creatinine levels were determined as described in Methods. p values are indicated. a,b,c designations are used to enable statistical comparisons between groups.

Administration of PrC-210 conferred significant reductions in both the creatine ($p = 0.032$) and BUN ($p = 0.046$) kidney damage markers.

3.2. Activated Caspase Levels in Post-Transplant Kidneys

Levels of activated caspase in BN kidney homogenates were significantly reduced in BN allograft kidneys that were not exposed to PrC-210 treatment during the 20 h following transplant (Figure 5). Perfusion of BN kidneys with PrC-210-containing UW Solution and transplant into Lewis rats that received systemic PrC-210 doses resulted in the same activated caspase activity to that seen in the "0 h" control kidneys.

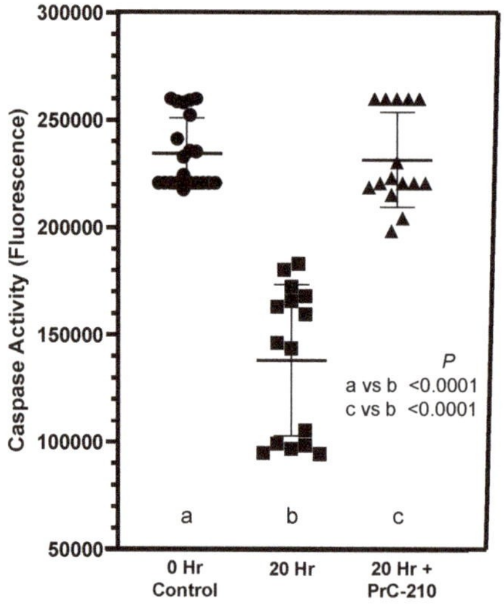

Figure 5. Caspase activity in post-transplant kidneys. Kidney supernatant activated caspase activity was measured enzymatically as described in Materials and Methods in a 60 min reaction. a,b,c designations are used to enable statistical comparisons between groups.

3.3. *Inflammatory Cytokine Levels Following BN Kidney Allograft*

In screening experiments, kidney homogenate supernates from 0 h controls and 20 h No Treatment kidneys were screened with the Proteome Profiler 29 cytokine array to detect altered, inflammation-associated, cytokine and chemokine expression levels 20 h post-transplant. As shown in the two Figure 6 microarray insets, changes were seen in TIMP-1, TNF-alpha and MIP-3a/CCL20. Individual ELISA plates were then used to quantify these changes in kidney homogenates and sera, now including rats treated with PrC-210 as well. Both TIMP-1 and TNF-alpha levels were increased 20 h post-transplant, and in both cases, their levels were decreased in the presence of PrC-210 (Figure 6A–C). MIP-3a/CCL20 levels were increased at 20 h, but they increased significantly higher in PrC-210-treated rats (Figure 6D).

3.4. *PrC-210 Protection of Rat Kidney Mitochondria*

Sustained mitochondrial function during and after kidney transplant is required for survival of the transplanted organ. ROS generated during transplant-associated ischemia, ischemia-reperfusion, and post-implant inflammation can all affect kidney mitochondrial performance and survival. Because of the important mitochondrial role, we isolated mitochondria from rat kidneys and determined whether PrC-210 at achievable pharmacologic concentrations could protect these organelles from an ROS insult.

In Figure 7, purified rat kidney mitochondria incubated with an •OH generator [23]. Following the brief reaction, we saw significant ROS fragmentation of rat kidney mitochondrial DNA. Following the ROS insult, an aliquot of the mitochondria was solubilized in SDS-containing gel-loading buffer, and mitochondrial DNA was separated and "sized" using agarose gel chromatography (Figure 7). The ROS insult clearly reduced the mean size of the rat kidney mitochondrial DNA (lane b), and addition of PrC-210 to the mitochondria prevented the DNA breakage (lanes c–g) in a PrC-210 concentration-dependent manner.

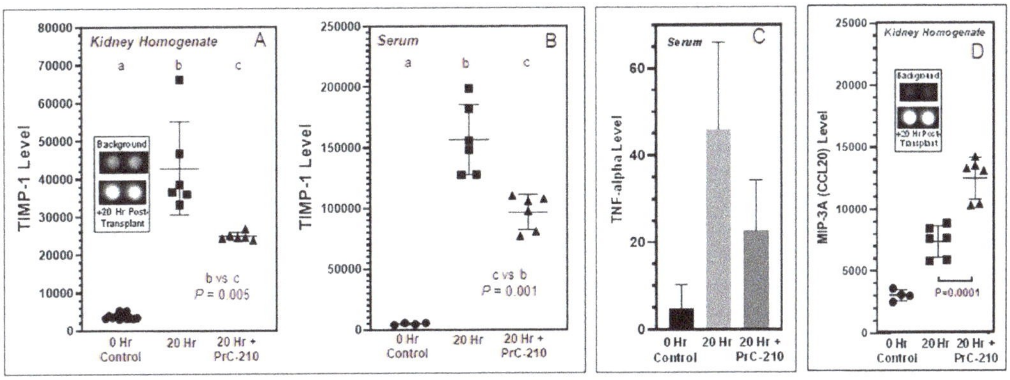

Figure 6. Effect of PrC-210 administration on serum and kidney homogenate cytokine levels 20 h after transplant of the BN kidney into a LEW rat. (**A**,**B**) TIMP-1, (**C**) TNF-alpha and (**D**) MIP-3A/CCL20 levels were determined by ELISA analysis. Serum collection and transplanted BN kidney harvest and homogenization were done as described in Methods. *p* values are indicated. a,b,c designations are used to enable statistical comparisons between groups.

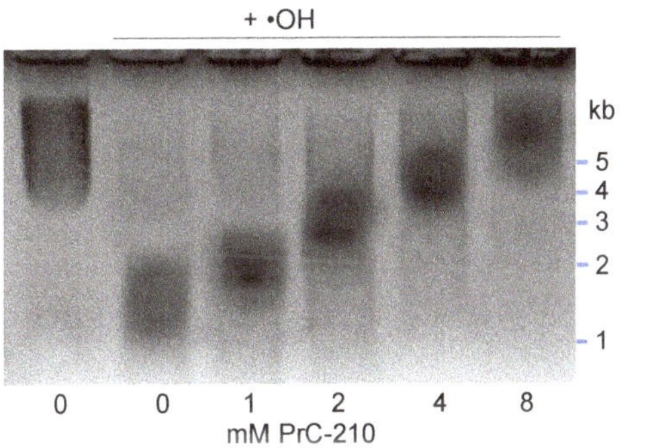

Figure 7. Suppression of hydroxyl radical-induced rat kidney mitochondrial DNA fragmentation by PrC-210 in a dose-dependent manner. Ten minutes before addition of the hydroxyl radical generator, PrC-210 was added to the incubations at the indicated concentrations. Following 20 min incubation at 37 °C, a reaction aliquot (10 µL) was mixed with 0.1% SDS loading dye and heated to 60 °C (1 min). Electrophoresis was performed, and gels were stained with ethidium bromide.

4. Discussion

Allograft kidney transplantation, which triggers innate host cellular- and antibody-mediated rejection of the kidney, is a major contributor to short and long-term kidney damage during transplant, and the associated Delayed Graft Function seen in up to 50% of transplanted kidneys. We undertook this study to determine whether PrC-210 would be effective in suppressing the severity of the damage induced following allograft kidney transplant in a rat model that largely eliminates transplant ischemic time and its associated oxidative stress. Our assumption was that this approach should allow us to see the impact of PrC-210 on the post-transplantation inflammation insult with minimal ischemia interference.

The increase in TNF-alpha and substantial mononuclear infiltration demonstrate that allograft kidney transplantation induces pronounced acute inflammation in the 20 h after transplantation, and this was correlated with the damage of the kidney tubular cells seen in the kidney histology (Banff Scores). TNF-alpha is mainly produced by activated macrophages and is a cell signaling protein involved in acute inflammation. It is closely associated with the pathogenesis of acute and chronic allograft rejection [24].

In contrast to the above findings in non-treated BN kidneys, PrC-210 given as part of the UW Solution and administered systemically in the post-transplant rats reduced both TNF-alpha level and kidney infiltration by mononuclear cells, which are both signs of reduced acute inflammation. PrC-210 reduced the kidney damage as seen in the histological kidney injury scores (Banff Scores) to untreated background levels and lowered levels of both kidney pathology functional scores, creatinine and BUN.

Inflammation in untreated BN kidneys was associated with an increase in both TIMP-1 and MIP-3A/CCL20. By comparison, we saw that PrC-210 treatment significantly reduced the TIMP-1 level and significantly increased the MIP-3A/CCL20 level.

Tissue inhibitor of metalloproteinase-1 (TIMP-1) is an important regulator of extracellular matrix (ECM) synthesis and degradation. Excess ECM accumulation is the main pathological mechanism of fibrosis development during and after acute kidney injury. There is essentially no expression of TIMP-1 in normal kidney tissue [25], an observation which is corroborated in our Figure 6A,B, but TIMP-1 is known to be expressed in injured kidneys, mainly in renal tubular epithelial cells, renal tubular basement membrane and the cytoplasm of interstitial cells. Increased TIMP-1 expression was positively correlated with the simultaneous deterioration of renal function [26]. Rats treated with PrC-210 showed a profound reduction in TIMP-1 levels ($p = 0.001$), both in kidney homogenate and plasma; this implies that PrC-210 exerts a strong protective effect against transplantation-induced reorganization of the kidney extracellular matrix.

The chemokine MIP-3a/CCL20 activates the CCR6 receptor, which is expressed especially on regulatory T-cells (Tregs). CCL20 is expressed by tubular endothelial and interstitial cells and is also upregulated in kidneys with acute kidney injury. The CCL20–CCR6 pathway plays a vital role in Treg-mediated T-cell recruitment to the kidney, and Tregs have been described to have a positive role in kidney repair, transplant tolerance, and kidney survival. Both antibody blocking of the CCL20–CCR6 pathway, as well as the use of CCR6-deficient mice in acute kidney injury experiments, were shown to increase the severity of kidney failure and mortality [27]. This suggests, that clinically, CCL20–CCR6 pathway enhancement and Treg activation may be a possible therapeutic route to limit acute and chronic kidney injury [28]. In our study (Figure 6D), the MIP-3a/CCL20 level was significantly higher in PrC-210-treated rats than in untreated rats. We speculate that this is one of the reasons for both the (i) significantly lower recruitment of mononuclear cells to kidneys (Figure 3C) and (ii) the significantly reduced kidney damage (Figure 2) in the PrC-210-treated rats.

Normal kidney mitochondrial function, and importantly, insults to it during the kidney storage, implant, and post-implant inflammation steps are significant determinants of ROS injury, and kidney failure during transplant. It was thus significant that PrC-210 was shown to confer complete suppression of mitochondrial DNA fragmentation (Figure 7) at concentrations (2–4 mM) that have been achieved in the plasma of both mice and rats that were given either intraperitoneal or oral systemic 0.5 MTD doses of PrC-210 that were tolerated with no detectable toxicities [29].

In our earlier kidney transplant-related studies [16,17], we saw substantial increases in activated caspase in kidneys exposed to "cold ischemia" and "ischemia-reperfusion" injury. These ischemia-induced insults to the kidneys were reduced to background by treatment with PrC-210 (Figure 8). In the studies of this manuscript (Figure 5), in which cold ischemia and ischemia-reperfusion were essentially eliminated by immediate transplant, there was no increase in activated caspase in transplanted kidneys. Rather, activated caspase was significantly reduced at +20 h in "No Drug Treatment" controls, and PrC-210 treatment completely eliminated this caspase reduction in +20 h rats and kept the caspase level stable. Our interpretation of these interesting results is that absent any significant ischemia-induced free radical insult through ROS and RNS to the post-transplant kidneys, there is no associated cell death and apoptosis markers like activated caspases. Rather, in these allograft kidney transplants, inflammatory signals from newly expressed cyto- and chemokines now regulate cell metabolism, which includes influencing the apoptosis pathway. The literature describes that overexpression of TIMP-1 leads to suppression of apoptosis [26]. Our caspase results (Figure 5) support this described TIMP-1 effect, and they imply that TIMP-1 is important in regulating the pathophysiology of cell damage after kidney allograft transplantation. In corroboration of the earlier PrC-210 suppression of TIMP-1 expression (Figure 6A,B), PrC-210 treatment completely ablated the caspase change, keeping the caspase levels at the same level seen in the control "0 h" kidneys. Because the reduced PrC-210 serum TIMP-1 levels at +20 h (Figure 6B) accurately reflect the significant suppression of allograft: (i) apoptosis (Figure 5), (ii) histologic pathology (Figures 2 and 3) and (iii) inflammatory cell infiltration (Figure 3), we expect that monitoring serum TIMP-1 levels in human kidney allograft recipients will be a logical way to monitor PrC-210 clinical efficacy in future clinical trials.

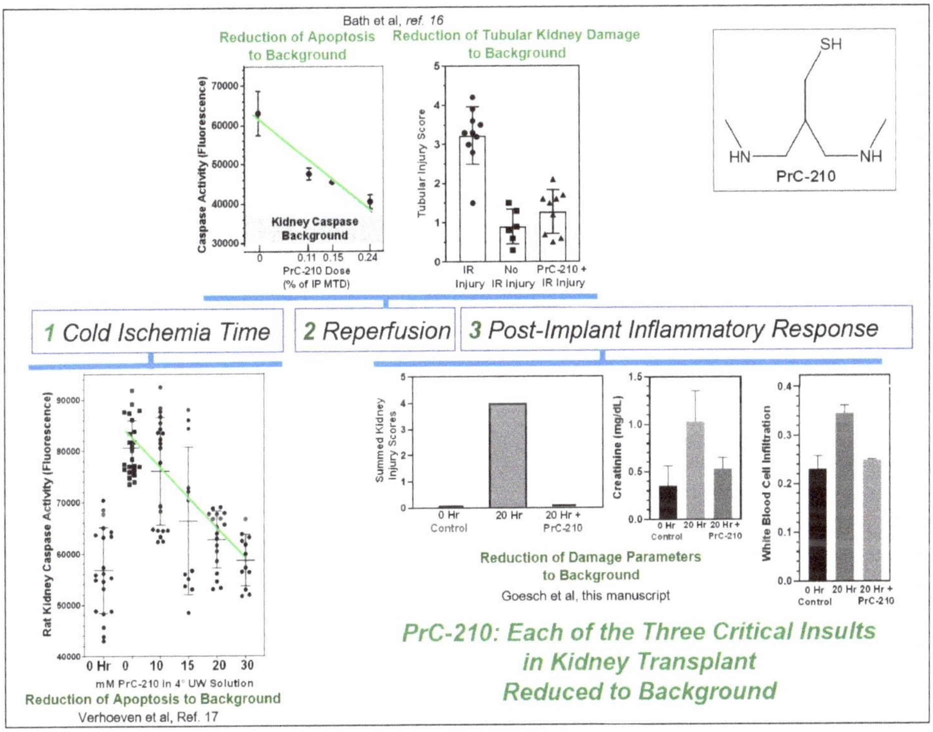

Figure 8. Schematic summarizing PrC-210 efficacy in suppressing each of the three organ insults that occur during kidney transplantation.

In our work to date [16,17], we have shown that PrC-210 is able to protect transplanted kidneys against both the cold-ischemia and ischemia-reperfusion insults. In this study, we now see that PrC-210 also protects allograft kidneys from the non-ischemia inflammatory insults that occur after kidney implant. PrC-210 significantly reduces levels of acute inflammatory cytokines, such as TNF-alpha, and suppresses expression of the TIMP-1 chemokine. Both of these events, and potentially, further supported by additional CCL20 expression, would be expected to: (i) reduce allograft kidney damage, (ii) suppress T-cell recruitment to the kidney, and (iii) suppress activation of the innate and adaptive immune system. In Figure 8, we summarize these findings to support the role that we feel PrC-210 can play in human kidney transplantation; it suppresses: (i) cold ischemia reactive oxygen species (ROS) and reactive nitrogen species (RNS) damage to background [17],

(ii) ischemia-reperfusion ROS damage to background [16], and (iii) allograft inflammation damage substantially, in some cases, to background.

Since the primary PrC-210 mechanism of action for PrC-210 is scavenging oxygen and nitrogen free radicals, this implies that these free radicals are an important contributor to the kidney damage seen in non-ischemic conditions, i.e., the allograft-associated inflammation studied in this manuscript.

In summary, this suggests that PrC-210 could provide broadly applicable organ protection for many allograft transplantation conditions; it could protect transplanted kidneys during and after all stages of the transplantation process—from organ donation, through transportation, re-implantation and the post-operative inflammation—to minimize acute and chronic rejection.

Author Contributions: N.A.W., W.Z. (Weifeng Zeng), W.Z. (Weixiong Zhong), B.M.V. and M.M.C.G. participated in performance of the research and data analysis; T.R.G. and N.A.W. participated in research design and data analysis; W.E.F. and T.R.G. participated in research design, performance of the research, data analysis, and writing of the paper. All authors have read and agreed to the published version of the manuscript.

Funding: This research was funded by the UW Institute for Clinical and Translational Research and grant UL1TR002373 to UW ICTR from NIH/NCATS, American Society of Transplant Surgeons (133 AAA1552), American College of Surgeons (133 AAB2176), and grant support to W.E.F. (#R03CA176799).

Institutional Review Board Statement: This research was prospectively approved by School of Medicine and Public Health Institutional Animal Care and Use Committee at the University of Wisconsin (Animal Protocol #M005204).

Data Availability Statement: The study data are fully available within this manuscript.

Conflicts of Interest: The authors declare no conflict of interest.

References

1. Wang, H.; Naghavi, M.; Allen, C.; Barber, R.M.; Bhutta, Z.A.; Casey, D.C.; Charlson, F.J.; Chen, A.Z.; Coates, M.M.; Coggeshall, M.; et al. Global, regional, and national life expectancy, all-cause mortality, and cause-specific mortality for 249 causes of death, 1980–2015: A systematic analysis for the Global Burden of Disease Study 2015. *Lancet* **2016**, *388*, 1459–1544. [CrossRef]
2. Treat, E.; Chow, E.; Peipert, J.D.; Waterman, A.; Kwan, L.; Massie, A.B.; Thomas, A.G.; Bowring, M.G.; Leeser, D.; Flechner, S.; et al. Shipping living donor kidneys and transplant recipient outcomes. *Arab. Archaeol. Epigr.* **2017**, *18*, 632–641. [CrossRef] [PubMed]
3. Weight, S.C.; Bell, P.R.; Nicholson, M.L. Renal ischaemia-reperfusion injury. *Br. J. Surg.* **1996**, *83*, 162–170. [PubMed]
4. Sellarés, J.; De Freitas, D.G.; Mengel, M.; Reeve, J.; Einecke, G.; Sis, B.; Hidalgo, L.G.; Famulski, K.; Matas, A.; Halloran, P.F. Understanding the Causes of Kidney Transplant Failure: The Dominant Role of Antibody-Mediated Rejection and Nonadherence. *Am. J. Transplant.* **2012**, *12*, 388–399. [CrossRef]
5. Siedlecki, A.; Irish, W.; Brennan, D.C. Delayed Graft Function in the Kidney Transplant. *Arab. Archaeol. Epigr.* **2011**, *11*, 2279–2296. [CrossRef]
6. Hidalgo, L.G.; Campbell, P.M.; Sis, B.; Einecke, G.; Mengel, M.; Chang, J.; Sellares, J.; Reeve, J.; Halloran, P.F.; Hidalgo, L.G.; et al. De NovoDonor-Specific Antibody at the Time of Kidney Transplant Biopsy Associates with Microvascular Pathology and Late Graft Failure. *Arab. Archaeol. Epigr.* **2009**, *9*, 2532–2541. [CrossRef]
7. Loupy, A.; Hill, G.S.; Jordan, S.C. The impact of donor-specific anti-HLA antibodies on late kidney allograft failure. *Nat. Rev. Nephrol.* **2012**, *8*, 348–357. [CrossRef]
8. Huang, G.; Wilson, N.A.; Reese, S.R.; Jacobson, L.M.; Zhong, W.; Djamali, A. Characterization of Transfusion-Elicited Acute Antibody-Mediated Rejection in a Rat Model of Kidney Transplantation. *Arab. Archaeol. Epigr.* **2014**, *14*, 1061–1072. [CrossRef] [PubMed]
9. Jordan, S.C.; Pescovitz, M.D. Presensitization: The Problem and Its Management. *Clin. J. Am. Soc. Nephrol.* **2006**, *1*, 421–432. [CrossRef] [PubMed]
10. Stegall, M.; Gloor, J.; Winters, J.; Moore, S.; DeGoey, S. A Comparison of Plasmapheresis Versus High-Dose IVIG Desensitization in Renal Allograft Recipients with High Levels of Donor Specific Alloantibody. *Arab. Archaeol. Epigr.* **2006**, *6*, 346–351. [CrossRef] [PubMed]
11. Peebles, D.D.; Soref, C.M.; Fahl, W.E. ROS-scavenger and radioprotective efficacy of the new PrC-210 aminothiol. *Radiat. Res.* **2012**, *178*, 57–68. [CrossRef]
12. Soref, C.M.; Hacker, T.A.; Fahl, W.E. A New Orally Active, Aminothiol Radioprotector-Free of Nausea and Hypotension Side Effects at Its Highest Radioprotective Doses. *Int. J. Radiat. Oncol.* **2012**, *82*, e701–e707. [CrossRef]

13. Techapiesancharoenkij, N.; Fiala, J.L.; Navasumrit, P.; Croy, R.G.; Wogan, G.N.; Groopman, J.D. Sulforaphane, a cancer chemopreventive agent, induces pathways associated with membrane biosynthesis in response to tissue damage by af-latoxin B1. *Toxicol. Appl. Pharmacol.* **2015**, *282*, 52–60. [CrossRef] [PubMed]
14. Jermusek, F.; Benedict, C.; Dreischmeier, E.; Brand, M.; Uder, M.; Jeffery, J.J.; Ranallo, F.N.; Fahl, W.E. Significant Suppression of CT Radiation-Induced DNA Damage in Normal Human Cells by the PrC-210 Radioprotector. *Radiat. Res.* **2018**, *190*, 133–141. [CrossRef] [PubMed]
15. Hacker, T.A.; Diarra, G.; Fahl, B.L.; Back, S.; Kaufmann, E.; Fahl, W.E. Significant reduction of ischemia-reperfusion cell death in mouse myocardial infarcts using the immediate-acting PrC-210 ROS-scavenger. *Pharmacol. Res. Perspect.* **2019**, *7*, e00500. [CrossRef]
16. Bath, N.M.; Fahl, W.E.; Redfield, R.R. Significant reduction of murine renal ischemia-reperfusion cell death using the im-mediate-acting PrC-210 reactive oxygen species-scavenger. *Transplant. Direct.* **2019**, *5*, e549–e555. [CrossRef] [PubMed]
17. Verhoven, B.M.; Karim, A.S.; Bath, N.M.; Fahl, C.J.S.; Wilson, N.A.; Redfield, R.R.; Fahl, W.E. Significant Improvement in Rat Kidney Cold Storage Using UW Organ Preservation Solution Supplemented with the Immediate-Acting PrC-210 Free Radical Scavenger. *Transplant. Direct* **2020**, *6*, e578. [CrossRef]
18. Giese, A.P.J.; Guarnaschelli, J.G.; Ward, J.A.; Choo, D.I.; Riazuddin, S.; Ahmed, Z.M. Radioprotective Effect of Aminothiol PrC-210 on Irradiated Inner Ear of Guinea Pig. *PLoS ONE* **2015**, *10*, e0143606. [CrossRef]
19. Copp, R.R.; Peebles, D.D.; Fahl, W.E. Synthesis and growth regulatory activity of a prototype member of a new family of aminothiol radioprotectors. *Bioorganic Med. Chem. Lett.* **2011**, *21*, 7426–7430. [CrossRef]
20. Fahl, W.E.; Peebles, D.; Copp, R.R. Amino Thiol Compounds and Compositions for Use in Conjunction with Cancer Therapy. U.S. Patent 7,314,959, 12 April 2004.
21. Haas, M.; Sis, B.; Racusen, L.C.; Solez, K.; Glotz, D.; Colvin, R.B.; Castro, M.C.R.; David, D.S.R.; Davidneto, E.; Bagnasco, S.M.; et al. Banff 2013 Meeting Report: Inclusion of C4d-Negative Antibody-Mediated Rejection and Antibody-Associated Arterial Lesions. *Arab. Archaeol. Epigr.* **2014**, *14*, 272–283. [CrossRef]
22. Hruszkewycz, A.M. Evidence for mitochondrial DNA damage by lipid peroxidation. *Biochem. Biophys. Res. Commun.* **1988**, *153*, 191–197. [CrossRef]
23. Floyd, R.A.; Watson, J.J.; Wong, P.K. Sensitive assay of hydroxyl free radical formation utilizing high pressure liquid chromatography with electrochemical detection of phenol and salicylate hydroxylation products. *J. Biochem. Biophys. Methods* **1984**, *10*, 221–235. [CrossRef]
24. Sang-Kyu, P. Association between tumor necrosis factor-alpha (TNF-α) polymorphism (-308, G/A) and acute rejec-tion of solid organ allograft: A meta-analysis. *Int. J. Clin. Exp. Med.* **2016**, *9*, 17060–17068.
25. Qiang, Y. Expression of MMP-2 and TIMP-1 in renal tissue of patients with chronic active antibody-mediated renal graft rejection. *Diag. Pathol.* **2012**, *7*, 141–148.
26. Gangyong, L. Tissue Inhibitor of Metalloproteinase-1 inhibits apoptosis of human breast epithelial cells. *Cancer Res.* **1999**, *59*, 6267–6275.
27. González-Guerrero, C.; Morgado-Pascual, J.L.; Cannata-Ortiz, P.; Ramos-Barron, M.A.; Gómez-Alamillo, C.; Arias, M.; Mezzano, S.; Egido, J.; Ruiz-Ortega, M.; Ortiz, A.; et al. CCL20 blockade increases the severity of nephrotoxic folic acid-induced acute kidney injury. *J. Pathol.* **2018**, *246*, 191–204. [CrossRef] [PubMed]
28. Min, H. Regulatory T cells in kidney disease. and transplantation. *Kidney Int. J.* **2016**, *90*, 502–514.
29. Dreischmeier, E.; Fahl, W.E. Determination of Plasma Levels of the Active Thiol Form of the Direct-Acting PrC-210 ROS-Scavenger Using a Fluorescence-Based Assay. *Anal. Biochem.* **2021**, *616*, 114100. [CrossRef]

biomolecules

MDPI

Review

Mitochondrial Redox Signaling and Oxidative Stress in Kidney Diseases

Ana Karina Aranda-Rivera [1,2], Alfredo Cruz-Gregorio [3], Omar Emiliano Aparicio-Trejo [4] and José Pedraza-Chaverri [1,*]

1 Laboratorio F-315, Departamento de Biología, Facultad de Química, Universidad Nacional Autónoma de México, Mexico City 04510, Mexico; anaaranda025@gmail.com
2 Posgrado en Ciencias Biológicas, Universidad Nacional Autónoma de México, Ciudad Universitaria, Mexico City 04510, Mexico
3 Laboratorio F-225, Departamento de Biología, Facultad de Química, Universidad Nacional Autónoma de México, Mexico City 04510, Mexico; cruzgalfredo@gmail.com
4 Departmento de Fisiopatología Cardio-Renal, Instituto Nacional de Cardiología "Ignacio Chávez", Mexico City 14080, Mexico; emilianoaparicio91@gmail.com
* Correspondence: pedraza@unam.mx; Tel.: +52-555622-3878

Abstract: Mitochondria are essential organelles in physiology and kidney diseases, because they produce cellular energy required to perform their function. During mitochondrial metabolism, reactive oxygen species (ROS) are produced. ROS function as secondary messengers, inducing redox-sensitive post-translational modifications (PTM) in proteins and activating or deactivating different cell signaling pathways. However, in kidney diseases, ROS overproduction causes oxidative stress (OS), inducing mitochondrial dysfunction and altering its metabolism and dynamics. The latter processes are closely related to changes in the cell redox-sensitive signaling pathways, causing inflammation and apoptosis cell death. Although mitochondrial metabolism, ROS production, and OS have been studied in kidney diseases, the role of redox signaling pathways in mitochondria has not been addressed. This review focuses on altering the metabolism and dynamics of mitochondria through the dysregulation of redox-sensitive signaling pathways in kidney diseases.

Keywords: acute kidney injury (AKI); chronic kidney disease (CKD); tricarboxylic acid (TCA) cycle; mitochondrial metabolism; mitochondrial redox signaling; mitochondrial proteins; oxidative phosphorylation (OXPHOS); fatty acid (FA) β-oxidation; mitochondrial dynamics; biogenesis; mitophagy

Citation: Aranda-Rivera, A.K.; Cruz-Gregorio, A.; Aparicio-Trejo, O.E.; Pedraza-Chaverri, J. Mitochondrial Redox Signaling and Oxidative Stress in Kidney Diseases. *Biomolecules* **2021**, *11*, 1144. https://doi.org/10.3390/biom11081144

Academic Editor: Liang-Jun Yan

Received: 13 July 2021
Accepted: 1 August 2021
Published: 3 August 2021

1. Introduction

Kidney diseases are a severe health problem that causes high economic costs worldwide in medical attention, emergency, therapies, among others [1,2]. These are divided into acute kidney injury (AKI) and chronic kidney diseases (CKD). AKI encompasses a set of pathologies characterized by the rapid loss of renal function in a short period [3]. AKI is often caused by the use of chemotherapeutics agents such as cisplatin, episodes of renal ischemia/reperfusion (I/R), and exposure to contaminants [4]. AKI is associated with high morbidity and mortality, contributing to CKD development and affecting approximately between 7% and 12% of the world [5]. CKD cause renal fibrosis development [6–8]. The latter comprises an unsatisfactory repair process and is the consequence of severe and persistent damage that does not restore organ function [9]. Renal fibrosis, in turn, is one of the principal mechanisms involved in AKI to CKD transition [5].

Mitochondria are responsible for several cell functions such as cell growth, cell survival, and apoptosis induction, playing a significant role in kidney physiology and the development of kidney diseases. Mitochondria also coordinate the biosynthesis of lipids, amino acids, and nucleotides and bioenergetics processes such as tricarboxylic acid (TCA) cycles, electron transport systems (ETSs), and fatty acids (FA) β-oxidation [10]. During these processes, reactive oxygen species (ROS) are produced. Low levels of ROS are needed to regulate cellular signaling, but an excess of ROS induces oxidative stress (OS). OS causes oxidative damage in organelles including mitochondria and biomolecules, such as proteins, lipids, and deoxyribonucleic acid (DNA), which may be conducive to cell death. Indeed, OS is associated with AKI development and its transition to CKD, where mitochondria dysfunction is the principal characteristic of both [11,12]. Although mitochondrial metabolism, ROS production, and OS have been studied in kidney diseases, the role of redox signaling pathways in renal mitochondria impairment is not well understood. This review focuses on altering the metabolism and dynamics of mitochondria through the dysregulation of redox-sensitive signaling pathways in kidney diseases.

2. Redox-Sensitive Signaling in Kidney Diseases

ROS at low levels act as secondary messengers, activating signaling pathways and cellular enzymes and regulating several cellular processes such as cell proliferation, survival, and growth [13]. Among the plethora of ROS, hydrogen peroxide (H_2O_2) and nitric oxide ($^{\bullet}NO$) are the central redox signaling agents [13–16]. These ROS induce oxidative post-translational modifications (Ox-PTMs) in proteins that contain redox-sensitive amino acid residues, regulating their structure, localization, and function [17,18]. These redox-sensitive amino acids are arginine (Arg), cysteine (Cys), histidine (His), lysine (Lys), methionine (Met), proline (Pro), threonine (Thr), and tyrosine (Tyr) [19,20]. However, Cys and Met are most prone to be attacked by ROS. These amino acids contain oxidizable sulfur groups on the side chains, of which the oxidation states depend on the redox microenvironment changes [13]. Thus, the protein function is regulated by the oxidate states of these amino acids [21,22].

Cys residues perform structural functions, such as the assembly of iron-sulfur (Fe-S) groups, heme prosthetic groups, and zinc finger motifs and are essential in the active sites of enzymes. The sulfur in Cys is a large, polarizable, electron-rich atom. These characteristics give it high reactivity and the ability to adopt multiple oxidation states. In addition, pKa influences the formation of the nucleophilic thiolate anion (S^-). Because the surrounding milieu influences pKa, the thiol (SH) protonation form of Cys depends on the cellular microenvironment. For example, at physiological pH (pH 7.3), the pKa of Cys is 8.3, which means that Cys is in a protonated and less reactive state [23]. However, if the cellular microenvironment becomes alkaline (pH < 7.8), Cys can adopt a deprotonated state S^-. Note that positively charged amino acids adjacent to Cys can substantially decrease the pKa of SH. Furthermore, the presence of other contiguous positive charges, such as the positive partial charge of a dipole, can also reduce the pKa of Cys, making them more reactive [24]. On the other hand, since Cys oxidation reactions are predominantly bimolecular nucleophilic substitution (S_N2) reactions in the protonated form, steric hindrance can prevent Cys oxidation [25]. In this way, in the steric hindrance, the surrounding amino acids and redox microenvironment will determine whether a Cys will be reactive to undergo redox modification, dictating selectivity for SH modification. Therefore, there is a wide range of Ox-PTMs that also depend on the present electrophiles. For example, H_2O_2 oxidizes SH, with an oxidate state of −2, in an oxidative microenvironment, producing sulfenic acid (R–SOH) with an oxidate state of 0 [26]. R-SOH can react with proximal SH groups to form disulfide bonds (S–S), with a −1 oxidate state or s-glutathionylation (R-SSG) by reacting with glutathione (GSH) (Figure 1) [27,28]. This reaction is reversed by glutaredoxin (Grx). However, under OS, R-SOH forms sulfinic acid (R–SO_2H) with a +2 oxidate state or sulfonic acid (R–SO_3H) with a +4 oxidate state. The latter is not enzymatically reversible [29]. R–SOH can also be condensed with another R–SOH to form thiosulfinate [R–S(O)–S–R′]. R–S(O)–S–R′, in turn, reacts with amine or amide groups to form sulfenamide (R–SN–R′) [30]. Additionally, Cys can suffer from other modifications such as S-nitrosylation and persulfonation. Regarding S-nitrosylation, it comprises the $^{\bullet}NO$ covalent link to the thyil radical (R–$S^{\bullet}$) to form S-nitrosothiol (R–SNO), [31] and in persulfonation, the hydrogen sulfide (H_2S) reacts with R–$S^{\bullet}$ to form persulfide (R–S–SH) [20].

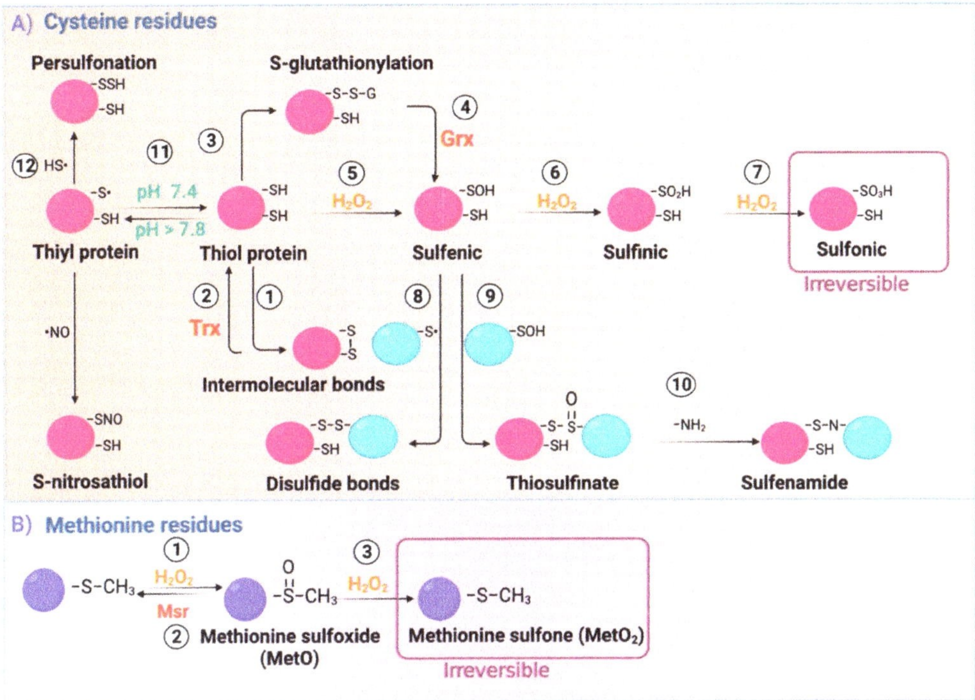

Figure 1. Reactive oxygen species (ROS) induce modifications in cysteine (Cys) and methionine (Met) residues. (**A**) ROS oxidize Cys residues in its thiol form, named thiol (SH) protein, forming (1) intermolecular bonds, (2) reversed by thioredoxin (Trx), or (3) S-glutathionylation (R–SSG). R–SSG can be reversed by (4) the glutaredoxin (Grx) enzyme, acquiring a sulfenic (R–SOH) form. Moreover, (5) R–SOH is formed through oxidation by H_2O_2 from SH. If oxidation continues, R–SOH forms (6) sulfinic (R–SO_2H) or (7) sulfonic (R–SO_3H); the last is irreversible. Furthermore, R–SOH can condense with another R-SOH to form (8) disulfide bonds (R–S–S–R′) or (9) thiosulfinate [R–S(O)–S-R]. The latter can react with amide groups (–NH_2), forming (10) sulfenamide (R–SN–R′). At the alkaline cellular microenvironment (pH > 7.8), R-SH proteins are deprotonated, forming (11) thiyl proteins (–S•). The latter can react with hydrogen sulfide (H_2S) to form (12) persulfide (RSSH) by persulfination (R–S–S–H). –S• can also react with nitrosothiol (•NO), forming S-nitrosothiol (RSNO). (**B**) Met residues are oxidized to form (1) methionine sulfoxide [MetO (R–$SOCH_3$)], reversed by (2) methionine sulfoxide reductase (Msr). However, if oxidation persists, MetO is further oxidized by H_2O_2 to (3) methionine sulfone [$MetO_2$ (RSO_2CH_3)], which is irreversible. H_2O_2: hydrogen peroxide; $HS^•$: hydrosulfide radical. Created with BioRender.com.

On the other hand, Met residues are oxidated by H_2O_2, generating methionine sulfoxide [MetO (R–$SOCH_3$)] [32]. This oxidation is reversed by methionine sulfoxide reductase (Msr). However, if OS persists, MetO is transformed into methionine sulfone [$MetO_2$ (RSO_2CH_3)], an enzymatically irreversible product [33].

The alkalinization of the mitochondrial matrix is due to pumping protons from the matrix into the intermembrane space (IMS). It has a tremendous impact on the potential reactivity of Cys, because alkalinization induces Cys protein residues to exist in an ionized state, which gives them a higher reactivity for ROS oxidation. Thus, mitochondria harbor a unique environment that promotes Cys modification.

As mentioned, there are Ox-PTMs classified as enzymatically reversible and irreversible. Ox-PTMs are reversible by thioredoxin (Trx), peroxiredoxin (Prx), Grx, glutathione peroxidase (GPx), Msr isoform A (MsrA), or the GSH [34,35]. Since reversible Ox-PTMs have enzymatic systems that remove oxidations, they are considered part of cellular signaling processes [34]. In contrast, irreversible ROS modifications are not considered part of cell signaling, because they do not have enzymatic reduction mechanisms [18].

Regarding cysteine persulfidation, it is a reversible Ox-PTM modification of SH to RSSH, which can be formed by reacting with H_2S, more precisely HS^-, and oxidized protein thiols, reaction between inorganic polysulfides and protein thiolates, and radical reaction by other reactive sulfur species [24]. Persulfidation regulates fine tune protein function, localization and interaction in cells [36]. It also modulates biological processes, including autophagy, cellular metabolism, inflammation, cell cycle, and cell death under physiological and pathological contexts [37]. It has been shown that the Cys' architecture and spatial arrangement determine if the residue can be persulfonated and the effect on protein function. Allosteric impediments protect Cys residues from being oxidized under SOH, preventing permanent damage and preserving protein function. Likewise, in a reduced environment, this Ox-PTM is reversed [38]. Additionally, persulfonation depends on ROS levels in the cells. The latter is supported, because the transient ROS increase might augment oxidized Cys forms (SOH and S–S), highly reactive to H_2S [39]. Thus, the reversibility of this Ox-PTM relies on Cys residues and the cellular microenvironment.

Ox-PTMs inactivate numerous proteins that contain Cys groups, such as protein tyrosine phosphatases (PTPs). The inactivation of PTPs prevents the deactivation of protein tyrosine kinases (PTKs), maintaining the activation of cell signaling pathways, such as mitogen-activated protein kinases (MAPK), promoting cell proliferation [40]. In addition, ROS such as H_2O_2 have also been shown to promote Cys oxidation and the formation of S–S in proteins such as growth factor receptors (GFRs), inducing their activation and cell growth. Since oO-PTMs regulate cellular functions involving cell proliferation, growth, migration, differentiation, and death, the concentration of ROS must be balanced to maintain cell homeostasis and prevent kidney diseases [13].

In kidney physiology, 25% of mitochondrial Cys in proteins can suffer nitrosylation, and around 70% of these proteins depend on the activity of endothelial nitric oxide synthase (eNOS) [41]. These nitrosylations are protective in the kidneys. For instance, Zhou et al. [42] reported that in I/R, the denitrosylase enzyme aldo-keto reductase family 1 member A1 (AKR1A1) is associate with kidney damage. In contrast, its deletion is protective in this organ. The authors also found that S-nitroso-coenzyme A (CoA) reductase induces pyruvate kinase isoform M2 (PKM2) nitrosylation, inhibiting its activity. This nitrosylation favors the pentose phosphatase pathway instead of glycolysis. The latter reduces equivalents increase, attributed to ROS detoxification, which has a protector effect during I/R [42]. S-glutathionylation is also a protective Ox-PTM in renal mitochondria [43]. In the kidneys, the reduction of mitochondrial protein S-glutathionylation induced by folic acid is associated with acute kidney damage [44]. The latter has been demonstrated, because of glutathionylated proteins decrease and the levels of GSH and the activity of the enzymes involved in S-glutathionylation in the mitochondria 24 h (h) after treatment with folic acid [44]. In the cisplatin-induced AKI model, the reduction of Met is a protective Ox-PTM, since MsrA deficiency exacerbates cisplatin-induced damage by increasing mitochondrial susceptibility [45]. Furthermore, mice $MsrA^{-/-}$ are more sensible to kidney injury induced by I/R [46]. The deficiency of MsrA, cystathionine-β-synthase (CBS), and cystathionine-γ-lyase (CSE), which are enzymes involved in the transsulfuration pathway, decreases homocysteine and H_2S [46]. Low levels of H_2S have also been reported after unilateral ureteral obstruction (UUO) due to CBS and CSE decrease, inducing superoxide anion radical ($O_2^{\bullet-}$) and H_2O_2 production. The latter promotes the augment of oxidative damage markers such as malondialdehyde (MDA) and 4-hydroxynonenal (4-HNE) [47]. The MsrA deficiency in this model contributes to fibrosis development by increasing collagen deposition and augmenting fibrosis markers [48]. Therefore, MsrA regulates the Met metabolism and the production of H_2S in the kidney [49].

Indeed, H_2S regulates several kidney biological processes. CSE and CBS are expressed in the kidney, producing H_2S that controls sodium reabsorption and glomerular filtration [50]. In addition, CSE is commonly expressed in endothelial and mesangial cells and podocytes [51]. In kidney diseases, H_2S ameliorates renal damage [52]. It has been shown that H_2S attenuates kidney injury during I/R and diabetic kidney disease (DKD). In animal models of I/R, H_2S donors administration before or after I/R ameliorates renal damage by decreasing OS, inflammation, and apoptosis [53]. Furthermore, in patients, it has been found that CSE messenger RNA (mRNA) expression positively correlates with glomerular filtration rate recovery after two weeks of kidney transplantation [54], suggesting a protective role of H_2S. Supporting the latter, CBS and CSE levels and H_2S production are low in animal models suffering I/R [53]. Regarding DKD, patients with type 2 diabetes show lowed H_2S plasma levels than non-diabetic people [55]. In UUO, the low production of H_2S is also due to the fact that CBS levels are decreased [56]. In this model, H_2S decrease is related to fibrosis and inflammation development. Following this, in vitro, sodium hydrosulfide (NaHS) treatment reduces fibrosis and inflammation by inhibiting transforming growth factor-beta 1 (TGFβ1) [56]. Together, these data support the relevance of H_2S signaling in kidney injury.

Cys persulfidation is an essential process in kidney physiology and disease, protecting against kidney injury [24]. Persulfidation in Cys 105 of mitochondrial glyceraldehyde 3-phosphate dehydrogenase (GAPDH) elevates its enzymatic activity [57]. In contrast, the persulfidation in Cys 156 or Cys 152 induces enzymatic deactivation [58]. Moreover, the persulfidation in Keap 1 Cys 150 promotes nuclear factor erythroid 2–related factor 2 (Nrf2) release and translocation to the nucleus, activating its genes targets [59]. This mechanism is critical for regulating redox homeostasis in the kidney. In renal pathologies, Nrf2 is deactivated, leading to OS increase [60,61]. In the kidney, persulfides regulate blood pressure and sodium reabsorption. In line with this, in tubular epithelial cells, the persulfidation of epidermal growth factor receptor (EGFR) (Cys 797 and 798) results in the loss of function of sodium/potassium-adenosine triphosphatase (Na^+/K^+-ATPase) [62]. Cys persulfide is produced by CSE and CSB, which are depleted in kidney diseases. However, a previous study showed that mice lacking CSE and CBS have significant levels of Cys persulfide proteins, which suggested the possibility of alternative processes [63–65]. Indeed, cysteinyl-transfer RNA (tRNA) synthetases (CARSs) act as cysteine persulfide synthases in vivo [66]. Moreover, CARSs are also involved in the regulation of mitochondrial bioenergetics (oxygen consumption and membrane potential) and dynamics (dynamin-related protein 1 (Drp1)) [66]. Thus, this mechanism may be active in kidney diseases, which deserves investigation. The latter might suggest therapeutic targeting OS-induced mitochondrial dysfunction in renal pathologies.

3. Ox-PTMs Regulate Manganese Superoxide Dismutase (Mn-SOD) in Kidney Injury

Mn-SOD is in the mitochondrial matrix, while copper/zinc-SOD (Cu/Zn-SOD) is in the space of the inner mitochondrial membrane (IMM) and IMS. These two enzymes catalyze the dismutation of $O_2^{\bullet-}$ to H_2O_2 in the mitochondrial matrix and IMS, respectively [67]. This dismutation is crucial to avoid the $O_2^{\bullet-}$-induced ferric iron (Fe^{3+}) to ferrous iron (Fe^{2+}) reduction in Fe-S clusters of critical enzymes such as aconitase (Acn). The latter leads to the release of Fe^{2+} and the inactivation of these enzymes [68]. During Fenton/Haber–Weiss reaction, Fe^{2+} reacts with H_2O_2 to produce a highly reactive ROS, hydroxyl radical ($^\bullet OH$), so the regulation of $O_2^{\bullet-}$ is essential to maintain mitochondrial homeostasis. Moreover, $O_2^{\bullet-}$ overproduction, directly and indirectly, leads to the inactivation of Mn-SOD, promoting mitochondrial dysfunction. For instance, $O_2^{\bullet-}$ can react with $^\bullet NO$ to produce peroxynitrite ($ONOO^-$), and $ONOO^-$ can induce Mn-SOD deactivation via nitration of Tyr34 residue in its active site [69].

Mn-SOD can also be S-glutathionylated in Cys 196, avoiding irreversible oxidation of SH. Renal I/R injury induces $O_2^{\bullet-}$ and $ONOO^-$ production [70], increasing nitration levels of mitochondrial proteins such as Mn-SOD and cytochrome c (cyt c), inactivating them and inducing OS and mitochondrial dysfunction [71–74]. Likewise, in folic-acid-induced renal damage, Mn-SOD activity reduction in isolated mitochondria is related to decreased mitochondrial S-glutathionylation [44], making it more susceptible to nitration. Moreover, in human renal transplants and experimental rat models of chronic renal nephropathy, there are elevated levels of $ONOO^-$ [72].

In DKD, mitochondrial OS reduces Mn-SOD enzyme activity due to Tyr nitration of this enzyme. Interestingly, the use of resveratrol, a potent antioxidant, reduces OS and Tyr nitration of Mn-SOD, preserving Mn-SOD activity. Moreover, kidneys of mice treated with streptozotocin to induced diabetic nephropathy (DN) show nitration in Mn-SOD Tyr 34, which results in a decrease of Mn-SOD. However, antagonists of thromboxane A2 receptors reduce diabetes-induced renal injury, which is associated with Mn-SOS Tyr nitration reduction [75].

In models of hypertension-related kidney injury, where hypertension is induced by angiotensin II (Ang II), the production of $O_2^{\bullet-}$ is promoted through the activation of nicotinamide adenine dinucleotide phosphate (NADPH) oxidases (NOXs) [76]. In these models, Mn-SOD activity deactivation is also associated with Tyr nitration, inducing OS [77]. Moreover, spontaneously hypertensive rats treated with N(G)-nitro-L-arginine-methyl ester (L-NAME), a potent nitric oxide synthase (NOS) inhibitor, reduce nitration of Mn-SOD, preserving its activity [78].

OS closely regulates aging-related kidney dysfunction, primarily by mitochondrial ROS (mtROS). A protein closely associated with aging is Klotho. This protein induces the activation of transcription factors such as forkhead box O (FoxO) that cause the expression of antioxidant enzymes such as Mn-SOD. Interestingly, Klotho$^{-/-}$ mouse models show high nitrotyrosine levels in Mn-SOD [79].

4. Crosstalk between NOXs and Mitochondria in Kidney Diseases

Mitochondria and NOXs are the primary ROS sources in the kidney. The NOXs family consists of seven isoforms, being NOX1, NOX2, NOX4, and NOX5, the most expressed in renal cells [80]. The $O_2^{\bullet-}$ production of NOX1 and NOX2 needs to assemble membrane subunit p22-phox and the cytosolic subunits p47- and p67-phox and ras-related C3 botulinum toxin substrate 1 (Rac1), while NOX4, abundantly expressed in mitochondrial membranes of renal cells, does not require cytosolic subunits and produces H_2O_2 [81–83].

The levels of NOXs augment in several AKI and CKD models, inducing ROS overproduction [84]. In the folic acid model, NOXs-linked ROS overproduction (in the kidney cortex, proximal tubules (PT), and distal tubules (DT)) is related to mtROS enhancement, because these two sources establish a pathological circle of ROS production, favoring AKI to CKD transition [44]. In the 5/6 nephrectomy model, mtROS also induce NOXs activation, increasing inflammation and fibrosis. Moreover, this mechanism contributes to fibrosis development in UUO [85]. Following the latter, several authors have established that in kidney pathologies, mtROS and NOXs-produced ROS increase mitochondrial damage and mitochondrial membrane potential depolarization ($\downarrow\Delta\Psi m$) (Figure 2) [86,87]. Ang II with the angiotensin type 1 receptor (ATR-1) also participates in the crosstalk between NOXs (NOX2 and NOX4) and mitochondria [88]. In addition to Ang II, in DN, the interaction of advanced glycation end products (AGE), produced by high glucose levels, activates NOXs by AGE receptor (RAGE) to generate ROS production [89]. Protein kinase C (PKC) epsilon (PKC-ε) also activates NOXs by inducing the phosphorylation of p47-phox, triggering ROS production [90]. NOXs-induced ROS cause the opening of a mitochondrial adenosine triphosphate (ATP)-sensitive potassium (K) channel (mt-K_{ATP}), triggering $\downarrow\Delta\Psi m$ [91]. In AKI and CKD models, the use of the NOXs inhibitor, apocynin, decreases mtROS production [92,93], supporting the idea that crosstalk between NOXs and mitochondria is involved.

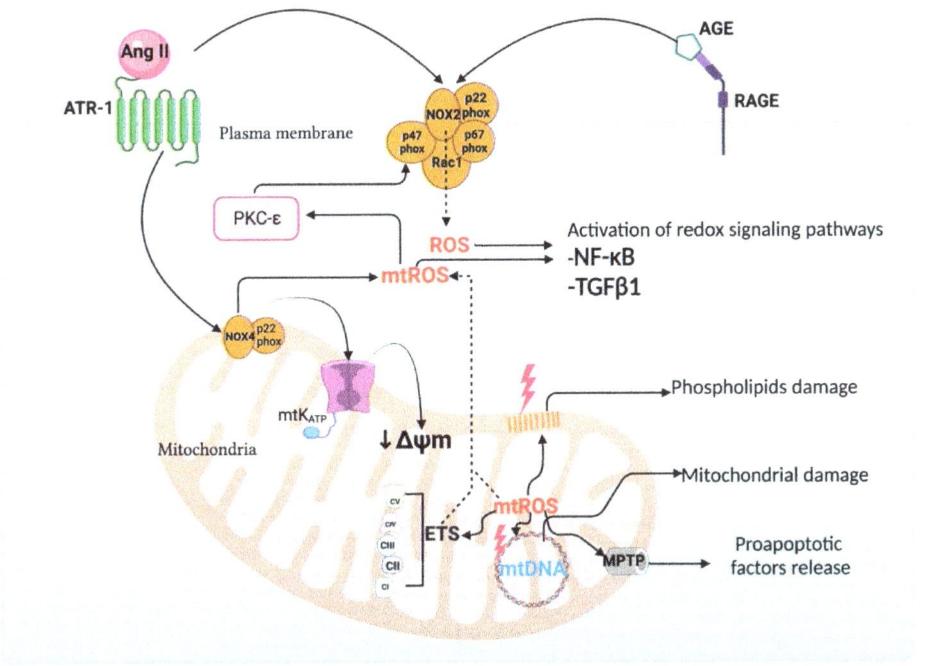

Figure 2. Crosstalk between nicotinamide adenine dinucleotide phosphate (NADPH) oxidases (NOXs) and mitochondria in the kidney. Angiotensin II (Ang II) binds to the angiotensin type 1 receptor (ATR-1), which activates NOX2 and NOX4. In addition, the binding of advanced glycation end products (AGEs) to the receptor for advanced glycation end products (RAGE) induces the activation of NOX2. Furthermore, protein kinase C (PKC) epsilon (PKC-ε), activated by mitochondrial ROS (mtROS), activates NOX2 through the p47-phox subunit, inducing ROS production. NOXs-induced ROS promote the phosphorylation and opening of an mitochondrial adenosine triphosphate (ATP)-sensitive potassium K channel (mt-K_{ATP}), decreasing mitochondrial membrane potential depolarization (↓ΔΨm). ROS and mtROS activate the redox signaling pathways: transforming growth factor-beta 1 (TGFβ1) and nuclear factor kappa-light-chain-enhancer of activated B cells (NF-κB). mtROS also affect mitochondrial function by inducing damage to mitochondrial deoxyribonucleic acid (DNA) (mtDNA) and phospholipids that, in the last instance, generates mitochondrial dysfunction. The mtROS also favor the mitochondrial permeability transition pore (MPTP) opening, inducing the release of proapoptotic factors into the cytosol. p67-phox: subunit from NOX2; p22-phox: subunit from NOX2 and NOX4; Rac1: Ras-related C3 botulinum toxin substrate 1; ETS: electron transport system; CI: complex I; CII: complex II; CIII: complex III; CIV: complex IV; CV: complex V. Created with BioRender.com.

The binding of Ang II and AGEs to their receptors activates TGFβ1 and nuclear factor kappa-light-chain-enhancer of activated B cells (NF-κB) redox-sensitive signaling pathways (Figure 2) [94]. Furthermore, mtROS can stimulate TGFβ1 through the upregulation of Smad 2/3, inducing its nuclear translocation. The latter is supported by the fact that the mitochondrial-targeting antioxidants coenzyme Q (mitoQ) and 2-(2,2,6,6-tetramethylpiperidin-1-oxyl-4-ylamino)-2-oxoethyl)triphenylphosphonium chloride (mitoTEMPO) prevents the activation of TGFβ1 along with the transcription of *TGFβ1* genes [95]. mtROS also activate NF-κB by inducing monocyte/macrophage infiltration, increasing interstitial inflammation in UUO kidneys, and treating the antioxidant curcumin to decrease them and preventing interstitial inflammation [96]. mtROS also activate NF-κB in macrophages. In this sense, Herb et al. [97] demonstrated that in macrophages, mtROS, and not ROS produced by NOX2, activate NF-κB by deactivating the regulatory subunit of inhibitor IKK complex (IKKγ) through the disulfide linkage formation. However, in kidney diseases, this mechanism has not been investigated.

TGFβ1 and NF-κB are also localized in mitochondria, regulating mitochondrial proteins [98]. Moreover, the mitochondrial localization of both proteins might indicate that inflammation and fibrosis processes are regulated by mtROS production. Following the latter, in DN, hyperglycemia triggers the Smad4 translocation into mitochondria. This translocation reduces oxidative phosphorylation (OXPHOS), inducing inflammation, fibrosis, and podocyte injury [98].

The production of mtROS and ROS produced by NOXs also damages phospholipids and mitochondrial DNA (mtDNA). mtDNA is particularly susceptible to ROS, because it does not contain histones to protect it, causing DNA integrity loss and resulting in the acquisition of mutations [99]. mtROS also favor the opening of mitochondrial permeability transition pores (MPTPs) into the cytosol [86]. On the other hand, mtROS induce phospholipids oxidation, principally cardiolipin, leading to the ↓ΔΨm, MPTP opening, and ETS activity reduction [100]. Thus, the pathological crosstalk between NOXs and mitochondrial induces ROS, affecting mitochondrial metabolism and biomolecules integrity. The latter induces mitochondrial impairment, favoring the AKI to CKD transition.

5. Mitochondrial Metabolism, ROS, and OS in Kidney Diseases

The kidneys remove waste from the blood, reabsorb nutrients, regulate electrolyte balance, maintain acid-base homeostasis and regulate blood pressure [10]. These processes require high amounts of energy, which come from OXPHOS or anaerobic glycolysis, depending on the region of the kidneys. For example, the renal cortex uses OXPHOS and low amounts of glucose, while the renal medulla uses glycolysis and lactate. Therefore, the medulla necessarily uses anaerobic glycolysis due to low oxygen levels. In contrast, the renal cortex uses OXPHOS, fed mainly by FA oxidation [10,101].

A significant number of mitochondria in renal cells is located in PT, the most metabolically active [10]. OXPHOS is the principal mechanism to produce ATP in renal proximal tubular cells (RPTCs). Ninety percentage of ATP is required to reabsorption of glucose, ions, and nutrients through the sodium-potassium ATP pump (Na^+/K^+ ATPase) [102]. PT uses FA, such as palmitate, through FA β-oxidation to produce high ATP levels (Figure 3) [103]. Therefore, RPTCs have high levels of carnitine *O*-palmitoyl transferase I (CPT I) isoforms A (CPT IA) and B (CPT IB), and carnitine *O*-palmitoyl transferase II (CPT II) [104].

5.1. *FA β-Oxidation Dysfunction in Kidney Diseases*

The impairment of β-oxidation has been reported in AKI and CKD. In patients and animal models, mRNAs along with β-oxidation proteins levels are decreased [105–107]. In the folic-acid-induced AKI model, ATP production is reduced, and mitochondria are uncoupling due to β-oxidation dysfunction [108]. Moreover, in maleic-acid (MA)-induced AKI, the FA β-oxidation-linked oxygen consumption rate (OCR) is diminished [109]. On the other hand, the transcriptomic analysis showed that the acyl-CoA dehydrogenase family member 10 (ACAD10) is downregulated in DN [110]. Moreover, in the 5/6 nephrectomy model, medium-chain acetyl dehydrogenase (MCAD) decreases at twenty-eight days [107]. Thus, the decrease in these enzymes is related to FA β-oxidation deregulation in AKI and CKD. Following the latter, FA β-oxidation has been evaluated in a course temporal in 5/6 nephrectomy, which decreases from early times [111]. Consequently, the reduction of FA β-oxidation causes intrarenal lipids accumulation, inducing lipotoxicity and impairing renal function [112]. In line with this, Nishi et al. [113] demonstrated that lipid accumulation is evident in tubular epithelial cells (TECs). Lipid deposition increases according to kidney lesion, suggesting a metabolic reprogramming that shifts β-oxidation to lipid synthesis [114,115]. According to the latter, UUO increases triglycerides synthesis from one day after obstruction [116]. Triglycerides increase is partly due to the overexpression of the transporter of the long-chain FA cluster of differentiation 36 (CD36) in PT [7,117]. CD36 also promotes signaling pathway activation such as epithelial–mesenchymal transition (EMT), inflammation, and others, leading to fibrosis. In this context, $CD36^{-/-}$ mice subjected to UUO show less fibrosis than sham groups [118]. However, Kang et al. [106] showed that mice that overexpress CD36 show the accumulation of lipids but low expression levels of fibrotic markers. Therefore, the authors hypothesized that impaired FA β-oxidation is sufficient to induce the development of fibrosis and lipid accumulation is the only consequence of this dysfunction [106]. Thus, it has been suggested that defective FA β-oxidation is one of the principal mechanisms associated with fibrosis development [103,106].

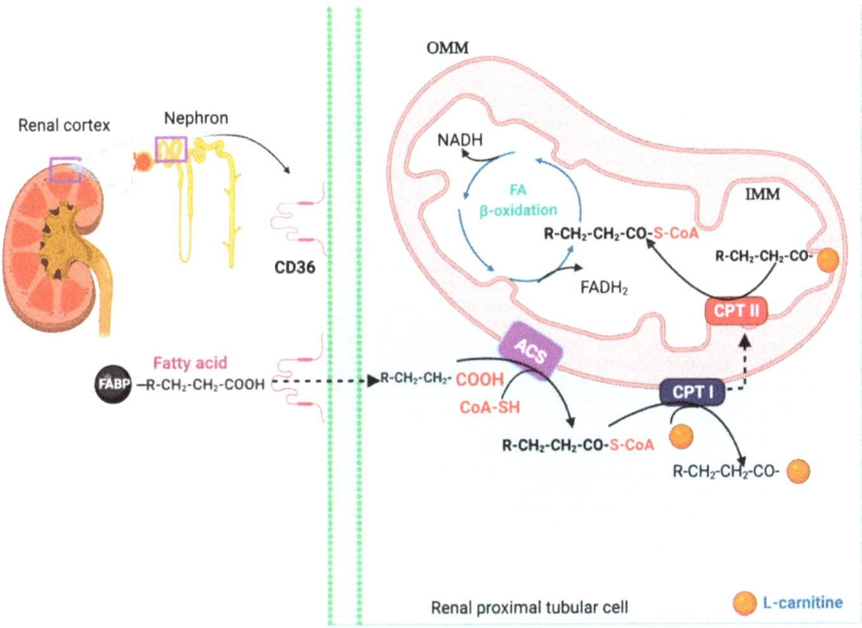

Figure 3. Renal proximal tubule (RPTC) cells use fatty acids (FA) β-oxidation to produce adenosine triphosphate (ATP). In the kidney, the proximal tubules of the nephrons of the renal cortex use fatty acids (FA) as the primary source of energy. FA bound to the fatty acid-binding protein (FABP) enter the RPTC through the cluster of differentiation 36 (CD36). In the cytosol, acyl-coenzyme A (CoA) synthetase (ACS) (attached to the outer mitochondrial membrane (OMM)) activates FA by the addition of acetyl-CoA (CoA-SH). The latter allows FA to enter the OMM through carnitine *O*-palmitoyl transferase I (CPT I). CPTI exchanges acetyl-CoA for L-carnitine. In turn, FA goes to the inner mitochondrial membrane (IMM). In the IMM, carnitine *O*-palmitoyl transferase II (CPT II) removes the carnitine group and adds acetyl-CoA (CoA-SH). The latter allows FA to enter the mitochondrial matrix. Fatty acyl-CoA undergoes β-oxidation, generating nicotinamide adenine dinucleotide phosphate (NADH) and flavin adenine dinucleotide ($FADH_2$). Created with Biorender.com.

The alterations in CPT I levels also contribute to FA β-oxidation impairment. For example, modifications in transporters of plasma acylcarnitine have been reported in CKD [119]. In this context, Prieto-Carrasco et al. [120] showed CPT I levels decreased in a temporal course from 2 to 28 days after nephrectomy, associated with progressive impairment in mitochondrial β-oxidation. The authors concluded that the decrease in CPT I favors mitochondrial β-oxidation impairment and the subsequence fibrosis development. The latter is supported by the fact that patients with CKD show a correlation between low CPT IA levels and fibrosis [121]. In this sense, the transgenic mouse models overexpressing CPT IA are able to restore oxidative metabolism, avoiding fibrosis development in UUO, folic acid nephropathy, and adenine-induced nephrotoxicity [121]. In addition, CPT IA overexpression also reduces fibrosis by decreasing TGFβ1 levels [121].

In summary, defective FA β-oxidation is observed in kidney diseases from early times, promoted through decreased mRNA expression and downregulation in the activity and levels of the proteins involved in this process and ETS activity reduction (discussed below).

Later, the overexpression of CD36 contributes to lipid accumulation and the activation of mechanisms that lead to the fibrotic process. However, other factors contribute to the impairment of β-oxidation.

Oxidation and OS Production and in Kidney Diseases

Renal pathologies cause a disturbance in mitochondria homeostasis, affecting mitochondrial metabolism, which leads to AKI to CKD transition. OS might cause these alterations in mitochondrial metabolism produced during ETS, β-oxidation, and Krebs cycle activity [103]. For instance, Kowaltowski's group [122,123] demonstrated that H_2O_2 is produced during the first step of FA β-oxidation, catalyzed by a very long-chain acyl-CoA dehydrogenase (VLCAD) enzyme in liver mitochondria. If H_2O_2 produced is not degraded, it could induce mitochondria-decoupling electron leakage during OXPHOS. Even β-oxidation may be damaged. The latter has been supported by Aparicio-Trejo et al. [44], showing that folic acid causes damage to mitochondria and decreasing the OXPHOS associated with FA β-oxidation is due to ROS overproduction. The use of the antioxidant N-acetylcysteine (NAC) prevents the reduction in OXPHOS capacity associated with FA β-oxidation impairment from 2 to 28 days of the administration, avoiding CKD transition [108]. In accordance, Briones-Herrera et al. [109] showed that MA, another inductor of OS, decreases β-oxidation and the use of antioxidant sulforaphane (SF) prevents this decrease [109].

On the other hand, Tan et al. [110] showed by transcriptomic analysis in diabetic mice that mitochondrial FA β-oxidation is downregulated. This downregulation is attributed to the overexpression of the C5 substrate of the complement system receptor 1 (C5aR1). Thus, C5aR1 is implicated in lipids metabolism in diabetes [124]. C5aR1 is also upregulated in kidney diseases, producing FA β-oxidation impairment in DN [110]. In addition, C5aR1 upregulation disrupts mitochondrial respiration, generating high levels of ROS. These results showed that C5aR1-induced ROS overproduction alters FA metabolism in DN.

During mitochondria decoupling, electron leakage from ETS occurs, which induces the reduction of oxygen (O_2) to the radical $O_2^{\bullet-}$, a type of ROS that triggers the production of other ROS such as H_2O_2 [125]. High levels of $^{\bullet}OH$ and $ONOO^-$ induce OS and significantly oxidative damage of proteins, lipids, and DNA. The oxidation of lipids produces highly reactive molecules such as MDA and 4-HNE as products of chain lipid peroxidation that also induce mtDNA damage (Figure 4) [68]. Forty-eight hours after cisplatin treatment, induced AKI, 4-HNE, and MDA levels increase along with GPx4 levels decrease in the renal cortex, indicating lipid membrane peroxidation [126]. Furthermore, in MA-induced Fanconi syndrome, 24 h after injection with MA, the mitochondrial levels of 4-HNE are elevated [109]. Additionally, GPx activity decreases, favoring H_2O_2 accumulation and mitochondrial lipidic peroxidation [109]. In folic-acid-induced AKI, 24 h after folic acid administration, mitochondrial MDA and 4-HNE levels increase [44]. Both OS markers also increase in nephrectomy models [127,128]. Together, these results show that mitochondria suffer lipid peroxidation in AKI and AKI to CKD transition.

In kidney diseases, the uptake of lipids by CD36, along with the dysfunction of FA β-oxidation, causes lipids accumulation in lipid droplets (LDs), inducing ROS overproduction (Figure 4) [129,130]. Since ROS and their products induce severe cell damage, a cellular balance of ROS is needed. This balance is performed by different antioxidants that include enzymatic and non-enzymatic antioxidants [131]. The reduction of the antioxidant system has been widely reported in kidney diseases [11,132]. In renal ischemia and nephrotoxicity, catalase (CAT), SOD, and glutathione S-transferase (GST) levels are depleted [70,133]. Moreover, in cisplatin-induced AKI, mitochondrial GSH and NADPH levels are decreased [134]. In the 5/6 nephrectomy, the activities of CAT, SOD, GPx, GR, and GST fall at 20 h in glomeruli, PT, and DT [128]. Moreover, early after UUO, these enzymes' mRNA and protein levels decrease, while oxidative markers increase [135]. Indeed, the decrease in antioxidant-system-induced OS has been suggested as a factor to induce the AKI to CKD transition.

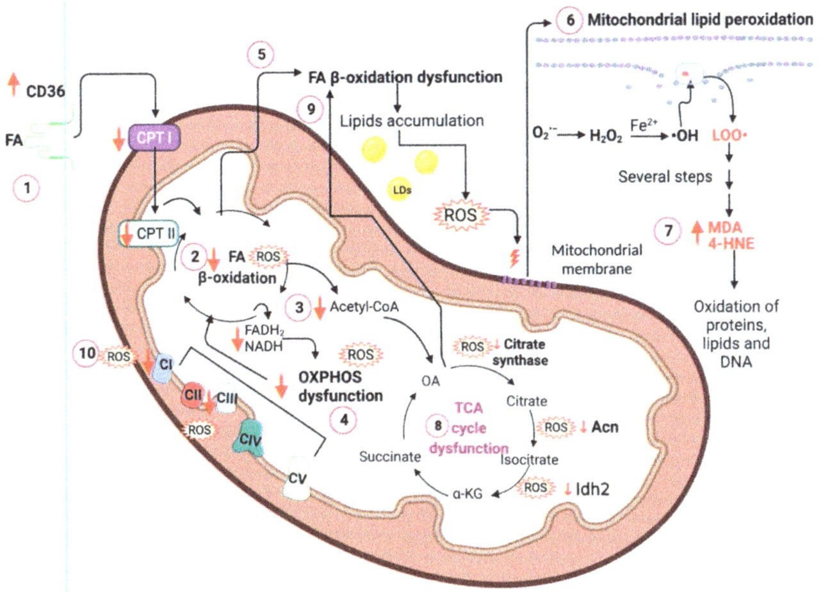

Figure 4. ROS deregulate mitochondrial metabolism in kidney diseases. (1) In renal damage, the cluster of differentiation 36 (CD36) overexpression causes a high fatty acids (FA) uptake. In addition, carnitine *O*-palmitoyl transferase I (CPT I) and carnitine *O*-palmitoyl transferase II (CPT II) are decreased. (2) ROS cause FA β-oxidation decrease, inducing (3) a tricarboxylic acid (TCA) cycle and (4) oxidative phosphorylation (OXPHOS) capacity reduction. The decrease in β-oxidation also (5) induces the accumulation of lipids and a further ROS overproduction. The latter (6) damages mitochondrial membranes by inducing mitochondrial lipid peroxidation, (7) forming the products malondialdehyde (MDA) and 4-hydroxynonenal (4-HNE). These products are highly reactive and damage other lipids, proteins, and mitochondrial DNA (mtDNA). On the other hand, (8) ROS downregulate aconitase (Acn), citrate synthase, and isocitrate dehydrogenase isoform 2 (Idh2), inducing TCA cycle dysfunction. Moreover, (9) TCA cycle impairment induces FA β-oxidation dysfunction. (10) ROS decrease CI and CIII activities, inducing β-oxidation dysfunction [111]. $O_2^{\bullet-}$, superoxide anion radical; H_2O_2, hydrogen peroxide; •OH, hydroxyl radical; LOO•, lipid peroxyl radical; LDs, lipid droplets; OA, oxalacetate; α-KG, alpha-ketoglutarate; DNA, deoxyribonucleic acid; NADH, nicotinamide adenine dinucleotide phosphate; $FADH_2$, flavin adenine dinucleotide; CoA, coenzyme A; CII, complex II; CIV, complex IV; CV, complex V. Created with BioRender.com.

The decreases in acetyl-CoA induced by FA β-oxidation impairment reduce the TCA cycle capacity. Interestingly, in AKI and CKD models, the reduction in TCA cycle enzymes is observed, before FA β-oxidation dysfunction occurs, suggesting this point as the start of the vicious cycle, which further increases the mitochondrial damage (Figure 4). In the next section, we will address the impact that ROS have on TCA cycle dysfunction.

5.2. TCA Cycle Redox-Sensitive Signaling Pathway in Kidney Diseases

The urinary excretion of the non-diabetic CKD patients shows low levels of TCA cycle metabolites (e.g., citrate, cis-aconitate, isocitrate, alpha-ketoglutarate (α-KG), and succinate) [136]. In addition, kidney biopsies have reduced aconitate, isocitrate, alpha-ketoglutarate dehydrogenase (α-KGDH), and succinate gene expression. These results show TCA cycle dysfunction [136]. In contrast, in a mouse model of DN, pyruvate, citrate, α-KGDH, and fumarate are upregulated [137]. Moreover, in UUO, succinate levels increase, attributed to TCA cycle dysfunction [138]. Note that the amount of the metabolites is tissue- and disease-dependent, so the identification of these metabolites could give advantages in the early detection of mitochondrial damage in these diseases.

TCA cycle dysfunction might be attributed to ROS alterations. In vitro studies have postulated that high glucose oxidation rates lead to the excessive production of electron donors from the TCA cycle. As a consequence, ETS becomes overloaded, promoting $O_2^{\bullet -}$ overproduction [139]. In line with this, podocytes treated with high glucose levels have high ROS levels, and the treatment with mitoTEMPO decreases them [140], suggesting that ROS are specifically delivered from mitochondria. Controversially, the determination of mtROS in the diabetic mouse model shows that it is reduced [141]. Further studies in vivo are needed to elucidate the mtROS overproduction-induced TCA cycle dysfunction in DN.

In the kidney, TCA cycle enzymes can be sulfenylated or S-glutathionylated. For example, Acn can be reversibly inactivated by the oxidation of the sulfhydryl group by $O_2^{\bullet -}$ and H_2O_2. However, if OS persists, Acn can be irreversibly deactivated by the disruption of the 4Fe-4S group [142]. In AKI induced by folic acid, mitochondrial Acn activity decreases, and the pre-treatment with NAC prevents it [44], suggesting that ROS promote the deactivation of Acn. In addition, the relation between Acn and citrate synthase diminishes, supporting the idea that decreasing in Acn activity is related to OS [44]. Moreover, Mapuskar et al. [143] reported that the persistent increase of $O_2^{\bullet -}$ decreases Acn and citrate synthase activity in cisplatin-induced kidney injury, of which the effects are ameliorated by SOD mimetic avasopasem manganese (GC4419) treatment. The authors reported that in the AKI phase, Acn and citrate synthase activities do not show changes, suggesting that high levels of ROS are required for their inactivation in this model [143]. The latter is demonstrated due to the fact that high levels of ROS are more evident in cisplatin-induced CKD [143].

In kidney pathologies, the levels of mitochondrial isocitrate dehydrogenase isoform 2 (Idh2) are decreased [144,145]. In the cisplatin model, Idh2 function is affected by decreased mitochondrial NADPH and GSH and increased H_2O_2 production [145]. Furthermore, Han et al. [144] showed that OS generated during I/R reduces Idh2 levels in kidney tubule cells from mice. Since S-glutathionylation deactivates Idh2, this Ox-PTM may be produced during OS under I/R conditions [146]. The deletion of the Idh2 (Idh2$^{-/-}$) gene in these mice exacerbates kidney tubule injury by increasing plasma creatinine and blood urea nitrogen (BUN) levels. In addition, OS increases the reduction of mitochondrial $NADP^+$ along with GST and GPx activities. In contrast, mitochondrial GSSG/GSH ratio augments. Idh2$^{-/-}$ mice show mitochondrial dysfunction and fragmentation, which induces apoptosis in kidney tubule cells [144]. After UUO, Idh2 decreases, and its deletion increases OS markers such as 4-HNE and H_2O_2 in mitochondrial fractions [147]. Additionally, inflammatory cell filtration was more evident in Idh2$^{-/-}$ than wild-type (WT) groups. Together, these results highlighted the importance of Idh2 in managing OS, and its deactivation exacerbates mitochondrial damage.

Note that the fact that ROS-induced TCA cycle dysfunction affects FA β-oxidation has been demonstrated, because TCA cycle impairment is observed early before FA β-oxidation damage in time course studies of the AKI to CKD transition, (Figure 4). In this regard, OXPHOS capacity is also decreased by ROS in early times, suggesting that both events are required to enhancement FA β-oxidation dysfunction [44,108,120].

5.3. OXPHOS Redox-Sensitive Signaling Pathway in Kidney Diseases

As mentioned above, kidney energy demand depends on OXPHOS, which in turn is regulated by Ox-PTMs. However, in the context of kidney diseases, it is poorly studied. OXPHOS capacity is downregulated in renal diseases, inducing ROS overproduction. In this regard, the production of mitochondrial H_2O_2 has been reported in kidney injury models, associated with $\downarrow\Delta\Psi m$, decreasing OXPHOS capacity [8,148]. In remnant kidney from 5/6 nephrectomy, OXPHOS linked to complex I (CI) feeding decrease in a temporal course of nephrectomy from 2 to 28 days [111]. Moreover, male Sprague Dawley rats subjected to nephrectomy showed ATPβ, NDUSF8, and cytochrome c oxidase subunit 1 (Cox I) reduction [8]. Consequently, CI, complex III (CIII) activities, and cyt c diminished, impairing mitochondrial function [8]. Avila-Rojas et al. [148] reported that potassium dichromate

($K_2Cr_2O_7$) decreases the CI + CII-linked S3 respiratory state. In addition, $\Delta\Psi m$ in CI + complex II (CII)-linked respiration and respiration associated with OXPHOS are reduced, suggesting that $K_2Cr_2O_7$ principally affects the synthesis of mitochondrial ATP [148]. Although redox signaling has not been investigated in previous studies, components of OXPHOS might be regulated by Ox-PTMs.

6. ROS Induce Uncoupling Proteins (UCPs) Dysregulation in Kidney Diseases

UCPs are proton transporters (H^+), which move H^+ from the IMM into the mitochondrial matrix. These transporters are localized in the IMM and dissipate the proton gradient from the mitochondrial matrix into the IMS [149]. mtROS induce UCP2 activation, decreasing the proton gradient and preventing mtROS overproduction [149].

It has been shown that UCP2 deletion aggravates tubular injury in the I/R model by inducing ROS overproduction, supporting the importance of these transporters in ROS dissipation [150]. Moreover, the UCP2 inhibition worsens the damage caused by lipopolysaccharide (LPS), increasing apoptosis in TECs [151]. In CKD, Jian et al. [152] showed that in renal tubular cells (RTCs), the expression of UCP2 is induced three days after obstruction and continues after seven days, avoiding UUO-induced fibrosis. It suggests that UCP2 is crucial to avert fibrosis development induced by ROS in UUO.

Although UCP1 is commonly found in mitochondria from brown adipose tissue, it is expressed in the kidney. For instance, in AKI models induced by cisplatin or I/R, Jia et al. [153] found that UCP1 is upregulated in renal TECs and its presence is related to OS suppression. Chouchani et al. [154] showed that mtROS alter the redox status of UCP1 by inducing its sulfenylation in Cys 253, promoting UCP1 activity.

7. Redox-Sensitive Signaling Controls Mitochondrial Dynamics, Biogenesis, and Mitophagy

Mitochondrial dynamic is the balance between mitochondrial fusion and fission, and it is involved in regulating mitochondrial metabolism and cell death. Likewise, the shape and morphology of mitochondria are regulated by the metabolite and ROS concentration concentrations [155]. The integral membrane guanosine triphosphatases (GTPases) perform mitochondrial fusion: mitofusin 1 (Mfn1), mitofusin 2 (Mfn2), and optical atrophy 1 (Opa1) [136]. Meanwhile, mitochondrial fission is mediated by Drp1, the mitochondrial fission factor (Mff), the adaptor protein fission protein 1 (Fis1), and the mitochondrial elongation factor 1 (Mief1) and 2 (Mief2). Chronic OS has been shown to induce mitochondrial fission [155]. However, H_2O_2 sublethal amounts, and acute OS cause mitochondrial hyperfusion [156]. In HeLa cells, mitochondrial hyperfusion has been associated with the S-glutathionylation of Mfn2 and the formation of Opa1 oligomers. Moreover, mitochondrial hyperfusion increases the resistance to cellular stress and cell death, since it promotes antioxidant defense enzymes activation [156]. In the next section, we analyze the current trends and paradigms of ROS-mediated signaling that form a link between redox-sensing elements and mitochondrial dynamics and their possible role in kidney diseases.

7.1. Redox-Sensitive Proteins Participating in Fission and Fusion

Redox signaling conveys external and internal signals between redox-sensitive receptors and the downstream effectors of fission machinery. Mitochondrial dynamics require the recruitment of proteins to mitochondria. Indeed, the importation of several proteins to mitochondria depends on proton electrochemical gradient H^+ created by ETS at the IMM, which is called the proton motive force (PMF) [157]. In addition, several redox-sensitive proteins are activated to induce proteins translocation from the cytosol. For instance, previous studies in HeLa cells have shown that MAPKs are involved in regulating the fission process in response to OS through Ras [158,159]. In line with this, the redox-sensitive extracellular regulated kinase 2 (ERK2) protein mediates the phosphorylation of Drp1 in Ser 616 to induce mitochondrial fragmentation [158]. In addition to ERK 2, PKC-delta (PKC-δ) promotes Drp1 phosphorylation in Ser 579 under OS [160]. Drp1 is also regulated by ROS. In this context, Kim et al. [161] showed that in vascular diseases related to diabetes and aging models, protein disulfide isomerase A1 (PDIA1) is depleted, inducing the sulfenylation of Drp1 in Cys 644. The latter leads to mitochondrial fragmentation, mtROS increase, and senescence induction. Furthermore, the authors demonstrated that the restoration of the PDIA1/Drp1 axis could be used as a therapeutic strategy to improve vascular diseases [161], suggesting that PDIA1 has a thiol reductase function for Drp1. Drp1 persulfidation was previously reported, inducing its inactivation. Persulfidation is carried out by CARS2, altering mitochondrial dynamics and favoring fusion [66].

In renal pathologies, the upregulation of Drp1 is related to OS conditions [44,147,148]. Likewise, the phosphorylation of ERK 1/2 increases along with Drp1 levels in response to OS in the I/R rat model [162]. Thus, ROS might induce the recruitment of Drp1 through ERK1/2 (Figure 5).

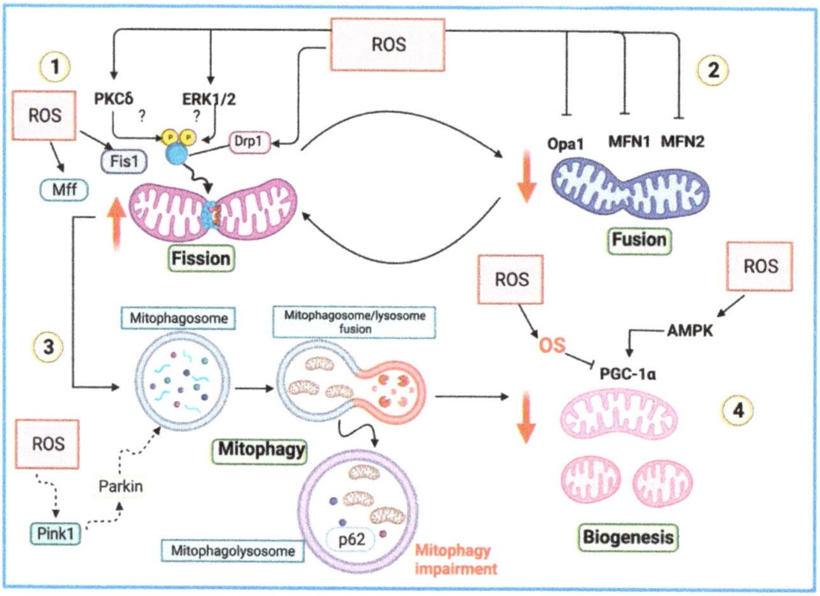

Figure 5. Redox-sensitive signaling regulates mitochondrial dynamics in kidney diseases. (1) ROS activate the redox-sensitive extracellular regulated kinase 2 (ERK2) protein and protein kinase C (PKC) isoform δ (PKC-δ). These proteins phosphorylate and activate dynamin-related protein 1 (Drp1), inducing its translocation to the OMM to triggering fission. Likewise, ROS overproduction upregulates Drp1, fission 1 (Fis1), and mitochondrial fission factor (Mff), inducing fission increase. (2) ROS also induce fusion decrease by downregulating optical atrophy 1 (Opa1) and mitofusin 1 (Mfn1) and 2 (Mfn2) proteins. (3) The augment of fission promotes mitophagy activation by inducing the translocation of phosphatase and tensin homolog (PTEN)-induced putative kinase 1 (Pink1) that in turn recruit to parkin in the OMM to induce mitophagosome formation. Mitophagosome fuses with lysosome to form mitophagolysome formation. However, mitophagy flux is impaired, inducing the accumulation of damaged mitochondria. (4) Low ROS levels induce mitochondrial biogenesis by activating the redox-sensitive adenosine monophosphate (AMP)-activated protein kinase (AMPK) and the peroxisome proliferator-activated receptor (PPAR) γ coactivator-1 alpha (PGC-1α). However, high levels of ROS producing oxidative stress (OS) induce mitochondrial biogenesis decrease, triggering mitochondrial mass decrease. Created with BioRender.com.

In kidney diseases, ROS overproduction induces mitochondrial proteins fusion decrease and fission increase. In line with this, in the $K_2Cr_2O_7$ rat model, Drp1 increase and curcumin treatment decrease it [148], suggesting that ROS induce Drp1 upregulation (Table 1). The latter is supported by the fact that ROS promote Drp-1 translocation to mitochondria. Drp1 is also recruited in RPTC treated with cisplatin, inducing mitochondrial fragmentation and ATP depletion [163]. Moreover, Drp1 is upregulated in Idh2$^{-/-}$ mice due to NADPH and GSH decrease, which increase mtROS. mtROS also downregulate Opa1 (Table 1) [44,144,145]. Thus, mtROS trigger fission increase and fusion decrease. Following the latter, Opa1 levels are diminished under H_2O_2 treatment, and the diminishment is even higher in Idh2 small interfering RNA (siRNA)-transfected mProx24 cells [144]. Mfn1 is another fusion protein decreased in AKI models (Table 1) [44,164].

Table 1. ROS regulate mitochondrial dynamics, biogenesis, and mitophagy in acute kidney injury (AKI) models.

AKI Model	In Vivo Model	Mitochondrial Dynamic Protein	Mechanism	References
Cisplatin-induced nephrotoxicity and I/R	C57BL/6 mice	↑Drp1	Drp1 translocates to the mitochondria in response to ROS overproduction.	Brooks et al. [163]
Maleate-induced nephrotoxicity	Male Wistar rats	↑Drp1, Fis1	Maleate-induced OS promotes mitochondrial fission by increasing Drp1 and Fis1.	Molina-Jijón et al. [84]
Cisplatin	C57BL/6 mice	↓Opa1, Mfn1 ↑Fis1 ↑Pink1, parkin	ROS and mtROS promote fission and decrease the mitochondrial fusion process.	Ortega-Domínguez et al. [164]
I/R	C57BL/6 mice	Drp1$^{-/-}$	The deletion of Drp1 improves mitochondrial function by decreasing mtROS.	Perry et al. [165]
Cisplatin	Female C57BL/6 Idh2$^{-/-}$ mice	↓Opa1 ↑Drp1	Idh2$^{-/-}$ decreases NADPH and GSH levels, inducing OS and triggering fission increase and fusion decrease.	Kong et al. [145]
I/R	Female C57BL/6 Idh2$^{-/-}$ mice	↑Drp1, Fis1 ↓Opa1	Idh2$^{-/-}$-induced mtROS, decreasing the levels of fusion proteins and augmenting fission proteins.	Han et al. [144]
Folic acid	Male Wistar rats	↑Fis1, Drp1 ↓Opa1, Mfn1 ↑Pink1, ↓LC3 ↓PGC-1α, ↓NRF1, NRF2	ROS overproduction increases fission and reduces the fusion process.	Aparicio-Trejo et al. [44]
Nephrotoxicity by $K_2Cr_2O_7$	Male Wistar rats	↑Drp1 ↓PGC-1α	ROS overproduction increases fission and reduces biogenesis.	Ávila-Rojas et al. [148]
MA: induced Fanconi syndrome	Male Wistar rats	↓TFAM, ↑Fis1, Drp1 ↑Parkin, p62, LC3-II	SF prevents mitochondrial fission increase and TFAM decrease and regulates mitophagy.	Briones-Herrera et al. [109]

Abbreviations: ↑: increase; ↓: decrease; I/R, ischemia/reperfusion; Drp1, dynamin-related protein 1; Fis1, fission 1; OS, oxidative stress; Opa1, optical atrophy 1; Mf1, mitofusin 1; Pink1, phosphatase and tensin homolog (PTEN)-induced putative kinase 1; Idh2, Isocitrate dehydrogenase isoform 2; $K_2Cr_2O_7$, potassium dichromate; PGC-1α, peroxisome proliferator-activated receptor (PPAR) γ coactivator-1 alpha; TFAM, transcription factor A; NRF1, nuclear respiratory factor 1; NRF2, nuclear respiratory factor 2; p62, sequestosome; LC3, microtubule-associated protein 1A/1B-light chain 3 phosphatidylethanolamine conjugate; NADPH, nicotinamide adenine dinucleotide phosphate; GSH, glutathione; mtROS, mitochondrial reactive oxygen species; MA, maleic acid; SF, sulforaphane.

The blockage of mitochondrial fission has been suggested as a strategy to ameliorates mitochondrial damage [166]. For instance, the loss of DRP1, six hours after suffering bilateral I/R, re-establish mitochondrial function due to the reduction of mtROS production associated with fission decrease (Table 1) [165]. In addition, mice Drp1$^{-/-}$ does not have tubulointerstitial fibrosis [165], suggesting that the Drp1 blocking might avoid the AKI to CKD transition. Furthermore, the specific inhibitor of Drp1, the mitochondrial division inhibitor 1 (mdivi-1), ameliorates I/R-mediated AKI (Table 1) [163]. However, in the case of UUO, the usage of mdivi-1 augments fibrosis [167], which correlates with midivi-1-treated human proximal tubular cells (HK2) under hypoxic conditions, showing fibrosis markers increase [167]. Therefore, Drp1 blocking might be employed in the case of AKI, but not in CKD.

The management of redox recovery homeostasis might be utilized as a strategy to improve mitochondrial dynamics. Consistent with this, the use of antioxidants that target mitochondria has been shown to enhance the homeostasis of mitochondrial dynamics [44,148]. Therefore, ROS regulate proteins involved in mitochondrial fission and fusion. According to the latter, it has been hypothesized that the treatment with sublethal amounts of H_2O_2 induces acute stress, promoting a hyperfused mitochondrial state [24,168]. In folic-acid-induced kidney injury, Drp1 and Fis1 increase, and the treatment with NAC decreases them [44]. In response to cisplatin or bilateral I/R, Drp1 is recruited to mitochondria, triggering apoptosis induction by delivering cyt c and decreasing the antiapoptotic protein B cell lymphoma 2 (Bcl-2) [163]. Moreover, the activation of peroxisome proliferator-activated receptor γ (PPARγ) stabilizes mitochondrial potential, reducing ROS. The latter results in decreasing Drp1, augmenting Mfn2, Opa1 and restoring mitochondrial dynamics [8], suggesting that the rescue of mitochondrial biogenesis might recover mitochondrial dynamics due to mtROS decrease. In CKD, the levels of fusion proteins (e.g., Mfn1, Mfn2, and Opa1) are also decreased, while fission proteins are increased (e.g., Drp1 and Mff) (Table 2) [8,167,169,170], attributed to mitochondrial OS increase.

Table 2. ROS regulate mitochondrial dynamics, biogenesis, and mitophagy in chronic kidney disease (CKD) models.

CKD Model	In Vivo Model	Mitochondrial Dynamic Proteins Alteration	Effects	References
DN	Male C57BL/6J mice	↓PGC-1α, AMPK	Reduced ROS levels decrease mitochondrial biogenesis.	Dugan et al. [141]
5/6 nephrectomy	Male Sprague-Dawley rats	↓Mfn2, Opa1 ↑Drp1 ↓PPARγ	The use of pioglitazone, a peroxisome proliferator-activated receptor γ (PPARγ) activator, decreases mtROS, improving mitochondrial dynamics.	Sun et al. [8]
5/6 nephrectomy	Male Wistar rats	↑Mfn1, Opa1 ↓Fis1, Drp1	ROS overproduction favors mitochondrial fusion.	Aparicio-Trejo et al. [128]
UUO	Male C57BL/6J mice	↑LC3, Pink1, parkin	ROS-induced senescence impairs mitophagy.	Liu et al. [169]
UUO	Male C57BL/6J mice	↓Drp1 ↑LC3, Pink1, parkin	mtROS recruit Drp1 to the OMM, regulating mitophagy parkin-dependent.	Li et al. [167]
5/6 Nephrectomy	Male Wistar rats	↓NRF1, NRF2, TFAM PGC-1α, PPARα ↓Mfn2, Opa1 ↑LC3, p62	Mitochondrial biogenesis and dynamics are altered temporal courses.	Prieto-Carrasco et al. [111]

Abbreviations: ↑: increase; ↓: decrease; DN, diabetic nephropathy; PPARα, peroxisome proliferator-activated receptor γ coactivator-1α; AMPK, adenosine monophosphate (AMP)-activated protein kinase; PPARγ, peroxisome proliferator-activated receptor γ; OMM, outer mitochondrial membrane; Drp1, dynamin-related protein 1; Fisl, fission 1; OS, oxidative stress; Opa1, optical atrophy 1; Mf1, mitofusin 1; Pink1, phosphatase and tensin homolog (PTEN)-induced putative kinase 1; PGC-1α, peroxisome proliferator-activated receptor (PPAR) γ coactivator-1 alpha; TFAM, transcription factor A; NRF1, nuclear respiratory factor 1; NRF2, nuclear respiratory factor 2; p62, sequestosome; LC3, microtubule-associated protein 1A/1B-light chain 3 phosphatidylethanolamine conjugate.

Ox-PTMs might regulate mitochondrial dynamics proteins; however, they are poorly studied in kidney disease. In line with this, S-nitrosylation in Drp1 Ser 644 induces Drp1 dimerization and augments its GTPase activity. Thus, NO promotes Drp1-induced mitochondrial fission [171]. Furthermore, OS can trigger mitochondrial hyperfusion [156]. Further studies are needed to clarify the role of ROS to induce Ox-PTMs regulation in mitochondrial dynamics and biogenesis in renal disease.

7.2. Redox-Sensitive Proteins Participating in Mitochondrial Biogenesis

Mitochondria biogenesis is triggered to increase the number and size of mitochondria [172]. PPARγ coactivator-1 alpha (PGC-1α) controls the biogenesis process through the transcription of nuclear respiratory factors 1 (NRF-1) and 2 (NRF-2), PPARs, transcription factor A (TFAM), estrogen, and estrogen-related receptors (ERRs), among others [173]. These proteins are redox-sensitive. For instance, the exposure to H_2O_2 in skeletal muscle cells for 24 h increases the activity of the PGC-1α promoter as well as mRNA expression. These effects are blocked with NAC treatment [174], showing that ROS mediate the activation of PGC-1α. However, the impact of ROS over PGC-1α depends on the concentration and cellular type. For example, low ROS levels lead to the reduced expression of PGC-1α, while high levels induce its transcription through redox-sensitive adenosine monophosphate (AMP)-activated protein kinase (AMPK) [174], which function as a cellular energy sensor by regulating mitochondrial biogenesis and maintaining redox homeostasis [175]. AMPK reduces ROS through PGC1-α, which induces the overexpression of CAT, Mn-SOD, UCP2, and nicotinamide adenine dinucleotide (NAD)-dependent deacetylase sirtuin-3 (SIRT3) [176]. Therefore, the activation of AMPK induces PGC-1α promoter activity and mRNA levels increase. PGC-1α is commonly downregulated in kidney diseases, leading to mitochondrial mass and metabolism decrease [10]. In the folic-acid-induced AKI model, PGC-1α, TFAM, NRF1, and NRF2 levels decrease 24 h after the treatment, and NAC treatment prevents this effect [44]. Moreover, male Wistar rats subject to nephrotoxicity by $K_2Cr_2O_7$ show a decrease in PGC-1α levels [148]. Both studies showed that the treatment with antioxidants (NAC and curcumin) upregulates biogenesis [47,152,169]. Importantly, NAC and curcumin antioxidants have shown mitochondria protection by promoting bioenergetics preservation and maintaining redox homeostasis to avoid mtROS [16,94]. In 5/6 nephrectomy-induced CKD, the levels of PGC-1α decrease in a temporal course from two days after the nephrectomy, causing reductions in NRF1 and NRF2 [120]. The treatment with pioglitazone, an antidiabetic drug, reduces mtROS, restoring mitochondrial biogenesis. Interestingly, the biogenesis decrease in DN is attributed to the low production of $O_2^{\bullet -}$ [141]. AMPK activity decreases in this model, and AMPK activation induces $O_2^{\bullet -}$ production, activating PGC-1α [141]. Therefore, in the DN mice model, low ROS levels are essential factors to trigger mitochondrial biogenesis.

7.3. Mitophagy, ROS, and OS in Kidney Diseases

The dysregulation of mitophagy has been previously reported in renal diseases [177,178]. At biological levels, ROS are involved in mitophagy regulation. Upon mitochondrial damage, depolarization, or mitochondrial OS, mitophagy is triggered. ROS promote the recruitment of phosphatase and tensin homolog (PTEN)-induced putative kinase 1 (Pink1) in the outer mitochondrial membrane (OMM) [179]. The latter leads to the recruitment and phosphorylation of parkin to begin mitophagy [180]. In the AKI models, Pink1 and parkin are upregulated after damage [44,109,164]. The upregulation of these proteins is related to ROS and mtROS increase. Furthermore, in CKD models, both proteins are augmented [120,167,169]. For instance, in the 5/6 nephrectomy model, levels of Pink1, parkin, microtubule-associated protein 1A/1B-light chain 3 phosphatidylethanolamine conjugate (LC3-II), and sequestosome (p62) increase in a temporal course way [120]. However, in both cases, mitophagy has been reported impaired. Although the levels of LC3-II increase, the accumulation of p62 is evident, suggesting dysfunctional mitophagy [120]. p62 accumulation is considered a marker of mitophagy malfunction. In kidney injury, mitophagosome accumulation is evident by increasing Pink1, parkin, and p62 [181]. Together, these data suggest that ROS induce Pink1 and parkin translocation to mitochondria, but the high ROS levels exacerbate mitophagy machinery, damaging it. Mitophagy damage is supported, because antioxidants can increase mitophagy flux in AKI and CKD models [44,181].

On the other hand, excessive autophagy has been described in models of UUO, which triggers endothelial dysfunction [182]. Consistent with this, Chen et al. [183] found that treatment with NaHS, an exogenous H_2S donor, in UUO mice decreases the expression levels of LC3-II/I, beclin-1, and AMPK proteins. On the contrary, the p62, CBS, and CSE levels increase compared to in the sham groups. Therefore, H_2S is considered a protective mechanism, because it moderates OS that promotes the dysregulation of autophagy.

8. Concluding Remarks

ROS are second messengers that modify redox sensitive proteins in mitochondria by inducing Ox-PTMs. Thus, low levels of ROS are necessary to render these modifications. However, in kidney diseases, ROS trigger mitochondrial dysfunction evident by alterations in FA β-oxidation, TCA cycle, OXPHOS, mitophagy, mitochondrial dynamics, and biogenesis.

Moreover, the crosstalk between NOXs and mitochondria generates ROS. The impairment in one of these elements can trigger an uncontrolled ROS production increase. Therefore, perturbations in mitochondrial redox homeostasis are common characteristics that allow the transition from AKI to CKD.

Author Contributions: Conceptualization, A.K.A.-R.; writing—original draft preparation, A.K.A.-R.; writing—review and editing, A.K.A.-R., A.C.-G., O.E.A.-T., and J.P.-C.; figures preparation, A.K.A.-R.; funding acquisition, J.P.-C. All authors have read and agreed to the published version of the manuscript.

Funding: This research was funded by Consejo Nacional de Ciencia y Tecnología (CONACYT; grant number: A1-S-7495) and by Dirección General de Asuntos del Personal Académico (DGAPA; grant numbers: IN202219 and IN200922).

Institutional Review Board Statement: Not applicable.

Informed Consent Statement: Not applicable.

Data Availability Statement: Not applicable.

Acknowledgments: A.K.A.-R. is a Ph.D. student from Posgrado en Ciencias Biológicas at the Universidad Nacional Autónoma de México and is recipient of a scholarship from CONACyT, México (CVU 818062). We gratefully acknowledge the Postdoctoral Grant's Program (POSTDOC) from the Dirección General de Asuntos Académicos (DGAPA), UNAM, for the postdoctoral fellow position to A.C.-G.

Conflicts of Interest: The authors declare no conflict of interest.

Abbreviations

↓ΔΨm	mitochondrial membrane potential depolarization
•NO	nitric oxide
•OH	hydroxyl radical
4-HNE	4-hydroxynonenal
α-KΓ	Alpha-ketoglutarate
α-KΓΔH	alpha-ketoglutarate dehydrogenase
ACAD10	acyl-CoA dehydrogenase family member 10

ACS	acyl-CoA synthetase
Acn	aconitase
AGE	advanced glycation end products
AKR1A1	aldo-keto reductase family 1 member A1
AKI	acute kidney injury
AMPK	adenosine monophosphate (AMP)-activated protein kinase
Ang II	angiotensin II
Arg	arginine
ATP	adenosine triphosphate
ATR-1	angiotensin type 1 receptor
Bcl-2	B cell lymphoma 2
BUN	blood urea nitrogen
CI	complex I
CII	complex II
CIII	complex III
CIV	complex IV
CV	complex V
C5aR1	C5 substrate of the complement system receptor 1
CAT	catalase
CBS	cystathionine β-synthase
CD36	cluster of differentiation-36
CKD	chronic kidney diseases
CoA	coenzyme A
Cox I	cytochrome c oxidase subunit 1
CPT I	carnitine *O*-palmitoyl transferase I
CPT II	carnitine *O*-palmitoyl transferase II
CPT IA	CPT I isoform A
CPT IB	CPT I isoform B
CSE	cystathionine γ-lyase
Cu/Zn-SOD	copper/zinc-SOD
Cys	cysteine
CARSs	cysteinyl-transfer RNA (tRNA) synthetases
Cyt c	cytochrome c
DKD	diabetic kidney disease
DN	diabetic nephropathy
DNA	deoxyribonucleic acid
Drp1	dynamin-related protein 1
DT	distal tubules
EGFR	epidermal growth factor receptor
EMT	epithelial–mesenchymal transition
eNOS	endothelial nitric oxide synthase
ERK2	extracellular regulated kinase 2
ERR	estrogen and estrogen-related receptors
ETS	electron transport system
FA	fatty acids
FABP	fatty-acid-binding protein
$FADH_2$	flavin adenine dinucleotide
Fe^{2+}	ferrous iron
Fe^{3+}	ferric iron
Fe-S	iron-sulfur
Fis1	fission protein 1
FoxO	forkhead box O
GC4419	avasopasem manganese
GFRs	growth factor receptors
GPx	glutathione peroxidase
Grx	glutaredoxin
GSH	glutathione
GTPases	guanosine triphosphatases
GST	glutathione S-transferase
h	hours

HK2	human proximal tubular cells
$HS^•$	hydrosulfide radical.
H_2O_2	hydrogen peroxide
H_2S	gydrogen sulfide
His	histidine
I/R	ischemia/reperfusion
Idh2	isocitrate dehydrogenase isoform 2
IKKγ	regulatory subunit of inhibitor IKK complex
IMM	inner mitochondrial membrane
IMS	intermembrane space
$K_2Cr_2O_7$	potassium dichromate
L-NAME	N(G)-nitro-L-arginine-methyl ester
LC3-II	microtubule-associated protein 1A/1B-light chain 3 phosphatidylethanolamine conjugate
LDs	lipid droplets
$LOO^•$	lipid peroxyl radical
LPS	lipopolysaccharides
Lys	lysine
MA	maleic acid
MAPK	mitogen-activated protein kinases
MCAD	medium-chain acetyl dehydrogenase
MDA	malondialdehyde
Met	methionine
MetO (R–$SOCH_3$)	methionine sulfoxide
$MetO_2$ (RSO_2CH_3)	methionine sulfone
Mff	mitochondrial fission factor
Mfn1	mitofusin 1
Mfn2	mitofusin 2
Mief1	mitochondrial elongation factor 1
Mief2	mitochondrial elongation factor 2
mitoTEMPO	2-(2,2,6,6-tetramethylpiperidin-1-oxyl-4-ylamino)-2-oxoethyl) triphenylphosphonium chloride
mitoQ	mitochondrial-targeting antioxidants coenzyme Q
Mn-SOD	manganese superoxide dismutase
mRNA	messenger RNA
Msr	methionine sulfoxide reductase
MsrA	Msr isoform A
MPTP	mitochondrial permeability transition pore
mt-K_{ATP}	mitochondrial ATP-sensitive potassium K channel
mtDNA	mitochondrial DNA
mtROS	mitochondrial ROS
Na^+/K^+ ATPase	sodium–potassium ATP pump
NAC	N-acetylcysteine
NAD	nicotinamide adenine dinucleotide
NADPH	nicotinamide adenine dinucleotide phosphate
NaHS	sodium hydrosulfide
NF-κB	nuclear factor kappa-light-chain-enhancer of activated B cells
NOS	nitric oxide synthase
NOXs	NADPH oxidases
NRF-1	nuclear respiratory factor 1
NRF-2	nuclear respiratory factor 2
Nrf2	nuclear factor erythroid 2–related factor 2
O_2	oxygen
$O_2^{•-}$	superoxide anion radical
OA	oxalacetate
OCR	oxygen consumption rate
OMM	outer mitochondrial membrane
$ONOO^-$	peroxynitrite

Opa1	optical atrophy 1
OS	oxidative stress
Ox-PTMs	oxidative post-translational modifications
OXPHOS	oxidative phosphorylation
p62	sequestosome
PDIA1	protein disulfide isomerase A1
PGC-1α	peroxisome proliferator-activated receptor (PPAR) γ coactivator-1 alpha
Pink1	phosphatase and tensin homolog (PTEN)-induced putative kinase 1
PKC	protein kinase C
PKC-ε	PKC-epsilon
PKC-δ	PKC-delta
PKM2	pyruvate kinase isoform M2
PMF	proton motive force
PPARs	peroxisome proliferator-activated receptor
Pro	proline
Prx	peroxiredoxin
PT	proximal tubules
PTKs	protein tyrosine kinases
PTM	post-translational modifications
PTPs	protein tyrosine phosphatases
R–S$^\bullet$	thyil radical
R–SNO	S-nitrosothiol
R–SOH	sulfenic acid
R–SO_2H	sulfinic acid
R–SO_3H	sulfonic acid
R–S(O)–S–R	thiosulfinate
R–SN–R′	sulfenamide
R–S–S–H	persulfonation
Rac1	Ras-related C3 botulinum toxin substrate 1
RAGE	AGE receptor
ROS	reactive oxygen species
RPTCs	renal proximal tubular cells
RTC	renal tubular cells
S^-	thiolate anion
S–S	disulfide bonds
SF	sulforaphane
SH	thiol
siRNA	small interfering RNA
SIRT3	nicotinamide adenine dinucleotide (NAD)-dependent deacetylase sirtuin-3
S_N2	bimolecular nucleophilic substitution
SSG	glutathionylation
TCA	tricarboxylic acid
TECs	tubular epithelial cells
TFAM	transcription factor A
TGFβ1	transforming growth factor-beta 1
Thr	threonine
Trx	thioredoxin
Tyr	tyrosine
UCPs	uncoupling proteins
UCP1	uncoupling protein 1
UCP2	uncoupling protein 2
UCPs	uncoupling proteins
UUO	unilateral ureteral obstruction
VLCAD	very long-chain acyl-CoA dehydrogenase
WT	wild type

References

1. Manns, B.; Hemmelgarn, B.; Tonelli, M.; Au, F.; So, H.; Weaver, R.; Quinn, A.E.; Klarenbach, S.; for Canadians Seeking Solutions and Innovations to Overcome Chronic Kidney Disease. The Cost of Care for People With Chronic Kidney Disease. *Can. J. Kidney Health Dis.* **2019**, *6*. [CrossRef]
2. Collister, D.; Pannu, N.; Ye, F.; James, M.; Hemmelgarn, B.; Chui, B.; Manns, B.; Klarenbach, S. Health Care Costs Associated with AKI. *Clin. J. Am. Soc. Nephrol.* **2017**, *12*, 1733–1743. [CrossRef] [PubMed]
3. Thomas, M.E.; Blaine, C.; Dawnay, A.; Devonald, M.A.J.; Ftouh, S.; Laing, C.; Latchem, S.; Lewington, A.; Milford, D.V.; Ostermann, M. The Definition of Acute Kidney Injury and Its Use in Practice. *Kidney Int.* **2015**, *87*, 62–73. [CrossRef] [PubMed]
4. Wang, Y.; Fang, Y.; Teng, J.; Ding, X. Acute Kidney Injury Epidemiology: From Recognition to Intervention. *Contrib. Nephrol.* **2016**, *187*, 1–8. [CrossRef] [PubMed]
5. Romagnani, P.; Remuzzi, G.; Glassock, R.; Levin, A.; Jager, K.J.; Tonelli, M.; Massy, Z.; Wanner, C.; Anders, H.-J. Chronic Kidney Disease. *Nat. Rev. Dis. Primers* **2017**, *3*, 17088. [CrossRef] [PubMed]
6. Martínez-Klimova, E.; Aparicio-Trejo, O.E.; Tapia, E.; Pedraza-Chaverri, J. Unilateral Ureteral Obstruction as a Model to Investigate Fibrosis-Attenuating Treatments. *Biomolecules* **2019**, *9*, 141. [CrossRef]
7. Susztak, K.; Ciccone, E.; McCue, P.; Sharma, K.; Böttinger, E.P. Multiple Metabolic Hits Converge on CD36 as Novel Mediator of Tubular Epithelial Apoptosis in Diabetic Nephropathy. *PLoS Med.* **2005**, *2*, e45. [CrossRef]
8. Sun, L.; Yuan, Q.; Xu, T.; Yao, L.; Feng, J.; Ma, J.; Wang, L.; Lu, C.; Wang, D. Pioglitazone Improves Mitochondrial Function in the Remnant Kidney and Protects against Renal Fibrosis in 5/6 Nephrectomized Rats. *Front. Pharmacol.* **2017**, *8*, 545. [CrossRef]
9. Hewitson, T.D.; Holt, S.G.; Smith, E.R. Progression of Tubulointerstitial Fibrosis and the Chronic Kidney Disease Phenotype—Role of Risk Factors and Epigenetics. *Front. Pharmacol.* **2017**, *8*, 520. [CrossRef]
10. Bhargava, P.; Schnellmann, R.G. Mitochondrial Energetics in the Kidney. *Nat. Rev. Nephrol.* **2017**, *13*, 629–646. [CrossRef]
11. Irazabal, M.V.; Torres, V.E. Reactive Oxygen Species and Redox Signaling in Chronic Kidney Disease. *Cells* **2020**, *9*, 1342. [CrossRef]
12. Pavlakou, P.; Liakopoulos, V.; Eleftheriadis, T.; Mitsis, M.; Dounousi, E. Oxidative Stress and Acute Kidney Injury in Critical Illness: Pathophysiologic Mechanisms—Biomarkers—Interventions, and Future Perspectives. *Oxidative Med. Cell. Longev.* **2017**, *2017*, 6193694. [CrossRef]
13. Holmström, K.M.; Finkel, T. Cellular Mechanisms and Physiological Consequences of Redox-Dependent Signalling. *Nat. Rev. Mol. Cell Biol.* **2014**, *15*, 411–421. [CrossRef]
14. Sies, H.; Jones, D.P. Reactive Oxygen Species (ROS) as Pleiotropic Physiological Signalling Agents. *Nat. Rev. Mol. Cell Biol.* **2020**, *21*, 363–383. [CrossRef]
15. Sies, H. Hydrogen Peroxide as a Central Redox Signaling Molecule in Physiological Oxidative Stress: Oxidative Eustress. *Redox Biol.* **2017**, *11*, 613–619. [CrossRef] [PubMed]
16. Cruz-Gregorio, A.; Aranda-Rivera, A.K. Redox-Sensitive Signalling Pathways Regulated by Human Papillomavirus in HPV-Related Cancers. *Rev. Med. Virol.* **2021**, e2230. [CrossRef]
17. D'Autréaux, B.; Toledano, M.B. ROS as Signalling Molecules: Mechanisms That Generate Specificity in ROS Homeostasis. *Nat. Rev. Mol. Cell Biol.* **2007**, *8*, 813–824. [CrossRef] [PubMed]
18. Wani, R.; Nagata, A.; Murray, B.W. Protein Redox Chemistry: Post-Translational Cysteine Modifications That Regulate Signal Transduction and Drug Pharmacology. *Front. Pharmacol.* **2014**, *5*, 224. [CrossRef]
19. Wong, C.M.; Cheema, A.K.; Zhang, L.; Suzuki, Y.J. Protein Carbonylation as a Novel Mechanism in Redox Signaling. *Circ. Res.* **2008**, *102*, 310–318. [CrossRef] [PubMed]
20. Anand, P.; Stamler, J.S. Enzymatic Mechanisms Regulating Protein S-Nitrosylation: Implications in Health and Disease. *J. Mol. Med.* **2012**, *90*, 233–244. [CrossRef]
21. Sies, H. Role of Metabolic H_2O_2 Generation: Redox Signaling and Oxidative Stress. *J. Biol. Chem.* **2014**, *289*, 8735–8741. [CrossRef] [PubMed]
22. Chao, C.C.; Ma, Y.S.; Stadtman, E.R. Modification of Protein Surface Hydrophobicity and Methionine Oxidation by Oxidative Systems. *Proc. Natl. Acad. Sci. USA* **1997**, *94*, 2969–2974. [CrossRef]
23. Winterbourn, C.C.; Hampton, M.B. Thiol Chemistry and Specificity in Redox Signaling. *Free Radic. Biol. Med.* **2008**, *45*, 549–561. [CrossRef]
24. Mailloux, R.J.; Jin, X.; Willmore, W.G. Redox Regulation of Mitochondrial Function with Emphasis on Cysteine Oxidation Reactions. *Redox Biol.* **2014**, *2*, 123–139. [CrossRef] [PubMed]
25. Lo Conte, M.; Carroll, K.S. The Redox Biochemistry of Protein Sulfenylation and Sulfinylation. *J. Biol. Chem.* **2013**, *288*, 26480–26488. [CrossRef] [PubMed]
26. Poole, L.B.; Nelson, K.J. Discovering Mechanisms of Signaling-Mediated Cysteine Oxidation. *Curr. Opin. Chem. Biol.* **2008**, *12*, 18–24. [CrossRef] [PubMed]
27. Hugo, M.; Turell, L.; Manta, B.; Botti, H.; Monteiro, G.; Netto, L.E.S.; Alvarez, B.; Radi, R.; Trujillo, M. Thiol and Sulfenic Acid Oxidation of AhpE, the One-Cysteine Peroxiredoxin from Mycobacterium Tuberculosis: Kinetics, Acidity Constants, and Conformational Dynamics. *Biochemistry* **2009**, *48*, 9416–9426. [CrossRef] [PubMed]
28. Zhang, J.; Ye, Z.-W.; Singh, S.; Townsend, D.M.; Tew, K.D. An Evolving Understanding of the S-Glutathionylation Cycle in Pathways of Redox Regulation. *Free Radic. Biol. Med.* **2018**, *120*, 204–216. [CrossRef]

29. Poole, L.B. The Basics of Thiols and Cysteines in Redox Biology and Chemistry. *Free Radic. Biol. Med.* **2015**, *80*, 148–157. [CrossRef]
30. Gupta, V.; Carroll, K.S. Sulfenic Acid Chemistry, Detection and Cellular Lifetime. *Biochim. Biophys. Acta Gen. Subj.* **2014**, *1840*, 847–875. [CrossRef] [PubMed]
31. Smith, P.K.; Krohn, R.I.; Hermanson, G.T.; Mallia, A.K.; Gartner, F.H.; Provenzano, M.D.; Fujimoto, E.K.; Goeke, N.M.; Olson, B.J.; Klenk, D.C. Measurement of Protein Using Bicinchoninic Acid. *Anal. Biochem.* **1985**, *150*, 76–85. [CrossRef]
32. Kaya, A.; Lee, B.C.; Gladyshev, V.N. Regulation of Protein Function by Reversible Methionine Oxidation and the Role of Selenoprotein MsrB1. *Antioxid. Redox Signal.* **2015**, *23*, 814–822. [CrossRef] [PubMed]
33. Stadtman, E.R.; Moskovitz, J.; Levine, R.L. Oxidation of Methionine Residues of Proteins: Biological Consequences. *Antioxid. Redox Signal.* **2003**, *5*, 577–582. [CrossRef]
34. García-Santamarina, S.; Boronat, S.; Hidalgo, E. Reversible Cysteine Oxidation in Hydrogen Peroxide Sensing and Signal Transduction. *Biochemistry* **2014**, *53*, 2560–2580. [CrossRef] [PubMed]
35. Matsuzawa, A. Thioredoxin and Redox Signaling: Roles of the Thioredoxin System in Control of Cell Fate. *Arch. Biochem. Biophys.* **2017**, *617*, 101–105. [CrossRef]
36. Roorda, M.; Miljkovic, J.L.; van Goor, H.; Henning, R.H.; Bouma, H.R. Spatiotemporal Regulation of Hydrogen Sulfide Signaling in the Kidney. *Redox Biol.* **2021**, *43*, 101961. [CrossRef]
37. Sawa, T.; Motohashi, H.; Ihara, H.; Akaike, T. Enzymatic Regulation and Biological Functions of Reactive Cysteine Persulfides and Polysulfides. *Biomolecules* **2020**, *10*, 1245. [CrossRef]
38. Paul, B.D.; Snyder, S.H. H_2S Signalling through Protein Sulfhydration and Beyond. *Nat. Rev. Mol. Cell Biol.* **2012**, *13*, 499–507. [CrossRef] [PubMed]
39. Nagy, P.; Winterbourn, C.C. Rapid Reaction of Hydrogen Sulfide with the Neutrophil Oxidant Hypochlorous Acid to Generate Polysulfides. *Chem. Res. Toxicol.* **2010**, *23*, 1541–1543. [CrossRef]
40. Frank, G.D.; Eguchi, S. Activation of Tyrosine Kinases by Reactive Oxygen Species in Vascular Smooth Muscle Cells: Significance and Involvement of EGF Receptor Transactivation by Angiotensin II. *Antioxid. Redox Signal.* **2003**, *5*, 771–780. [CrossRef]
41. Doulias, P.-T.; Tenopoulou, M.; Greene, J.L.; Raju, K.; Ischiropoulos, H. Nitric Oxide Regulates Mitochondrial Fatty Acid Metabolism Through Reversible Protein S-Nitrosylation. *Sci. Signal.* **2013**, *6*, rs1. [CrossRef]
42. Zhou, H.-L.; Zhang, R.; Anand, P.; Stomberski, C.T.; Qian, Z.; Hausladen, A.; Wang, L.; Rhee, E.P.; Parikh, S.M.; Karumanchi, S.A.; et al. Metabolic Reprogramming by the S-Nitroso-CoA Reductase System Protects against Kidney Injury. *Nature* **2019**, *565*, 96–100. [CrossRef] [PubMed]
43. Shang, Y.; Siow, Y.L.; Isaak, C.K.; O, K. Downregulation of Glutathione Biosynthesis Contributes to Oxidative Stress and Liver Dysfunction in Acute Kidney Injury. Available online: https://www.hindawi.com/journals/omcl/2016/9707292/ (accessed on 5 February 2021).
44. Aparicio-Trejo, O.E.; Reyes-Fermín, L.M.; Briones-Herrera, A.; Tapia, E.; León-Contreras, J.C.; Hernández-Pando, R.; Sánchez-Lozada, L.G.; Pedraza-Chaverri, J. Protective Effects of N-Acetyl-Cysteine in Mitochondria Bioenergetics, Oxidative Stress, Dynamics and S-Glutathionylation Alterations in Acute Kidney Damage Induced by Folic Acid. *Free Radic. Biol. Med.* **2019**, *130*, 379–396. [CrossRef]
45. Noh, M.R.; Kim, K.Y.; Han, S.J.; Kim, J.I.; Kim, H.-Y.; Park, K.M. Methionine Sulfoxide Reductase A Deficiency Exacerbates Cisplatin-Induced Nephrotoxicity via Increased Mitochondrial Damage and Renal Cell Death. *Antioxid. Redox Signal.* **2017**, *27*, 727–741. [CrossRef]
46. Kim, J.I.; Choi, S.H.; Jung, K.-J.; Lee, E.; Kim, H.-Y.; Park, K.M. Protective Role of Methionine Sulfoxide Reductase A against Ischemia/Reperfusion Injury in Mouse Kidney and Its Involvement in the Regulation of Trans-Sulfuration Pathway. *Antioxid. Redox Signal.* **2013**, *18*, 2241–2250. [CrossRef] [PubMed]
47. Jung, K.-J.; Jang, H.-S.; Kim, J.I.; Han, S.J.; Park, J.-W.; Park, K.M. Involvement of Hydrogen Sulfide and Homocysteine Transsulfuration Pathway in the Progression of Kidney Fibrosis after Ureteral Obstruction. *Biochim. Biophys. Acta Mol. Basis Dis.* **2013**, *1832*, 1989–1997. [CrossRef] [PubMed]
48. Kim, J.I.; Noh, M.R.; Kim, K.Y.; Jang, H.-S.; Kim, H.-Y.; Park, K.M. Methionine Sulfoxide Reductase A Deficiency Exacerbates Progression of Kidney Fibrosis Induced by Unilateral Ureteral Obstruction. *Free Radic. Biol. Med.* **2015**, *89*, 201–208. [CrossRef]
49. Lai, L.; Sun, J.; Tarafdar, S.; Liu, C.; Murphy, E.; Kim, G.; Levine, R.L. Loss of Methionine Sulfoxide Reductases Increases Resistance to Oxidative Stress. *Free Radic. Biol. Med.* **2019**, *145*, 374–384. [CrossRef] [PubMed]
50. Cao, X.; Bian, J.-S. The Role of Hydrogen Sulfide in Renal System. *Front. Pharmacol.* **2016**, *7*, 385. [CrossRef] [PubMed]
51. Kasinath, B.S.; Feliers, D.; Lee, H.J. Hydrogen Sulfide as a Regulatory Factor in Kidney Health and Disease. *Biochem. Pharmacol.* **2018**, *149*, 29–41. [CrossRef] [PubMed]
52. Ngowi, E.E.; Sarfraz, M.; Afzal, A.; Khan, N.H.; Khattak, S.; Zhang, X.; Li, T.; Duan, S.-F.; Ji, X.-Y.; Wu, D.-D. Roles of Hydrogen Sulfide Donors in Common Kidney Diseases. *Front. Pharmacol.* **2020**, *11*, 1706. [CrossRef]
53. Han, S.J.; Kim, J.I.; Park, J.-W.; Park, K.M. Hydrogen Sulfide Accelerates the Recovery of Kidney Tubules after Renal Ischemia/Reperfusion Injury. *Nephrol. Dial. Transplant.* **2015**, *30*, 1497–1506. [CrossRef] [PubMed]
54. Bos, E.M.; Wang, R.; Snijder, P.M.; Boersema, M.; Damman, J.; Fu, M.; Moser, J.; Hillebrands, J.-L.; Ploeg, R.J.; Yang, G.; et al. Cystathionine γ-Lyase Protects against Renal Ischemia/Reperfusion by Modulating Oxidative Stress. *J. Am. Soc. Nephrol.* **2013**, *24*, 759–770. [CrossRef]

55. Li, H.; Feng, S.-J.; Zhang, G.-Z.; Wang, S.-X. Correlation of Lower Concentrations of Hydrogen Sulfide with Atherosclerosis in Chronic Hemodialysis Patients with Diabetic Nephropathy. *Blood Purif.* **2014**, *38*, 188–194. [CrossRef]
56. Song, K.; Wang, F.; Li, Q.; Shi, Y.-B.; Zheng, H.-F.; Peng, H.; Shen, H.-Y.; Liu, C.-F.; Hu, L.-F. Hydrogen Sulfide Inhibits the Renal Fibrosis of Obstructive Nephropathy. *Kidney Int.* **2014**, *85*, 1318–1329. [CrossRef]
57. Mustafa, A.K.; Gadalla, M.M.; Sen, N.; Kim, S.; Mu, W.; Gazi, S.K.; Barrow, R.K.; Yang, G.; Wang, R.; Snyder, S.H. H_2S Signals through Protein S-Sulfhydration. *Sci. Signal.* **2009**, *2*, ra72. [CrossRef] [PubMed]
58. Jarosz, A.P.; Wei, W.; Gauld, J.W.; Auld, J.; Özcan, F.; Aslan, M.; Mutus, B. Glyceraldehyde 3-Phosphate Dehydrogenase (GAPDH) Is Inactivated by S-Sulfuration in Vitro. *Free Radic. Biol. Med.* **2015**, *89*, 512–521. [CrossRef] [PubMed]
59. Motohashi, H.; Yamamoto, M. Nrf2–Keap1 Defines a Physiologically Important Stress Response Mechanism. *Trends Mol. Med.* **2004**, *10*, 549–557. [CrossRef] [PubMed]
60. Bianco, M.; Lopes, J.A.; Beiral, H.J.V.; Filho, J.D.D.; Frankenfeld, S.P.; Fortunato, R.S.; Gattass, C.R.; Vieyra, A.; Takiya, C.M. The Contralateral Kidney Presents with Impaired Mitochondrial Functions and Disrupted Redox Homeostasis after 14 Days of Unilateral Ureteral Obstruction in Mice. *PLoS ONE* **2019**, *14*, e0218986. [CrossRef]
61. Kong, W.; Fu, J.; Liu, N.; Jiao, C.; Guo, G.; Luan, J.; Wang, H.; Yao, L.; Wang, L.; Yamamoto, M.; et al. Nrf2 Deficiency Promotes the Progression from Acute Tubular Damage to Chronic Renal Fibrosis Following Unilateral Ureteral Obstruction. *Nephrol. Dial. Transplant.* **2018**, *33*, 771–783. [CrossRef]
62. Ge, S.-N.; Zhao, M.-M.; Wu, D.-D.; Chen, Y.; Wang, Y.; Zhu, J.-H.; Cai, W.-J.; Zhu, Y.-Z.; Zhu, Y.-C. Hydrogen Sulfide Targets EGFR Cys797/Cys798 Residues to Induce Na^+/K^+-ATPase Endocytosis and Inhibition in Renal Tubular Epithelial Cells and Increase Sodium Excretion in Chronic Salt-Loaded Rats. *Antioxid. Redox Signal.* **2014**, *21*, 2061–2082. [CrossRef]
63. Shirozu, K.; Tokuda, K.; Marutani, E.; Lefer, D.; Wang, R.; Ichinose, F. Cystathionine γ-Lyase Deficiency Protects Mice from Galactosamine/Lipopolysaccharide-Induced Acute Liver Failure. *Antioxid. Redox Signal.* **2014**, *20*, 204–216. [CrossRef] [PubMed]
64. Yadav, P.K.; Martinov, M.; Vitvitsky, V.; Seravalli, J.; Wedmann, R.; Filipovic, M.R.; Banerjee, R. Biosynthesis and Reactivity of Cysteine Persulfides in Signaling. *J. Am. Chem. Soc.* **2016**, *138*, 289–299. [CrossRef] [PubMed]
65. Nakano, S.; Ishii, I.; Shinmura, K.; Tamaki, K.; Hishiki, T.; Akahoshi, N.; Ida, T.; Nakanishi, T.; Kamata, S.; Kumagai, Y.; et al. Hyperhomocysteinemia Abrogates Fasting-Induced Cardioprotection against Ischemia/Reperfusion by Limiting Bioavailability of Hydrogen Sulfide Anions. *J. Mol. Med.* **2015**, *93*, 879–889. [CrossRef]
66. Akaike, T.; Ida, T.; Wei, F.-Y.; Nishida, M.; Kumagai, Y.; Alam, M.M.; Ihara, H.; Sawa, T.; Matsunaga, T.; Kasamatsu, S.; et al. Cysteinyl-TRNA Synthetase Governs Cysteine Polysulfidation and Mitochondrial Bioenergetics. *Nat. Commun.* **2017**, *8*, 1177. [CrossRef] [PubMed]
67. Karnati, S.; Lüers, G.; Pfreimer, S.; Baumgart-Vogt, E. Mammalian SOD_2 Is Exclusively Located in Mitochondria and Not Present in Peroxisomes. *Histochem. Cell Biol.* **2013**, *140*, 105–117. [CrossRef]
68. Valko, M.; Leibfritz, D.; Moncol, J.; Cronin, M.T.D.; Mazur, M.; Telser, J. Free Radicals and Antioxidants in Normal Physiological Functions and Human Disease. *Int. J. Biochem. Cell Biol.* **2007**, *39*, 44–84. [CrossRef]
69. Yamakura, F.; Kawasaki, H. Post-Translational Modifications of Superoxide Dismutase. *Biochim. Biophys. Acta* **2010**, *1804*, 318–325. [CrossRef] [PubMed]
70. Walker, L.M.; Walker, P.D.; Imam, S.Z.; Ali, S.F.; Mayeux, P.R. Evidence for Peroxynitrite Formation in Renal Ischemia-Reperfusion Injury: Studies with the Inducible Nitric Oxide Synthase Inhibitor L-N(6)-(1-Iminoethyl)Lysine. *J. Pharmacol. Exp. Ther.* **2000**, *295*, 417–422. [PubMed]
71. MacMillan-Crow, L.A.; Crow, J.P.; Kerby, J.D.; Beckman, J.S.; Thompson, J.A. Nitration and Inactivation of Manganese Superoxide Dismutase in Chronic Rejection of Human Renal Allografts. *Proc. Natl. Acad. Sci. USA* **1996**, *93*, 11853–11858. [CrossRef]
72. MacMillan-Crow, L.A.; Cruthirds, D.L.; Ahki, K.M.; Sanders, P.W.; Thompson, J.A. Mitochondrial Tyrosine Nitration Precedes Chronic Allograft Nephropathy. *Free Radic. Biol. Med.* **2001**, *31*, 1603–1608. [CrossRef]
73. MacMillan-Crow, L.A.; Crow, J.P.; Thompson, J.A. Peroxynitrite-Mediated Inactivation of Manganese Superoxide Dismutase Involves Nitration and Oxidation of Critical Tyrosine Residues. *Biochemistry* **1998**, *37*, 1613–1622. [CrossRef]
74. Cruthirds, D.L.; Novak, L.; Akhi, K.M.; Sanders, P.W.; Thompson, J.A.; MacMillan-Crow, L.A. Mitochondrial Targets of Oxidative Stress during Renal Ischemia/Reperfusion. *Arch. Biochem. Biophys.* **2003**, *412*, 27–33. [CrossRef]
75. Xu, S.; Jiang, B.; Maitland, K.A.; Bayat, H.; Gu, J.; Nadler, J.L.; Corda, S.; Lavielle, G.; Verbeuren, T.J.; Zuccollo, A.; et al. The Thromboxane Receptor Antagonist S18886 Attenuates Renal Oxidant Stress and Proteinuria in Diabetic Apolipoprotein E-Deficient Mice. *Diabetes* **2006**, *55*, 110–119. [CrossRef]
76. Wang, H.D.; Xu, S.; Johns, D.G.; Du, Y.; Quinn, M.T.; Cayatte, A.J.; Cohen, R.A. Role of NADPH Oxidase in the Vascular Hypertrophic and Oxidative Stress Response to Angiotensin II in Mice. *Circ. Res.* **2001**, *88*, 947–953. [CrossRef] [PubMed]
77. Guo, W.; Adachi, T.; Matsui, R.; Xu, S.; Jiang, B.; Zou, M.-H.; Kirber, M.; Lieberthal, W.; Cohen, R.A. Quantitative Assessment of Tyrosine Nitration of Manganese Superoxide Dismutase in Angiotensin II-Infused Rat Kidney. *Am. J. Physiol. Heart Circ. Physiol.* **2003**, *285*, H1396–H1403. [CrossRef]
78. Majzunova, M.; Kvandova, M.; Berenyiova, A.; Balis, P.; Dovinova, I.; Cacanyiova, S. Chronic NOS Inhibition Affects Oxidative State and Antioxidant Response Differently in the Kidneys of Young Normotensive and Hypertensive Rats. *Oxidative Med. Cell. Longev.* **2019**, *2019*, 5349398. [CrossRef]

79. Kimura, T.; Shiizaki, K.; Akimoto, T.; Shinzato, T.; Shimizu, T.; Kurosawa, A.; Kubo, T.; Nanmoku, K.; Kuro-O, M.; Yagisawa, T. The Impact of Preserved Klotho Gene Expression on Antioxidative Stress Activity in Healthy Kidney. *Am. J. Physiol. Physiol.* **2018**, *315*, F345–F352. [CrossRef]
80. Bedard, K.; Krause, K.-H. The NOX Family of ROS-Generating NADPH Oxidases: Physiology and Pathophysiology. *Physiol. Rev.* **2007**, *87*, 245–313. [CrossRef]
81. Babelova, A.; Avaniadi, D.; Jung, O.; Fork, C.; Beckmann, J.; Kosowski, J.; Weissmann, N.; Anilkumar, N.; Shah, A.M.; Schaefer, L.; et al. Role of Nox4 in Murine Models of Kidney Disease. *Free Radic. Biol. Med.* **2012**, *53*, 842–853. [CrossRef]
82. Serrander, L.; Cartier, L.; Bedard, K.; Banfi, B.; Lardy, B.; Plastre, O.; Sienkiewicz, A.; Fórró, L.; Schlegel, W.; Krause, K.-H. NOX_4 Activity Is Determined by MRNA Levels and Reveals a Unique Pattern of ROS Generation. *Biochem. J.* **2007**, *406*, 105–114. [CrossRef]
83. Lee, H.; Jose, P.A. Coordinated Contribution of NADPH Oxidase- and Mitochondria-Derived Reactive Oxygen Species in Metabolic Syndrome and Its Implication in Renal Dysfunction. *Front. Pharmacol.* **2021**, *12*, 670076. [CrossRef]
84. Molina-Jijón, E.; Aparicio-Trejo, O.E.; Rodríguez-Muñoz, R.; León-Contreras, J.C.; Cárdenas-Aguayo, M.d.C.; Medina-Campos, O.N.; Tapia, E.; Sánchez-Lozada, L.G.; Hernández-Pando, R.; Reyes, J.L.; et al. The Nephroprotection Exerted by Curcumin in Maleate-Induced Renal Damage Is Associated with Decreased Mitochondrial Fission and Autophagy: Curcumin Decreases Maleate-Induced Renal Dysfunction. *BioFactors* **2016**, *42*, 686–702. [CrossRef]
85. Kim, S.; Kim, S.J.; Yoon, H.E.; Chung, S.; Choi, B.S.; Park, C.W.; Shin, S.J. Fimasartan, a Novel Angiotensin-Receptor Blocker, Protects against Renal Inflammation and Fibrosis in Mice with Unilateral Ureteral Obstruction: The Possible Role of Nrf2. *Int. J. Med. Sci.* **2015**, *12*, 891–904. [CrossRef]
86. Dikalov, S. Cross Talk between Mitochondria and NADPH Oxidases. *Free Radic. Biol. Med.* **2011**, *51*, 1289–1301. [CrossRef] [PubMed]
87. Wenzel, P.; Mollnau, H.; Oelze, M.; Schulz, E.; Wickramanayake, J.M.D.; Müller, J.; Schuhmacher, S.; Hortmann, M.; Baldus, S.; Gori, T.; et al. First Evidence for a Crosstalk Between Mitochondrial and NADPH Oxidase-Derived Reactive Oxygen Species in Nitroglycerin-Triggered Vascular Dysfunction. *Antioxid. Redox Signal.* **2008**, *10*, 1435–1448. [CrossRef] [PubMed]
88. Daiber, A. Redox Signaling (Cross-Talk) from and to Mitochondria Involves Mitochondrial Pores and Reactive Oxygen Species. *Biochim. Biophys. Acta Bioenerg.* **2010**, *1797*, 897–906. [CrossRef] [PubMed]
89. Hallan, S.; Sharma, K. The Role of Mitochondria in Diabetic Kidney Disease. *Curr. Diabetes Rep.* **2016**, *16*, 61. [CrossRef] [PubMed]
90. Rathore, R.; Zheng, Y.-M.; Niu, C.-F.; Liu, Q.-H.; Korde, A.; Ho, Y.-S.; Wang, Y.-X. Hypoxia Activates NADPH Oxidase to Increase [ROS]i and $[Ca^{2+}]$i through Mitochondrial ROS–PKCε Signaling Axis in Pulmonary Artery Smooth Muscle Cells. *Free Radic. Biol. Med.* **2008**, *45*, 1223–1231. [CrossRef] [PubMed]
91. Aparicio-Trejo, O.E.; Tapia, E.; Sánchez-Lozada, L.G.; Pedraza-Chaverri, J. Mitochondrial Bioenergetics, Redox State, Dynamics and Turnover Alterations in Renal Mass Reduction Models of Chronic Kidney Diseases and Their Possible Implications in the Progression of This Illness. *Pharmacol. Res.* **2018**, *135*, 1–11. [CrossRef]
92. Cheng, X.; Zheng, X.; Song, Y.; Qu, L.; Tang, J.; Meng, L.; Wang, Y. Apocynin Attenuates Renal Fibrosis via Inhibition of NOXs-ROS-ERK-Myofibroblast Accumulation in UUO Rats. *Free Radic. Res.* **2016**, *50*, 840–852. [CrossRef] [PubMed]
93. Asaba, K.; Tojo, A.; Onozato, M.L.; Goto, A.; Quinn, M.T.; Fujita, T.; Wilcox, C.S. Effects of NADPH Oxidase Inhibitor in Diabetic Nephropathy. *Kidney Int.* **2005**, *67*, 1890–1898. [CrossRef] [PubMed]
94. Aranda-Rivera, A.K.; Cruz-Gregorio, A.; Aparicio-Trejo, O.E.; Ortega-Lozano, A.J.; Pedraza-Chaverri, J. Redox Signaling Pathways in Unilateral Ureteral Obstruction (UUO)-Induced Renal Fibrosis. *Free Radic. Biol. Med.* **2021**, *172*, 65–81. [CrossRef] [PubMed]
95. Jain, M.; Rivera, S.; Monclus, E.A.; Synenki, L.; Zirk, A.; Eisenbart, J.; Feghali-Bostwick, C.; Mutlu, G.M.; Budinger, G.R.S.; Chandel, N.S. Mitochondrial Reactive Oxygen Species Regulate Transforming Growth Factor-β Signaling. *J. Biol. Chem.* **2013**, *288*, 770–777. [CrossRef] [PubMed]
96. Kuwabara, N.; Tamada, S.; Iwai, T.; Teramoto, K.; Kaneda, N.; Yukimura, T.; Nakatani, T.; Miura, K. Attenuation of Renal Fibrosis by Curcumin in Rat Obstructive Nephropathy. *Urology* **2006**, *67*, 440–446. [CrossRef]
97. Herb, M.; Gluschko, A.; Wiegmann, K.; Farid, A.; Wolf, A.; Utermöhlen, O.; Krut, O.; Krönke, M.; Schramm, M. Mitochondrial Reactive Oxygen Species Enable Proinflammatory Signaling through Disulfide Linkage of NEMO. *Sci. Signal.* **2019**, *12*, eaar5926. [CrossRef]
98. Li, J.; Sun, Y.B.Y.; Chen, W.; Fan, J.; Li, S.; Qu, X.; Chen, Q.; Chen, R.; Zhu, D.; Zhang, J.; et al. Smad4 Promotes Diabetic Nephropathy by Modulating Glycolysis and OXPHOS. *EMBO Rep.* **2020**, *21*, e48781. [CrossRef] [PubMed]
99. Hahn, A.; Zuryn, S. Mitochondrial Genome (MtDNA) Mutations That Generate Reactive Oxygen Species. *Antioxidants* **2019**, *8*, 392. [CrossRef]
100. Guo, C.; Sun, L.; Chen, X.; Zhang, D. Oxidative Stress, Mitochondrial Damage and Neurodegenerative Diseases. *Neural Regener. Res.* **2013**, *8*, 2003–2014. [CrossRef]
101. Alsahli, M.; Gerich, J.E. Renal Glucose Metabolism in Normal Physiological Conditions and in Diabetes. *Diabetes Res. Clin. Pract.* **2017**, *133*, 1–9. [CrossRef]
102. Pfaller, W.; Rittinger, M. Quantitative Morphology of the Rat Kidney. *Int. J. Biochem.* **1980**, *12*, 17–22. [CrossRef]
103. Simon, N.; Hertig, A. Alteration of Fatty Acid Oxidation in Tubular Epithelial Cells: From Acute Kidney Injury to Renal Fibrogenesis. *Front. Med.* **2015**, *2*, 52. [CrossRef] [PubMed]

104. Markwell, M.A.; McGroarty, E.J.; Bieber, L.L.; Tolbert, N.E. The Subcellular Distribution of Carnitine Acyltransferases in Mammalian Liver and Kidney. A New Peroxisomal Enzyme. *J. Biol. Chem.* **1973**, *248*, 3426–3432. [CrossRef]
105. Stadler, K.; Goldberg, I.J.; Susztak, K. The Evolving Understanding of the Contribution of Lipid Metabolism to Diabetic Kidney Disease. *Curr. Diabetes Rep.* **2015**, *15*, 40. [CrossRef]
106. Kang, H.M.; Ahn, S.H.; Choi, P.; Ko, Y.-A.; Han, S.H.; Chinga, F.; Park, A.S.D.; Tao, J.; Sharma, K.; Pullman, J.; et al. Defective Fatty Acid Oxidation in Renal Tubular Epithelial Cells Has a Key Role in Kidney Fibrosis Development. *Nat. Med.* **2015**, *21*, 37–46. [CrossRef] [PubMed]
107. Fedorova, L.V.; Tamirisa, A.; Kennedy, D.J.; Haller, S.T.; Budnyy, G.; Shapiro, J.I.; Malhotra, D. Mitochondrial Impairment in the Five-Sixth Nephrectomy Model of Chronic Renal Failure: Proteomic Approach. *BMC Nephrol.* **2013**, *14*, 209. [CrossRef]
108. Aparicio-Trejo, O.E.; Avila-Rojas, S.H.; Tapia, E.; Rojas-Morales, P.; León-Contreras, J.C.; Martínez-Klimova, E.; Hernández-Pando, R.; Sánchez-Lozada, L.G.; Pedraza-Chaverri, J. Chronic Impairment of Mitochondrial Bioenergetics and β-Oxidation Promotes Experimental AKI-to-CKD Transition Induced by Folic Acid. *Free Radic. Biol. Med.* **2020**, *154*, 18–32. [CrossRef] [PubMed]
109. Briones-Herrera, A.; Ramírez-Camacho, I.; Zazueta, C.; Tapia, E.; Pedraza-Chaverri, J. Altered Proximal Tubule Fatty Acid Utilization, Mitophagy, Fission and Supercomplexes Arrangement in Experimental Fanconi Syndrome Are Ameliorated by Sulforaphane-Induced Mitochondrial Biogenesis. *Free Radic. Biol. Med.* **2020**, *153*, 54–70. [CrossRef] [PubMed]
110. Tan, S.M.; Ziemann, M.; Thallas-Bonke, V.; Snelson, M.; Kumar, V.; Laskowski, A.; Nguyen, T.-V.; Huynh, K.; Clarke, M.V.; Libianto, R.; et al. Complement C5a Induces Renal Injury in Diabetic Kidney Disease by Disrupting Mitochondrial Metabolic Agility. *Diabetes* **2020**, *69*, 83–98. [CrossRef] [PubMed]
111. Aparicio-Trejo, O.E.; Rojas-Morales, P.; Avila-Rojas, S.H.; León-Contreras, J.C.; Hernández-Pando, R.; Jiménez-Uribe, A.P.; Prieto-Carrasco, R.; Sánchez-Lozada, L.G.; Pedraza-Chaverri, J.; Tapia, E. Temporal Alterations in Mitochondrial β-Oxidation and Oxidative Stress Aggravate Chronic Kidney Disease Development in 5/6 Nephrectomy Induced Renal Damage. *Int. J. Mol. Sci.* **2020**, *21*, 6512. [CrossRef]
112. Gai, Z.; Wang, T.; Visentin, M.; Kullak-Ublick, G.A.; Fu, X.; Wang, Z. Lipid Accumulation and Chronic Kidney Disease. *Nutrients* **2019**, *11*, 722. [CrossRef] [PubMed]
113. Nishi, H.; Higashihara, T.; Inagi, R. Lipotoxicity in Kidney, Heart, and Skeletal Muscle Dysfunction. *Nutrients* **2019**, *11*, 1664. [CrossRef] [PubMed]
114. Herman-Edelstein, M.; Scherzer, P.; Tobar, A.; Levi, M.; Gafter, U. Altered Renal Lipid Metabolism and Renal Lipid Accumulation in Human Diabetic Nephropathy. *J. Lipid Res.* **2014**, *55*, 561–572. [CrossRef]
115. Declèves, A.-E.; Zolkipli, Z.; Satriano, J.; Wang, L.; Nakayama, T.; Rogac, M.; Le, T.P.; Nortier, J.L.; Farquhar, M.G.; Naviaux, R.K.; et al. Regulation of Lipid Accumulation by AMP-Activated Kinase [Corrected] in High Fat Diet-Induced Kidney Injury. *Kidney Int.* **2014**, *85*, 611–623. [CrossRef]
116. Moosavi, S.M.S.; Ashtiyani, S.C.; Hosseinkhani, S.; Shirazi, M. Comparison of the Effects of L: -Carnitine and Alpha-Tocopherol on Acute Ureteral Obstruction-Induced Renal Oxidative Imbalance and Altered Energy Metabolism in Rats. *Urol. Res.* **2010**, *38*, 187–194. [CrossRef]
117. Bobulescu, I.A. Renal Lipid Metabolism and Lipotoxicity. *Curr. Opin. Nephrol. Hypertens.* **2010**, *19*, 393–402. [CrossRef]
118. Okamura, D.M.; Pennathur, S.; Pasichnyk, K.; López-Guisa, J.M.; Collins, S.; Febbraio, M.; Heinecke, J.; Eddy, A.A. CD36 Regulates Oxidative Stress and Inflammation in Hypercholesterolemic CKD. *J. Am. Soc. Nephrol.* **2009**, *20*, 495–505. [CrossRef]
119. Liu, J.-J.; Ghosh, S.; Kovalik, J.-P.; Ching, J.; Choi, H.W.; Tavintharan, S.; Ong, C.N.; Sum, C.F.; Summers, S.A.; Tai, E.S.; et al. Profiling of Plasma Metabolites Suggests Altered Mitochondrial Fuel Usage and Remodeling of Sphingolipid Metabolism in Individuals With Type 2 Diabetes and Kidney Disease. *Kidney Int. Rep.* **2017**, *2*, 470–480. [CrossRef] [PubMed]
120. Prieto-Carrasco, R.; García-Arroyo, F.E.; Aparicio-Trejo, O.E.; Rojas-Morales, P.; León-Contreras, J.C.; Hernández-Pando, R.; Sánchez-Lozada, L.G.; Tapia, E.; Pedraza-Chaverri, J. Progressive Reduction in Mitochondrial Mass Is Triggered by Alterations in Mitochondrial Biogenesis and Dynamics in Chronic Kidney Disease Induced by 5/6 Nephrectomy. *Biology* **2021**, *10*, 349. [CrossRef] [PubMed]
121. Miguel, V.; Tituaña, J.; Herrero, J.I.; Herrero, L.; Serra, D.; Cuevas, P.; Barbas, C.; Puyol, D.R.; Márquez-Expósito, L.; Ruiz-Ortega, M.; et al. Renal Tubule Cpt1a Overexpression Protects from Kidney Fibrosis by Restoring Mitochondrial Homeostasis. *J. Clin. Investig.* **2021**, *131*. [CrossRef]
122. Cardoso, A.R.; Kakimoto, P.A.H.B.; Kowaltowski, A.J. Diet-Sensitive Sources of Reactive Oxygen Species in Liver Mitochondria: Role of Very Long Chain Acyl-CoA Dehydrogenases. *PLoS ONE* **2013**, *8*, e77088. [CrossRef] [PubMed]
123. Kakimoto, P.A.H.B.; Tamaki, F.K.; Cardoso, A.R.; Marana, S.R.; Kowaltowski, A.J. H_2O_2 Release from the Very Long Chain Acyl-CoA Dehydrogenase. *Redox Biol.* **2015**, *4*, 375–380. [CrossRef] [PubMed]
124. Walport, M.J. Complement. Available online: https://www.nejm.org/doi/10.1056/NEJM200104053441406 (accessed on 18 June 2021).
125. Wallace, D.C. Mitochondria and Cancer. *Nat. Rev. Cancer* **2012**, *12*, 685–698. [CrossRef]
126. Hu, Z.; Zhang, H.; Yi, B.; Yang, S.; Liu, J.; Hu, J.; Wang, J.; Cao, K.; Zhang, W. VDR Activation Attenuate Cisplatin Induced AKI by Inhibiting Ferroptosis. *Cell Death Dis.* **2020**, *11*, 73. [CrossRef]
127. Deng, B.; Yang, W.; Wang, D.; Cheng, L.; Bu, L.; Rao, J.; Zhang, J.; Xie, J.; Zhang, B. Peptide DR8 Suppresses Epithelial-to-Mesenchymal Transition via the TGF-β/MAPK Signaling Pathway in Renal Fibrosis. *Life Sci.* **2020**, *261*, 118465. [CrossRef]

128. Aparicio-Trejo, O.E.; Tapia, E.; Molina-Jijón, E.; Medina-Campos, O.N.; Macías-Ruvalcaba, N.A.; León-Contreras, J.C.; Hernández-Pando, R.; García-Arroyo, F.E.; Cristóbal, M.; Sánchez-Lozada, L.G.; et al. Curcumin Prevents Mitochondrial Dynamics Disturbances in Early 5/6 Nephrectomy: Relation to Oxidative Stress and Mitochondrial Bioenergetics. *Biofactors* **2017**, *43*, 293–310. [CrossRef]
129. Ge, M.; Fontanesi, F.; Merscher, S.; Fornoni, A. The Vicious Cycle of Renal Lipotoxicity and Mitochondrial Dysfunction. *Front. Physiol.* **2020**, *11*, 732. [CrossRef]
130. Martínez-Klimova, E.; Aparicio-Trejo, O.E.; Gómez-Sierra, T.; Jiménez-Uribe, A.P.; Bellido, B.; Pedraza-Chaverri, J. Mitochondrial Dysfunction and Endoplasmic Reticulum Stress in the Promotion of Fibrosis in Obstructive Nephropathy Induced by Unilateral Ureteral Obstruction. *BioFactors* **2020**, *46*, 716–733. [CrossRef] [PubMed]
131. Cruz-Gregorio, A.; Manzo-Merino, J.; Lizano, M. Cellular Redox, Cancer and Human Papillomavirus. *Virus Res.* **2018**, *246*, 35–45. [CrossRef]
132. Gyurászová, M.; Gurecká, R.; Bábíčková, J.; Tóthová, Ľ. Oxidative Stress in the Pathophysiology of Kidney Disease: Implications for Noninvasive Monitoring and Identification of Biomarkers. *Oxidative Med. Cell. Longev.* **2020**, *2020*, 5478708. [CrossRef]
133. Ma, N.; Wei, W.; Fan, X.; Ci, X. Farrerol Attenuates Cisplatin-Induced Nephrotoxicity by Inhibiting the Reactive Oxygen Species-Mediated Oxidation, Inflammation, and Apoptotic Signaling Pathways. *Front. Physiol.* **2019**, *10*, 1419. [CrossRef] [PubMed]
134. Santos, N.A.G.; Catão, C.S.; Martins, N.M.; Curti, C.; Bianchi, M.L.P.; Santos, A.C. Cisplatin-Induced Nephrotoxicity Is Associated with Oxidative Stress, Redox State Unbalance, Impairment of Energetic Metabolism and Apoptosis in Rat Kidney Mitochondria. *Arch. Toxicol.* **2007**, *81*, 495–504. [CrossRef]
135. Kaeidi, A.; Taghipour, Z.; Allahtavakoli, M.; Fatemi, I.; Hakimizadeh, E.; Hassanshahi, J. Ameliorating Effect of Troxerutin in Unilateral Ureteral Obstruction Induced Renal Oxidative Stress, Inflammation, and Apoptosis in Male Rats. *Naunyn Schmiedebergs Arch. Pharmacol.* **2020**, *393*, 879–888. [CrossRef]
136. Hallan, S.; Afkarian, M.; Zelnick, L.R.; Kestenbaum, B.; Sharma, S.; Saito, R.; Darshi, M.; Barding, G.; Raftery, D.; Ju, W.; et al. Metabolomics and Gene Expression Analysis Reveal Down-Regulation of the Citric Acid (TCA) Cycle in Non-Diabetic CKD Patients. *EBioMedicine* **2017**, *26*, 68–77. [CrossRef]
137. Kim, N.H.; Hyeon, J.S.; Kim, N.H.; Cho, A.; Lee, G.; Jang, S.Y.; Kim, M.-K.; Lee, E.Y.; Chung, C.H.; Ha, H.; et al. Metabolic Changes in Urine and Serum during Progression of Diabetic Kidney Disease in a Mouse Model. *Arch. Biochem. Biophys.* **2018**, *646*, 90–97. [CrossRef]
138. Liu, H.; Li, W.; He, Q.; Xue, J.; Wang, J.; Xiong, C.; Pu, X.; Nie, Z. Mass Spectrometry Imaging of Kidney Tissue Sections of Rat Subjected to Unilateral Ureteral Obstruction. *Sci. Rep.* **2017**, *7*, 41954. [CrossRef] [PubMed]
139. Nishikawa, T.; Edelstein, D.; Du, X.L.; Yamagishi, S.; Matsumura, T.; Kaneda, Y.; Yorek, M.A.; Beebe, D.; Oates, P.J.; Hammes, H.-P.; et al. Normalizing Mitochondrial Superoxide Production Blocks Three Pathways of Hyperglycaemic Damage. *Nature* **2000**, *404*, 787–790. [CrossRef]
140. Chen, J.; Chen, J.-K.; Harris, R.C. EGF Receptor Deletion in Podocytes Attenuates Diabetic Nephropathy. *J. Am. Soc. Nephrol.* **2015**, *26*, 1115–1125. [CrossRef]
141. Dugan, L.L.; You, Y.-H.; Ali, S.S.; Diamond-Stanic, M.; Miyamoto, S.; DeCleves, A.-E.; Andreyev, A.; Quach, T.; Ly, S.; Shekhtman, G.; et al. AMPK Dysregulation Promotes Diabetes-Related Reduction of Superoxide and Mitochondrial Function. Available online: https://www.jci.org/articles/view/66218/pdf (accessed on 17 June 2021).
142. Mailloux, R.J.; Hamel, R.; Appanna, V.D. Aluminum Toxicity Elicits a Dysfunctional TCA Cycle and Succinate Accumulation in Hepatocytes. *J. Biochem. Mol. Toxicol.* **2006**, *20*, 198–208. [CrossRef]
143. Mapuskar, K.A.; Wen, H.; Holanda, D.G.; Rastogi, P.; Steinbach, E.; Han, R.; Coleman, M.C.; Attanasio, M.; Riley, D.P.; Spitz, D.R.; et al. Persistent Increase in Mitochondrial Superoxide Mediates Cisplatin-Induced Chronic Kidney Disease. *Redox Biol.* **2019**, *20*, 98–106. [CrossRef] [PubMed]
144. Han, S.J.; Jang, H.-S.; Noh, M.R.; Kim, J.; Kong, M.J.; Kim, J.I.; Park, J.-W.; Park, K.M. Mitochondrial NADP+-Dependent Isocitrate Dehydrogenase Deficiency Exacerbates Mitochondrial and Cell Damage after Kidney Ischemia-Reperfusion Injury. *J. Am. Soc. Nephrol.* **2017**, *28*, 1200–1215. [CrossRef]
145. Kong, M.J.; Han, S.J.; Kim, J.I.; Park, J.-W.; Park, K.M. Mitochondrial NADP + -Dependent Isocitrate Dehydrogenase Deficiency Increases Cisplatin-Induced Oxidative Damage in the Kidney Tubule Cells. *Cell Death Dis.* **2018**, *9*, 716. [CrossRef]
146. Kil, I.S.; Park, J.-W. Regulation of Mitochondrial NADP+-Dependent Isocitrate Dehydrogenase Activity by Glutathionylation. *J. Biol. Chem.* **2005**, *280*, 10846–10854. [CrossRef]
147. Kim, J.I.; Noh, M.R.; Yoon, G.-E.; Jang, H.-S.; Kong, M.J.; Park, K.M. IDH2 Gene Deficiency Accelerates Unilateral Ureteral Obstruction-Induced Kidney Inflammation through Oxidative Stress and Activation of Macrophages. *Korean J. Physiol. Pharmacol.* **2021**, *25*, 139–146. [CrossRef] [PubMed]
148. Avila-Rojas, S.H.; Aparicio-Trejo, O.E.; Briones-Herrera, A.; Medina-Campos, O.N.; Reyes-Fermín, L.M.; Martínez-Klimova, E.; León-Contreras, J.C.; Hernández-Pando, R.; Tapia, E.; Pedraza-Chaverri, J. Alterations in Mitochondrial Homeostasis in a Potassium Dichromate Model of Acute Kidney Injury and Their Mitigation by Curcumin. *Food Chem. Toxicol.* **2020**, *145*, 111774. [CrossRef]
149. Mailloux, R.J.; Harper, M.-E. Uncoupling Proteins and the Control of Mitochondrial Reactive Oxygen Species Production. *Free Radic. Biol. Med.* **2011**, *51*, 1106–1115. [CrossRef] [PubMed]

150. Zhou, Y.; Cai, T.; Xu, J.; Jiang, L.; Wu, J.; Sun, Q.; Zen, K.; Yang, J. UCP2 Attenuates Apoptosis of Tubular Epithelial Cells in Renal Ischemia-Reperfusion Injury. *Am. J. Physiol. Physiol.* **2017**, *313*, F926–F937. [CrossRef]
151. Zhong, X.; He, J.; Zhang, X.; Li, C.; Tian, X.; Xia, W.; Gan, H.; Xia, Y. UCP2 Alleviates Tubular Epithelial Cell Apoptosis in Lipopolysaccharide-Induced Acute Kidney Injury by Decreasing ROS Production. *Biomed. Pharmacother.* **2019**, *115*, 108914. [CrossRef] [PubMed]
152. Jiang, L.; Qiu, W.; Zhou, Y.; Wen, P.; Fang, L.; Cao, H.; Zen, K.; He, W.; Zhang, C.; Dai, C.; et al. A MicroRNA-30e/Mitochondrial Uncoupling Protein 2 Axis Mediates TGF-B1-Induced Tubular Epithelial Cell Extracellular Matrix Production and Kidney Fibrosis. *Kidney Int.* **2013**, *84*, 285–296. [CrossRef]
153. Jia, P.; Wu, X.; Pan, T.; Xu, S.; Hu, J.; Ding, X. Uncoupling Protein 1 Inhibits Mitochondrial Reactive Oxygen Species Generation and Alleviates Acute Kidney Injury. *EBioMedicine* **2019**, *49*, 331–340. [CrossRef]
154. Chouchani, E.T.; Kazak, L.; Jedrychowski, M.P.; Lu, G.Z.; Erickson, B.K.; Szpyt, J.; Pierce, K.A.; Laznik-Bogoslavski, D.; Vetrivelan, R.; Clish, C.B.; et al. Mitochondrial ROS Regulate Thermogenic Energy Expenditure and Sulfenylation of UCP1. *Nature* **2016**, *532*, 112–116. [CrossRef]
155. Liesa, M.; Shirihai, O.S. Mitochondrial Dynamics in the Regulation of Nutrient Utilization and Energy Expenditure. *Cell Metab.* **2013**, *17*, 491–506. [CrossRef] [PubMed]
156. Shutt, T.; Geoffrion, M.; Milne, R.; McBride, H.M. The Intracellular Redox State Is a Core Determinant of Mitochondrial Fusion. *EMBO Rep.* **2012**, *13*, 909–915. [CrossRef]
157. Hernansanz-Agustín, P.; Enríquez, J.A. Generation of Reactive Oxygen Species by Mitochondria. *Antioxidants* **2021**, *10*, 415. [CrossRef]
158. Kashatus, J.A.; Nascimento, A.; Myers, L.J.; Sher, A.; Byrne, F.L.; Hoehn, K.L.; Counter, C.M.; Kashatus, D.F. Erk2 Phosphorylation of Drp1 Promotes Mitochondrial Fission and MAPK-Driven Tumor Growth. *Mol. Cell* **2015**, *57*, 537–551. [CrossRef]
159. Taguchi, N.; Ishihara, N.; Jofuku, A.; Oka, T.; Mihara, K. Mitotic Phosphorylation of Dynamin-Related GTPase Drp1 Participates in Mitochondrial Fission. *J. Biol. Chem.* **2007**, *282*, 11521–11529. [CrossRef] [PubMed]
160. Qi, X.; Disatnik, M.-H.; Shen, N.; Sobel, R.A.; Mochly-Rosen, D. Aberrant Mitochondrial Fission in Neurons Induced by Protein Kinase C{delta} under Oxidative Stress Conditions in Vivo. *Mol. Biol. Cell* **2011**, *22*, 256–265. [CrossRef]
161. Kim, Y.-M.; Youn, S.-W.; Sudhahar, V.; Das, A.; Chandhri, R.; Cuervo Grajal, H.; Kweon, J.; Leanhart, S.; He, L.; Toth, P.T.; et al. Redox Regulation of Mitochondrial Fission Protein Drp1 by Protein Disulfide Isomerase Limits Endothelial Senescence. *Cell Rep.* **2018**, *23*, 3565–3578. [CrossRef]
162. Rojas-Morales, P.; León-Contreras, J.C.; Granados-Pineda, J.; Hernández-Pando, R.; Gonzaga, G.; Sánchez-Lozada, L.G.; Osorio-Alonso, H.; Pedraza-Chaverri, J.; Tapia, E. Protection against Renal Ischemia and Reperfusion Injury by Short-Term Time-Restricted Feeding Involves the Mitochondrial Unfolded Protein Response. *Free Radic. Biol. Med.* **2020**, *154*, 75–83. [CrossRef]
163. Brooks, C.; Wei, Q.; Cho, S.-G.; Dong, Z. Regulation of Mitochondrial Dynamics in Acute Kidney Injury in Cell Culture and Rodent Models. *J. Clin. Investig.* **2009**, *119*, 1275–1285. [CrossRef]
164. Ortega-Domínguez, B.; Aparicio-Trejo, O.E.; García-Arroyo, F.E.; León-Contreras, J.C.; Tapia, E.; Molina-Jijón, E.; Hernández-Pando, R.; Sánchez-Lozada, L.G.; Barrera-Oviedo, D.; Pedraza-Chaverri, J. Curcumin Prevents Cisplatin-Induced Renal Alterations in Mitochondrial Bioenergetics and Dynamic. *Food Chem. Toxicol.* **2017**, *107*, 373–385. [CrossRef]
165. Perry, H.M.; Huang, L.; Wilson, R.J.; Bajwa, A.; Sesaki, H.; Yan, Z.; Rosin, D.L.; Kashatus, D.F.; Okusa, M.D. Dynamin-Related Protein 1 Deficiency Promotes Recovery from AKI. *J. Am. Soc. Nephrol.* **2018**, *29*, 194–206. [CrossRef]
166. Bhatia, D.; Capili, A.; Choi, M.E. Mitochondrial Dysfunction in Kidney Injury, Inflammation, and Disease: Potential Therapeutic Approaches. *Kidney Res. Clin. Pract.* **2020**, *39*, 244–258. [CrossRef]
167. Li, S.; Lin, Q.; Shao, X.; Zhu, X.; Wu, J.; Wu, B.; Zhang, M.; Zhou, W.; Zhou, Y.; Jin, H.; et al. Drp1-Regulated PARK2-Dependent Mitophagy Protects against Renal Fibrosis in Unilateral Ureteral Obstruction. *Free Radic. Biol. Med.* **2020**, *152*, 632–649. [CrossRef]
168. Willems, P.H.G.M.; Rossignol, R.; Dieteren, C.E.J.; Murphy, M.P.; Koopman, W.J.H. Redox Homeostasis and Mitochondrial Dynamics. *Cell Metab.* **2015**, *22*, 207–218. [CrossRef]
169. Liu, T.; Yang, Q.; Zhang, X.; Qin, R.; Shan, W.; Zhang, H.; Chen, X. Quercetin Alleviates Kidney Fibrosis by Reducing Renal Tubular Epithelial Cell Senescence through the SIRT1/PINK1/Mitophagy Axis. *Life Sci.* **2020**, *257*, 118116. [CrossRef]
170. Sang, X.-Y.; Xiao, J.-J.; Liu, Q.; Zhu, R.; Dai, J.-J.; Zhang, C.; Yu, H.; Yang, S.-J.; Zhang, B.-F. Regulators of Calcineurin 1 Deficiency Attenuates Tubulointerstitial Fibrosis through Improving Mitochondrial Fitness. *FASEB J.* **2020**, *34*. [CrossRef]
171. Barsoum, M.J.; Yuan, H.; Gerencser, A.A.; Liot, G.; Kushnareva, Y.; Gräber, S.; Kovacs, I.; Lee, W.D.; Waggoner, J.; Cui, J.; et al. Nitric Oxide-Induced Mitochondrial Fission Is Regulated by Dynamin-Related GTPases in Neurons. *EMBO J.* **2006**, *25*, 3900–3911. [CrossRef]
172. Villena, J.A. New Insights into PGC-1 Coactivators: Redefining Their Role in the Regulation of Mitochondrial Function and Beyond. *FEBS J.* **2015**, *282*, 647–672. [CrossRef]
173. Hock, M.B.; Kralli, A. Transcriptional Control of Mitochondrial Biogenesis and Function. *Annu. Rev. Physiol.* **2009**, *71*, 177–203. [CrossRef]
174. Irrcher, I.; Ljubicic, V.; Hood, D.A. Interactions between ROS and AMP Kinase Activity in the Regulation of PGC-1alpha Transcription in Skeletal Muscle Cells. *Am. J. Physiol. Cell Physiol.* **2009**, *296*, C116–C123. [CrossRef] [PubMed]
175. Yan, Y.; Zhou, X.E.; Xu, H.E.; Melcher, K. Structure and Physiological Regulation of AMPK. *Int. J. Mol. Sci.* **2018**, *19*, 3534. [CrossRef]

176. Rabinovitch, R.C.; Samborska, B.; Faubert, B.; Ma, E.H.; Gravel, S.-P.; Andrzejewski, S.; Raissi, T.C.; Pause, A.; St-Pierre, J.; Jones, R.G. AMPK Maintains Cellular Metabolic Homeostasis through Regulation of Mitochondrial Reactive Oxygen Species. *Cell Rep.* **2017**, *21*, 1–9. [CrossRef] [PubMed]
177. Bhatia, D.; Choi, M.E. The Emerging Role of Mitophagy in Kidney Diseases. *J. Life Sci.* **2019**, *1*, 13–22. [CrossRef]
178. Wang, Y.; Cai, J.; Tang, C.; Dong, Z. Mitophagy in Acute Kidney Injury and Kidney Repair. *Cells* **2020**, *9*, 338. [CrossRef] [PubMed]
179. Xiao, B.; Goh, J.-Y.; Xiao, L.; Xian, H.; Lim, K.-L.; Liou, Y.-C. Reactive Oxygen Species Trigger Parkin/PINK1 Pathway-Dependent Mitophagy by Inducing Mitochondrial Recruitment of Parkin. *J. Biol. Chem.* **2017**, *292*, 16697–16708. [CrossRef] [PubMed]
180. Kondapalli, C.; Kazlauskaite, A.; Zhang, N.; Woodroof, H.I.; Campbell, D.G.; Gourlay, R.; Burchell, L.; Walden, H.; Macartney, T.J.; Deak, M.; et al. PINK1 Is Activated by Mitochondrial Membrane Potential Depolarization and Stimulates Parkin E3 Ligase Activity by Phosphorylating Serine 65. *Open Biol.* **2012**, *2*, 120080. [CrossRef]
181. Reyes-Fermín, L.M.; Avila-Rojas, S.H.; Aparicio-Trejo, O.E.; Tapia, E.; Rivero, I.; Pedraza-Chaverri, J. The Protective Effect of Alpha-Mangostin against Cisplatin-Induced Cell Death in LLC-PK1 Cells Is Associated to Mitochondrial Function Preservation. *Antioxidants* **2019**, *8*, 133. [CrossRef]
182. Li, L.; Zepeda-Orozco, D.; Black, R.; Lin, F. Autophagy Is a Component of Epithelial Cell Fate in Obstructive Uropathy. *Am. J. Pathol.* **2010**, *176*, 1767–1778. [CrossRef]
183. Chen, Q.; Yu, S.; Zhang, K.; Zhang, Z.; Li, C.; Gao, B.; Zhang, W.; Wang, Y. Exogenous H_2S Inhibits Autophagy in Unilateral Ureteral Obstruction Mouse Renal Tubule Cells by Regulating the ROS-AMPK Signaling Pathway. *Cell. Physiol. Biochem.* **2018**, *49*, 2200–2213. [CrossRef] [PubMed]

MDPI
St. Alban-Anlage 66
4052 Basel
Switzerland
Tel. +41 61 683 77 34
Fax +41 61 302 89 18
www.mdpi.com

Biomolecules Editorial Office
E-mail: biomolecules@mdpi.com
www.mdpi.com/journal/biomolecules

www.ingramcontent.com/pod-product-compliance
Lightning Source LLC
Chambersburg PA
CBHW061516210926
51850CB00008B/1934
9783036537573